Lecture Notes in Statistics

Edited by S. Fienberg, J. Olkin, and N. Wermuth

93

Bernard Van Cutsem (Editor)

Classification and Dissimilarity Analysis

Springer-Verlag
New York Berlin Heidelberg London Paris
Tokyo Hong Kong Barcelona Budapest

Bernard Van Cutsem
Laboratoire de Modélisation et Calcul
Institut IMAG
Université Joseph Fourier Grenoble I
B.P. 53
F-38041 Grenoble Cedex 9
France

Library of Congress Cataloging-in-Publication Data Available
Printed on acid-free paper.

© 1994 Springer-Verlag New York, Inc.
All rights reserved. This work may not be translated or copied in whole or in part without the written permission of the publisher (Springer-Verlag New York, Inc., 175 Fifth Avenue, New York, NY 10010, USA), except for brief excerpts in connection with reviews or scholarly analysis. Use in connection with any form of information storage and retrieval, electronic adaptation, computer software, or by similar or dissimilar methodology now known or hereafter developed is forbidden.
The use of general descriptive names, trade names, trademarks, etc., in this publication, even if the former are not especially identified, is not to be taken as a sign that such names, as understood by the Trade Marks and Merchandise Marks Act, may accordingly be used freely by anyone.

Camera ready copy provided by the author.
Printed and bound by Braun-Brumfield, Ann Arbor, MI.
Printed in the United States of America.

9 8 7 6 5 4 3 2 1

ISBN 0-387-94400-1 Springer-Verlag New York Berlin Heidelberg

Preface

During the years 1983–1987, the French Centre National de la Recherche Scientifique organised a research group on the subject "Analyse des Données" with a view to establish an overview of French contributions to that part of Statistics. Inside this group some people, mainly the contributors of the present publication, constituted in 1985–1987 a subgroup, extended to some foreign participants, interested in the theoretical aspects of Automatic Classification methods.

In the following years, this small group continued to meet and work exchanging ideas in an informal way. Its interest was concentrated on the basic notions of Automatic Classification. Gradually a central interest emerged around the ideas of similarities and dissimilarities and became what is known as the "Analysis of Dissimilarity".

During the years 1990-1992, with financial supports from the French Ministère de la Recherche et de la Technologie and from the Warwick University Research and Innovation Fund, a "Réseau Européen de Laboratoires" was created on the theme "Structures Métriques pour l'Analyse des Dissimilarités". This framework allowed the extension of the initial group to researchers from The Netherlands and Portugal. The main activity of this Réseau was the organisation in Rennes, France, in June 1992, of DISTANCIA'92 an International Meeting on "Analysis of Distances".

This publication contains the main contributions of the initial group in the field of the theoretical bases of the classification theory. Most of them have been presented in (often invited) communications at various specialised meetings. We hope that their presentation in a single volume will be of interest to many people concerned with various aspects of Classification.

Each chapter is self-contained and can be read independently of the others. However, Chapter 2 is important as it contains the main definitions in the domain of Dissimilarity Analysis.

The references of each chapter have been collated into a single unified reference list at the end of the book.

I would like to thank the Editorial Staff of Springer-Verlag for the help in bringing this book to publication.

Bernard Van Cutsem

Ordonner, classer les gènes, en faire l'inventaire ? Mais pour quoi faire, alors qu'on ne connaît pas encore leur fonction ? Il faut penser d'abord, ranger ensuite, laissaient entendre les théoriciens qui, dédaignant l'histoire des sciences, oubliaient qu'ordonner aide à penser, et que classer, c'est déjà théoriser. Et pourtant ! Peut-on imaginer que Newton ait pu énoncer les lois de la gravition universelle sans tenir compte de l'inventaire des planètes établi dès la Grèce antique ? La classification des espèces par le Suédois Linné au XVIIIe siècle n'a-t-elle pas frayé la voie à la théorie de l'évolution de Darwin ? Et la classification des éléments périodiques par le Russe Mendeleïev, à la fin du XIXe siècle, préparé le terrain pour la compréhension de la radioactivité au XXe ?

Ce que les biologistes n'avaient pas bien appréhendé, c'est qu'en cherchant à classer les gènes on découvrirait des fonctions nouvelles, des propriétés qui ouvrent des perspectives sur l'inconnu. Nous, les classificateurs de gènes, les baliseurs de génomes, les routiers et les cantonniers de l'ADN, apportions aux théoriciens des problèmes nouveaux, en un mot, de quoi alimenter leurs théories.

......

Nous faisons de la science, mais nous n'avons pas d'hypothèses. Nous pratiquons l'inverse du schéma admis : ce n'est plus l'hypothèse qui fonde la recherche, c'est la recherche qui se met en quête d'hypothèses nouvelles.

......

C'est ce type d'audace qui caractérisa tous les grands faiseurs d'inventaires de la nature, ceux des planètes, des espèces, des éléments chimiques, à la manière des Copernic, Linné et autres Mendeleïev. Et leurs inventaires fécondèrent ces moissons extraordinaires d'hypothèses et de théories qui frayèrent de nouveaux accès aux secrets de la nature.

Daniel COHEN *Les gènes de l'espoir*, Robert Laffont , Paris, 1993 (p 57 et 85-86)

Contents

Preface v
List of contributors xiii

Chapter 1

Introduction 1
Bernard Van Cutsem

    1.1 Classification in the history of Science 1
    1.2 Dissimilarity analysis 2
    1.3 Organisation of this publication 3
    1.4 References 4

Chapter 2

The partial order by inclusion of the principal classes of dissimilarity on a finite set, and some of their basic properties 5
Frank Critchley and Bernard Fichet

    2.1 Introduction 5
        2.1.1 What is dissimilarity analysis? 5
        2.1.2 What's in this chapter? 8
    2.2 Preliminaries 9
        2.2.1 The vector space $\mathcal{D}$ generated by the dissimilarities 9
        2.2.2 Some elementary dissimilarities 11
        2.2.3 Some general types of dissimilarities 12
        2.2.4 Stability under increasing transformations 13
        2.2.5 The quotient space of an even dissimilarity 14
        2.2.6 Embeddability in a metric space 15
        2.2.7 The set of all semi-distances 16
        2.2.8 Stability under replication 17

2.3 The general structures of dissimilarity data analysis and their geometrical and topological nature 18
   2.3.1 Euclidean semi-distances 18
   2.3.2 Semi-distances of L_1-type 21
   2.3.3 Hypermetric semi-distances 23
   2.3.4 Quasi-hypermetric dissimilarities 25
   2.3.5 Ultrametric semi-distances 26
   2.3.6 Tree semi-distances 28
   2.3.7 Star semi-distances 33
   2.3.8 Robinsonian dissimilarities 34
   2.3.9 Strongly-Robinsonian dissimilarities 38
2.4 Inclusions 39
   2.4.1 Some immediate inclusions 39
   2.4.2 Other inclusions 41
2.5 The convex hulls 46
2.6 When are the inclusions strict? 48
2.7 The inclusions shown are exhaustive 52
2.8 Discussion 53
   2.8.1 Further mathematical study 54
   2.8.2 Extensions to other types of data 56
   2.8.3 Connections with neighbouring disciplines 57
   2.8.4 The future of dissimilarity analysis 57
Acknowledgements 58
References 58

Chapter 3

Similarity functions 67
Serge Joly and Georges Le Calvé

3.1 Introduction 67
3.2 Definitions. Examples 68
   3.2.1 Definitions 68
   3.2.2 Examples 69
      3.2.2.1 Linear function 69
      3.2.2.2 Homographic function 69
      3.2.2.3 Quadratic function 70

      3.2.2.4 Exponential function 70
      3.2.2.5 Circular function 70
      3.2.2.6 Graphical representations 71
  3.3 The $W^M(D^p)$ forms 72
    3.3.1 Definitions and properties 72
    3.3.2 The $W^M(D^2)$ form 74
      Torgerson form 75
    3.3.3 Transformations of D 77
      D^α and the Euclidean distances 78
  3.4 The $W^M(D)$ form 79
    3.4.1 Geometrical interpretations and properties 79
    3.4.2 About metric projection 81
    3.4.3 $W^M(D)$ and "$\mathcal{M}^1$-type" distance 82
  Appendix: Some indices of dissimilarity for categorical variables 84
  References 86

Chapter 4

An order-theoretic unification and generalisation of certain fundamental bijections in mathematical classification. I 87

Frank Critchley and Bernard Van Cutsem

  4.1 Introduction and overview 87
  4.2 A few notes on ordered sets 89
    4.2.1 Introduction 89
    4.2.2 Duality and order-isomorphisms 90
    4.2.3 Semi-lattices and lattices 91
    4.2.4 Residual and residuated mappings 92
  4.3 Predissimilarities 93
  4.4 Bijections 95
    4.4.0 Overview 95
    4.4.1 Indexed hierarchies (S finite, $L = \mathbb{R}_+$) 95
    4.4.2 Dendrograms (S finite, $L = \mathbb{R}_+$) 97
    4.4.3 Numerically stratified clusterings (S finite, $L = \mathbb{R}_+$) 98
    4.4.4 Indexed regular generalised hierarchies (S arbitrary, $L = \mathbb{R}_+$) . . 98
    4.4.5 Generalised dendrograms (S arbitrary, $L = \mathbb{R}_+$) 99
    4.4.6 Prefilters (S arbitrary, $L = \mathbb{R}_+$) 100

4.4.6 Residual maps (S finite, L obeys LMIN and JSL) 101
4.5 The unifying and generalising result 102
4.6 Further properties of an ordered set 104
4.7 Stratifications and generalised stratifications 105
4.8 Residual maps 108
4.9 On the associated residuated maps 110
4.10 Some applications to mathematical classification 111
Acknowledgements 113
Appendix A : Proofs 114
References 118

Chapter 5
An order-theoretic unification and generalisation of certain fundamental bijections in mathematical classification. II 121
Frank Critchley and Bernard Van Cutsem

5.1 Introduction and overview 121
5.2 The case $E = A \times B$ of theorem 4.5.1 122
5.3 Other aspects of the case $E = A \times B$ 124
    5.3.1 Duality 124
    5.3.2 Multiway data 126
    5.3.3 Residual maps 126
5.4 Prefilters 127
5.5 Ultrametrics and reflexive level foliations 127
    5.5.1 The main result 127
    5.5.2 Remarks on Theorem 5.5.1 128
    5.5.3 Variants of Theorem 5.5.1 130
5.6 On generalisations of indexed hierarchies 131
    5.6.1 Introduction 131
    5.6.2 Benzécri structures 131
    5.6.3 Special cases of Benzécri structures 134
    5.6.4 The condition LSU 135
5.7 Benzécri structures 136
5.8 Subdominants 138
Acknowledgements 139
Appendix B: Proofs 140

References 147

Chapter 6

The residuation model for the ordinal construction of dissimilarities and other valued objects 149

Bruno Leclerc

    6.1 Introduction 149

    6.2 Residuated mappings and closure operators 151

        6.2.1 Residuated and residual mappings 151

        6.2.2 Closure and anticlosure operators 153

    6.3 Lattices of objects and lattices of values 154

        6.3.1 Lattices 154

        6.3.2 Distributivity 154

        6.3.3 Lattices of objects: ten examples 155

        6.3.4 Lattices of values 158

    6.4 Valued objects 159

        6.4.1 Consequences of the lattice structures hypothesis 159

        6.4.2 Valued objects: definitions and examples 160

    6.5 Lattices of valued objects 164

    6.6 Notes and conclusions 167

    Acknowledgements 169

    References 169

Chapter 7

On exchangeability-based equivalence relations induced by strongly Robinson and, in particular, by quadripolar Robinson dissimilarity matrices 173

Frank Critchley

    7.1 Overview 173

        7.1.1 Preamble 173

        7.1.2 Quadripolar, Robinson and strongly Robinson matrices 174

        7.1.3 Plan and principal results 176

    7.2 Preliminaries 177

    7.3 Quadripolar Robinson matrices of order four 179

    7.4 Equivalence relations induced by strongly Robinson matrices 181

        7.4.1 Exchangeability and connectedness 181

7.4.2 Internal evenness 184
  7.4.3 Logical relationships 185
 7.5 Reduced forms 187
  7.5.1 External evenness 187
  7.5.2 Properties of reduced forms 189
 7.6 Limiting r-forms of strongly Robinson matrices 191
 7.7 Limiting r-forms of quadripolar Robinson matrices 193
 References 197

Chapter 8
Dimensionality problems in L_1-norm representations 201
Bernard Fichet

 8.1 Introduction 201
 8.2 Preliminaries and notations 202
  8.2.1 Dissimilarities 202
  8.2.2 Some notations 202
  8.2.3 Some characterizations 203
 8.3 Dimensionality for semi-distances of L_p-type 204
 8.4 Dimensionality for semi-distances of L_1-type 207
 8.5 Numerical characterizations of semi-distances of L_1-type 209
  8.5.1 Solving the general problem 210
  8.5.2 Reducing the problem 211
  8.5.3 Approximations 212
   8.5.3.1 Least absolute deviations approximations 212
   8.5.3.2 Least squares approximation 213
   8.5.3.3 The additive constants 213
 8.6 Appendices 215
  8.6.1 Appendix 1 215
  8.6.2 Appendix 2 216
  8.6.3 Appendix 3 216
  8.6.4 Appendix 4 218
 References 222

Unified reference list **225**

List of contributors

Frank Critchley. University of Birmingham.
School of Mathematics and Statistics,
Edgbaston, B15 2TT BIRMINGHAM, U.K.

Bernard Fichet. Université d'Aix-Marseille II.
Laboratoire de Biomathématiques, Faculté de Médecine,
27, boulevard Jean Moulin, 13385 MARSEILLE Cedex, France

Serge Joly. Université de Haute Bretagne, Rennes II.
Laboratoire Analyse et Traitement des Données,
6, avenue Gaston Berger, 35043 RENNES Cedex, France

Georges Le Calvé. Université de Haute Bretagne, Rennes II.
Laboratoire Analyse et Traitement des Données,
6, avenue Gaston Berger, 35043 RENNES Cedex, France

Bruno Leclerc. Ecole des Hautes Etudes en Sciences Sociales, Paris.
Centre d'Analyse et de Mathématiques Sociales,
54, boulevard Raspail, 75270 PARIS Cedex06, France

Bernard Van Cutsem. Université Joseph Fourier, Grenoble I.
Laboratoire Modélisation et Calcul, CNRS URA 397,
B.P. 53, 38041 GRENOBLE Cedex9, France

Chapter 1.
Introduction[*]

Bernard Van Cutsem[†]

1.1. Classification in the history of Science

Classifying objects according to their likeness seems to have been, for all time, a step in the human process of acquiring knowledge. We could possibly find the beginning of this process as early as infancy, if we admit that the brain of a young child learns to distinguish categories of objects or persons (distinguishing one animal from another, his parents from strangers, ...), or of situations (what is allowed from what is not) proceeding by analogies. Historically the scientific process, even when it is purely descriptive, works in the same way. We dominate well a domain when all its notions are classified and categorised. Such classifications show some relationships from which the exploration yields, in a second stage, improvements in the knowledge of the domain. Even far back in time, we find texts and authors who have shown organisation and clustering particularly in descriptions in the areas of botany or geology (Aristotle). Closer to us, the scientists of the "siècle des lumières" and their heirs introduced some famous classifications (A.L. de Jussieu, G. Cuvier, G. L. Buffon, C. von Linneaus, ...). In the same way, we can find, outside natural science, various works in classification in the exact sciences (D.I. Mendeleïev, ...), in astronomy or in linguistics (see Marcotorchino (1991) for an interesting historical presentation). The criteria of classification used are generally empirical ones : mammals are separated from non mammals, vertebrates from non vertebrates, ... In a modern context, this is equivalent to deriving a hierarchical clustering of objects (we use this generic name for individuals, or statistical units, Operational Taxonomic Units (OTU), ...) using some variables which are introduced one after another in a given order. The last clusters of objects are divided according to the next variable in an ordered sequence of variables.

Modern clustering may be now considered as a part of Statistics, or at least as a main tool of Exploratory Data Analysis. As a matter of fact, under these terms, all the methods are put together to answer questions such as "Look at the data to see what patterns they may suggest" or more simply "to see what kind of simple (geometrical or relational) structures they have". So classification is close to factorial analysis methods, and to multi-dimensional scaling methods. Clustering is mainly used to verify the homogeneity or the non-homogeneity of the data (for example choosing between a mono- or a pluri-modal probablistic model). These are the basic methods essential for the practitioner examining new data. The basic information for such an analysis is generally contained in an Objects × Variables array. The objects are described by the rows of this

---

[*] *In* Van Cutsem, B. (Ed.), (1994) *Classification and Dissimilarity Analysis*, Lecture Notes in Statistics, Springer-Verlag, New York.

[†] Université Joseph Fourier, Grenoble, France.

array that is, by their profile (e.g. the set of values of the variables for these objects) and are represented by a point in a relevant space, the product of the spaces in which the variables take their values. The likeness between two objects is characterised by the likeness of their profiles in this representation space. If we define a measure of the proximity in this space, we get an Objects × Objects array which contains in a quantitative form information on the proximities. Measures of proximity are not easily to define, particularly when non numerical variables appear. Moreover they are not necessarily a distance in a mathematical sense. We denote this absence of structure of distance by the use of dissimilarities (for measuring difference), or of similarities (for measuring likeness). Of course it is possible that, in some cases, the data are directly given as an Objects × Objects array of dissimilarities or of similarities.

1.2. Dissimilarity analysis

These historical early beginnings we briefly recalled above show that the start of Classification was illustrated by many successes due to the ability of the practitioners. As Marcotorchino (1991) noticed: "Quel que soit le terrain applicatif, si l'intuition et la persévérance des praticiens ont quelquefois suffi pour mener à bien des tâches d'envergure, il est vite apparu que le recours à des techniques plus élaborées était inéluctable. L'outil mathématique s'est naturellement imposé comme interface entre les préoccupations des taxinomistes et les méthodes de classifications à vue pour donner naissance à de véritables algorithmes de traitement.", this progress was only possible with the help of more powerful algorithmic and theoretical tools. Actually, the main obstacle was the material difficulty in the clustering of large sets of objects and we were very far from being able to experiment with new methods and even more so from comparing them. The arrival of computers allowed an extremely fast evolution and we could see within a few years a multiplication of publications and of applications in a many diversified areas: archaeology, medicine, biology, economics, social science, artificial intelligence, computer science, and so on. The imagination had no limits and many ideas leading to partitions, hierarchies of grouping of objects starting from a collection of variables were explored. Nevertheless, it was also clear that the combinatorial aspect of the problems was an impossible obstacle. The number of all the partitions or hierarchies of a set of n objects has such a high growth rate with n, that the search for an optimal solution for a given criteria was out of reach - and still is despite the enormous power of the more recent computers. Then it was important to look for heuristics or for approximate solutions in some more restricted feasible sets. Here also, under pressure from practitioners in every domain, the abundance of ideas and of publications has been, and still is enormous.

All this investigative work about classification methods has been followed by theoretical research. The comparison of methods (which methods are special cases of others?), the search for new models, the definition of a "classifiability" concept, the influence of the choice of the dissimilarity measure, the robustness of the methods, the optimality, the convergence and the efficacy of algorithms (in particular, the existence of a unique solution) have been relatively well explored domains.

Recently, a few researchers were interested in the basic notion of classification: the theoretical study of the set of all the dissimilarities. A dissimilarity of a finite set of objects I is a function defined on $I \times I$ with values in $\mathbb{R}_+$ which is symmetric, zero on the diagonal, and their set is obviously a cone in a space $\mathbb{R}^k$ with the relevant dimension. Questions such as "What are the relationships and the structures of the set of dissimilarities? What are its principal subsets? How are these subsets organised from the point of view of inclusion, density in the topological sense of the word, ...?" were approached and much work remains to be done in these directions.

1.3. Organisation of this publication

The following papers constitute a relatively coherent whole, including definitions and analyses of mathematical properties in Dissimilarity Analysis. It was first necessary to describe the different notions of distances and dissimilarities used in Classification.

This appears in the second and third chapters. Chapter 2, by Frank Critchley and Bernard Fichet, gives the definitions of the various kinds of dissimilarities, a description of their corresponding subsets in the cone of all dissimilarities, and an analysis of how these subsets fit together. Strict inclusions and density or closedness properties are characterised. Chapter 3, by Serge Joly and Georges Le Calvé, on the one hand analyses the transformations between similarities and dissimilarities, and on the other hand gives a description of Euclidian and City-Block distances.

Such a panorama of the set of dissimilarities makes it clear that some similar notions which appear in different contexts can be unified and extended in a more general framework. The contributions of the theory exceed a good understanding of the links between a few tools as, together with the definition of the well adapted framework for their study, they open new perspectives and allow some generalisations which in turn offer solutions to some new problems. Two examples will illustrate this process. It is well known that ultrametrics and dissimilarities on a finite set of objects have representations by structures such as dendrograms, generalised dendrograms and indexed hierarchical classifications. These representations are essential to decant information and have been established for more than twenty years. The first example concerns the introduction by M. Janowitz of dissimilarities with values in an ordered set instead of the nonnegative real numbers $\mathbb{R}_+$. The aim was to see which were the algebraic structures of $\mathbb{R}_+$ necessary for the representation of dissimilarities as generalised dendrograms. M. Janowitz introduced a new type of representation with residual functions and high lighted the main role played by the duality between residuated and residual functions.

Bruno Leclerc emphasizes, in Chapter 6, a mathematical model in terms of ordered structures and presents some applications from this point of view illustrating the fundamental role of the ordered structures. Initiated with an æsthetic solicitation, this development now gives a theoretical framework well adapted to the definition of inferior maximum ultrametrics and superior minimal ones.

The second example is also illustrative. Frank Critchley and Bernard Van Cutsem establish, in Chapter 4 and 5, that all the known representations of the dissimilarities by dendrograms, generalised dendrograms or indexed hierarchical classifications are, in fact, the consequence of a simple theorem of representation of a function by its level sets. They then note that the known representations of dissimilarities can immediatly be extended to the case where the set of objects is non finite, the dissimilarities are nonsymmetric and, following M. Janowitz's idea, the dissimilarities take their values in an ordered set. This unification, and the generalisation it implies, introduces a good framework and gives the right tools to approach the practical problems coming from

- vector valued dissimilarities,
- comparing classifications,
- asymptotic studies when the number of objects increases to infinity,
- nonsymmetric dissimilarities,
- dissimilarities defined on three or more objects (multiway analysis), and so on.

This work ends with the study of inferior maximum ultrametrics for a given dissimilarity (existence, uniqueness, characterisation).

In Chapter 7, Frank Critchley discusses the properties of dissimilarities of a particular kind. These dissimilarities are assumed to be additive and compatible with an order on the objects. The conjunction of these two properties implies some interesting developments and some precise characterisations.

Finally, in Chapter 8, Bernard Fichet considers the special case of the cone of semi-distances of L_1-type. The dichotomies are used as a basic tool. The problem of the dimension and of the minimal dimension of the geometric representation of a semi-distance of L_1-type on a finite set of points is then investigated.

In brief, these mathematical developments present from a different point of view many aspects of Analysis of Dissimilarity and will enable in the near future the exploration of some new possibilities such as those already discussed. This is a work in Mathematical Classification, if we may use these words to denote the mathematics of Classification and of Analysis of Dissimilarity. Moreover, although precise applications do not appear in these studies, every researcher interested in Classification will be reading some of them between the lines in many of the notions we have considered.

We offer this work as a contribution to Analysis of Dissimilarity and, more generally, to Classification. Many of the results here have been presented in (often invited) papers at various specialised meetings. Putting them together in a single volume will, we hope, facilitate their diffusion to a wider public.

References

Marcotorchino, F. (1991), La classification mathématique aujourd'hui, *Publications Scientifiques et Techniques d'IBM France*, 2, pp. 35–93.

Chapter 2.
The partial order by inclusion of the principal classes of dissimilarity on a finite set, and some of their basic properties[*]

Frank Critchley[†]

Bernard Fichet[††]

2.1. Introduction

2.1.1. What is dissimilarity analysis?

We begin with a short overview of the whole domain of dissimilarity analysis. The contents of this particular paper are then summarised in Section 2.1.2.

A primary objective of many applications of dissimilarity analysis is to produce an appropriate visual display of complex information in order to arrive at a better understanding of it. Although not essential to this enterprise, such a graphical display may perhaps also then help to formulate an appropriate model, for example by suggesting certain hypotheses. Or it may serve to criticise an existing model, for example by identifying un-modelled subgroups or clusters of observations.

The corresponding array of data analysis methods largely rests on an analogy between dissimilarity and distance. Data dissimilarities between objects are represented by distances between points in some visualisable space. In classical multidimensional scaling, for example, these are distances between points in some low-dimensional Euclidean space. Again, in hierarchical cluster analysis, these are the ultrametric distances between the terminal nodes of a dendrogram.

Thus the data dissimilarity is represented by another dissimilarity which has a convenient visual form. Of course, it will not in general be possible to do this both directly and exactly. Rather, we often allow the data dissimilarity, or the representation dissimilarity, or both, to be transformed within some prescribed class of transformations. In any event, the (transformed) representation dissimilarity will in general only be an approximation in some sense to the (transformed) data dissimilarity. We may then define a method of dissimilarity analysis by answering the following key questions.

---

[*] *In* Van Cutsem, B. (Ed.), (1994) *Classification and Dissimilarity Analysis*, Lecture Notes in Statistics, Springer-Verlag, New York.
[†] University of Birmingham, U.K.
[††] Université d'Aix-Marseille II, France.

For this method:

(1) What form does the data dissimilarity take?

(2) What is the class of possible representations of the data dissimilarity?

(3) What transformations of the data dissimilarity and/or of its representation dissimilarity are permitted?

(4) What criterion is used to say how good is a particular (transformed) representation of the (transformed) data dissimilarity?

This produces a framework for dissimilarity analysis which consists in crossing all possible answers to these key questions.

The field of dissimilarity analysis is both large and expanding. In part its richness lies in the variety of possible answers to the above key questions, and in the many possibilities for combining answers to different questions. We concentrate here on the first two questions.

Distances naturally intervene in all domains of human enquiry, while dissimilarities are more general still. Thus, there is an abundance of data of the appropriate form for analysis in every domain from astronomy to zoology. Formally, a dissimilarity d on a set I is a function $d: I^2 \to \mathbb{R}$ that is symmetric, nonnegative and vanishes on the diagonal. That is,

$$d(i,j) = d(j,i) \geq d(i,i) = 0.$$

Intuitively, $d(i,j)$ measures how different i and j are in some sense. In practice, data dissimilarity measures may either be (1) observed directly or (2) calculated from other quantities. Some examples of these are, respectively:

(1.1) recommended road distances between towns in mainland Europe,

(1.2) first passage times between states in a reversible process,

(1.3) judged dissimilarities between stimuli in a psychometric experiment

and

(2.1) counts of the number of binary characteristics on which zoological taxonomic units differ,

(2.2) Mahalanobis distances between samples from a set of multivariate populations with a common covariance matrix

(2.3) geodesic distances between probability distributions in a differentiable manifold.

Usually a lot more is known about the properties of the resulting data dissimilarity in case (2). For example, in senses made exact below, (2.1) is of L_1-type while (2.2) is Euclidean. See in particular Fichet and Le Calvé (1984), Gower and Legendre (1986) and Theme 1 of the book edited by De Leeuw, Heiser, Meulman and Critchley (1986). This extra knowledge can then be exploited in the choice of the method of analysis and of its algorithmic implementation, and/or in the interpretation of the end results. On the other hand, the dependences among the $d(i,j)$ values usually take a simpler form in case (1). In particular, implicit assumptions of independence made in least-squares methods are typically more plausible in this case.

Answers to the second key question are limited only by human creative ability. As we discuss below, representations by Euclidean distances or by dendrograms have more recently been augmented by, amongst others, tree distances and pseudo-hierarchies. This abundance of data coupled with the utility of the resulting visual displays has produced a buoyant demand. And, as always, new problems encountered in new domains are the seeds of new research.

To take one example, the recent book of Crippen and Havel (1988) reveals a wide range of challenging mathematical problems that arise in applying distance geometry to molecular conformation problems. A central problem here is that of determining those conformations whose distances between atoms are nearest to those measured by nuclear magnetic resonance techniques. The continued prominence given to this area is due to the fundamental relations between the shape and function of a molecule. For example, the desire of genetic engineers to alter protein function has provided the latest impetus for determining protein conformation (Oxender and Fox, 1987).

However, this very profusion of methods of dissimilarity analysis poses certain problems: to the newcomer trying to find his way around the field, to the practitioner seeking a method well-suited to his particular problem, and even on occasion to the researcher wanting to know how his results fit into the wider knowledge-base of what is already established.

Again, notwithstanding the fundamental contributions that have already been made, there remains considerable scope for the further application of mathematics and, perhaps to a lesser extent, statistics in dissimilarity analysis. Balanced growth in any subject depends on a constructive interplay between its theoretical and practical aspects. In dissimilarity analysis, there is an abundance of empirical wisdom and heuristic algorithms but, at the same time, there are some glaring gaps in the theory. For example, the basic question of the existence of a unique solution remains open for a number of methods, not enough is yet known about the precise relationships between kindred methods, and there is a severe shortage of probability models for dissimilarity data which meet the often conflicting demands of realism and tractability.

For both these reasons we advocate the introduction of a systematic framework, such as that outlined above. This has two principal advantages. It places an order on the subject and encourages the application of mathematics and, where appropriate, statistics to it. This project has yet to be fully attempted. However, a start has been made in Critchley, Marriott and Salmon (1992) while the present paper offers some further contributions as we detail below. We underline two obvious disclaimers straight away. The above is not the only framework possible, nor is it necessarily the best for any particular question. And it is quite impossible for any treatment to be exhaustive.

2.1.2. What's in this chapter?

In this section we outline the contents of the present chapter and indicate how it contributes to the wider framework described above. This chapter is based upon Critchley and Fichet (1993).

We attempt here a systematic account of a wide variety of classes of representation of a single data dissimilarity on a finite set of objects. There are two equivalent ways to define many of these classes. We may either define a class as those dissimilarities admitting a certain visual form, and then characterise which dissimilarities fall in this class. Or, we may define the class as all those dissimilarities obeying certain conditions, and then deduce that these are precisely the dissimilarities admitting a certain visual form.

For each class we examine its principal topological and geometric properties. This is helpful from the point of view of approximating a given data dissimilarity by a member of the class. For example, the idea of minimising the distance from the data to a closed convex set of possible representations, or more generally of minimising a strictly convex function over this set, offers a unifying theme for a wide variety of methods in dissimilarity analysis, (see, for example, Critchley (1980)). At the same time, a single theorem guarantees the existence of a unique solution to any such problem. Moreover this solution can be characterised, a fact which can help both to determine its properties and to compute it in practice. A particularly nice example of this is minimising the Euclidean distance to a polyhedral closed convex cone. The unique solution corresponds to orthogonal projection onto the cone and, although it will not always be the fastest for any particular problem, there is a general algorithm which is guaranteed to converge in a finite number of iterations, (Fraser and Massam, 1989). In contrast, when least-squares projection methods are used for non-convex classes, such as the ultrametrics on more than two points, catastrophes in the formal sense can occur. That is, the solution set can be highly discontinuous as a function of the input data.

We also present the complete partial order by inclusion of all the classes of representation studied. These inclusions are of some interest in their own right. Moreover, they identify some methods of dissimilarity analysis as special cases of others. And, of course, they save time in the sense that, for example, a result established for all members of a class need not be separately established for any of its sub-classes.

The organisation of this paper is as follows. Section 2.2 deals with some necessary preliminaries. Section 2.3 discusses the principal types of dissimilarity, their characterisations and visual forms. The main geometric and topological properties of the sets of all dissimilarities of a given type are also given in Section 2.3. Inclusions between these sets are established in Section 2.4. The convex hulls of those sets not already convex are given in Section 2.5. In Section 2.6 it is shown that for each pairwise inclusion there is a smallest value of the cardinality n of I for which inclusion is strict. Each of these values is then given. Section 2.7 establishes that the inclusions presented are exhaustive: no other inclusions hold for all n. Some complements and extensions are noted in the final section.

Finally, we remark that this paper draws on a seminar by the first author, given initially at Université Paul Sabatier, Toulouse (April, 1986), and later developed on a number of occasions including the European Meeting of Statisticians in Thessaloniki (August, 1987) and the International Statistical Institute session held in Paris (August, 1989), and equally on the paper Fichet (1986) presented by the second author at the first world congress of the Bernoulli Society. Both authors have profited greatly from discussions with other members of the French CNRS coordinated research group "Analyse des Données et Informatique" and, more recently, with colleagues in the wider European network of laboratories "Metric structures for dissimilarity analysis". Members of these groups have been invited to give review talks on dissimilarity analysis on a number of occasions including a seminar at AFCET-B.U.R.O., Université Paris VI (1988) by G. Le Calvé, B. Van Cutsem and the second author, and at the international meeting DISTANCIA'92 on distance analysis held in Rennes (B. Van Cutsem and the first author).

2.2. Preliminaries

In this section we establish some basic notations, definitions and elementary results that will be used in the sequel.

2.2.1. The vector space $\mathcal{D}$ generated by the dissimilarities

Let I be a finite nonempty set with cardinality n. Let $\mathcal{D}(n)$ or simply $\mathcal{D}$ denote the set of all functions $d : I^2 \to \mathbb{R}$ that are symmetric ($d(i,j) = d(j,i)$) and vanish on the diagonal ($d(i,i) = 0$). Clearly, under the usual definitions of scalar multiplication and of addition of functions, $\mathcal{D}$ is a real vector space of dimension $\frac{1}{2}n(n-1)$.

Let d_0 denote the zero function ($d_0(i,j) = 0$) and, when $n > 1$, let $d^{\{i,j\}}$ ($i \neq j$) be defined by:

$$\forall (k,l) \in I^2, d^{\{i,j\}}(k,l) = \begin{cases} 1 & \text{if } \{k,l\} = \{i,j\} \\ 0 & \text{else} \end{cases}$$

Clearly, if $n = 1$, $\mathcal{D} = \{d_0\}$ while if $n > 1$, the $\frac{1}{2}n(n-1)$ functions $d^{\{i,j\}}$ form a canonical basis of $\mathcal{D}$ in which every $d \in \mathcal{D}$ can be expressed as:

$$d = \sum_{\{i,j\} \subseteq I} d(i,j) d^{\{i,j\}}$$

A member d of $\mathcal{D}$ is called a *dissimilarity* (on I) if it is nonnegative ($d(i,j) \geq 0$). Thus, the set $\mathcal{D}_+(n)$ or simply $\mathcal{D}_+$ of all dissimilarities is just the nonnegative orthant of $\mathcal{D}$ in the canonical basis. A *dissimilarity space* is just an ordered pair (I,d) with I as above and with $d \in \mathcal{D}_+$. A *subspace* of a dissimilarity space (I,d) is an ordered pair $(J, d|_J)$ where J is a nonempty subset of I and $d|_J$ is the restriction of d to J^2. Clearly such a subspace is a dissimilarity space in its own right.

Since $\mathcal{D}$ is of finite dimension, all norms on it are equivalent. The set $\mathcal{D}_+$, and any of its subsets introduced below, inherit their topological and geometric properties from $\mathcal{D}$ in the usual way.

Recall that:

(a) A subset K of $\mathcal{D}$ is called a *cone* if it is closed under nonnegative scalar multiplication. That is, if $\lambda K = K$ for all $\lambda \geq 0$.

(b) A nonempty cone $K \subseteq \mathcal{D}$ is called *pointed* if it contains no lines. That is, if $K \cap (-K) = \{d_0\}$.

(c) For any non zero $d \in \mathcal{D}$, $R(d) \equiv \{\lambda d \,|\, \lambda \geq 0\}$ is called the *ray* generated by d.

(d) A *face* of a convex set $C \subseteq \mathcal{D}$ is a convex subset C' of C such that every closed line segment in C with a relative interior point in C' has both endpoints in C'.

(e) An *extreme ray* of a convex cone $K \subseteq \mathcal{D}$ is a ray which is a face. That is, a ray of the form $R(d)$, d non zero, with

   (i) $R(d) \subseteq K$

and

   (ii) $\forall\, 0 < \lambda < 1,\ \forall\, k \in K,\ \forall\, k' \in K$,
   $$d = \lambda k + (1 - \lambda)\, k' \Rightarrow \{k, k'\} \subseteq R(d).$$

(f) A subset $P \subseteq \mathcal{D}$ is called *polyhedral* if it is the intersection of a finite number of closed half-spaces defined by affine hyperplanes. Thus, a polyhedron is necessarily closed and convex. Clearly, it is a cone if all the hyperplanes defining it pass through the origin.

(g) The convex hull of a set $S \subseteq \mathcal{D}$, denoted $\mathrm{conv}(S)$, is the intersection of all the convex sets in $\mathcal{D}$ containing S. Thus, $\mathrm{conv}(S)$ is the smallest convex set containing S.

(h) A convex cone $K \subseteq \mathcal{D}$ is polyhedral if and only if it is the convex hull of a finite number of rays. This intuitive but non-trivial result is proved, for example, in Rockafellar (1970, Theorem 19.1).

Clearly, $\mathcal{D}_+$ is a pointed cone in $\mathcal{D}$. The conic nature of $\mathcal{D}_+$ reflects the fact that there is no natural scale for dissimilarities (cf. the choice between miles and kilometres!). For essentially this same reason, every class of dissimilarities with which we shall be concerned is also a (necessarily pointed) cone.

When $n > 1$, we define $d_1 \in \mathcal{D}_+$ by $d_1(i,j) = 1$ whenever $i \neq j$. Thus d_1 corresponds to the discrete metric or, equivalently, to the unit regular simplex. Accordingly, we denote $\{\lambda d_1 \,|\, \lambda \geq 0\}$ by $\mathcal{D}_{rs}$. When $n = 1$, $\mathcal{D}_+ = \mathcal{D} = \{d_0\}$. When $n = 2$, $\mathcal{D}_+$ is just $\mathcal{D}_{rs}$, the ray generated by d_1. Thus primary interest centres upon the nontrivial cases where $n \geq 3$.

2.2.2. Some elementary dissimilarities

We introduce here the following elementary dissimilarities which play a fundamental role in parts of the sequel. Consider any equivalence relation on I. Equivalently, consider any partition of I into a certain number k ($1 \leq k \leq n$) of disjoint nonempty sets $J_1, \ldots, J_k$ which cover I. We define the corresponding *partition dissimilarity* $d_{J_1, \ldots, J_k}$ by:

$$\forall (i,j) \in I^2, d_{J_1, \ldots, J_k}(i,j) = \begin{cases} 1 & \text{if } k_i \neq k_j \\ 0 & \text{if } k_i = k_j \end{cases}$$

where $i \in J_{k_i}$ and $j \in J_{k_j}$. Of course, d_0 is the unique partition dissimilarity with $k = 1$. When $n > 1$, there is a unique partition dissimilarity with $k = n$, namely d_1.

A partition dissimilarity with $k = 2$ is called a *dichotomy*. We abbreviate d_{J,J^c} to d_J. Clearly, $d_J = d_{J^c}$. There are clearly $2^{n-1} - 1$ distinct dichotomies, and the cone which they generate is denoted $\mathcal{D}_{di}$. When $n > 2$, the *2-dichotomies* play an important role. These are the dichotomies in which at least one member of the partition contains exactly two elements which are said to generate it. The 2-dichotomy generated by $\{i,j\}$ is denoted $d_{\{i,j\}}$. For any $n > 4$, there are $\frac{1}{2}n(n-1)$ 2-dichotomies which form a basis of $\mathcal{D}$, alternative to the canonical basis. We have the following result (Fichet, (1987)).

Proposition 2.2.1. *Suppose $n > 4$. Then the 2-dichotomies form a basis of $\mathcal{D}$. In this basis any $d \in \mathcal{D}$ can be expressed as*

$$d = \sum_{\{i,j\} \subset I} \alpha(i,j) d_{\{i,j\}}$$

where

$$\alpha(i,j) = -\frac{1}{2}\left[d(i,j) - \frac{1}{n-4}\{d(i,\cdot) + d(j,\cdot)\} + \frac{1}{(n-2)(n-4)}d(\cdot,\cdot)\right]$$

and a dot denotes summation over an omitted index.

Proof. For any $\alpha \in \mathcal{D}$, define $d \in \mathcal{D}$ via $d = \sum \alpha(i,j)d_{\{i,j\}}$ where the summation is over all unordered pairs $\{i,j\}$ of distinct elements of I. Then:

$$\forall i \neq j, d(i,j) = \alpha(i,\cdot) + \alpha(j,\cdot) - 2\alpha(i,j).$$

Summing for fixed i over all $j \neq i$ gives:

$$d(i,\cdot) = \alpha(\cdot,\cdot) + (n-4)\alpha(i,\cdot).$$

Summing now over all i gives:

$$d(\cdot,\cdot) = 2(n-2)\alpha(\cdot,\cdot).$$

Thus, $d = d_0 \Rightarrow \alpha = d_0$ and so the 2-dichotomies are linearly independent. By dimensionality considerations, they therefore form a basis of $\mathcal{D}$. The explicit form of the coordinates $\alpha(i,j)$ of $d \in \mathcal{D}$ in this basis is now immediate. ∎

Although we do not use them here, it may be of interest to study other forms of partition dissimilarity, for example those with $2 < k < n$.

2.2.3. Some general types of dissimilarity

We introduce here three general types of dissimilarity. A dissimilarity d on I is said to be:

(a) *definite* (in French, proper) if $d(i,j) = 0 \Rightarrow i = j$,

(b) *even* (in French, semi-proper) if $d(i,j) = 0 \Rightarrow \forall k \in I, d(i,k) = d(j,k)$,

(c) a *semi-distance* if $d(i,j) \leq d(i,k) + d(j,k)$.

The following implications are obvious:

$$d \text{ definite} \Rightarrow d \text{ even} \Leftarrow d \text{ is a semi-distance}.$$

Trivially, a dissimilarity space (I,d) is a semi-metric space if and only if d is a semi-distance, in which case (I,d) is a metric space if and only if d is also definite. The subset of $\mathcal{D}_+$ comprising all dissimilarities that are definite (respectively even or semi-distances) is denoted $\mathcal{D}_+^+$ (respectively, $\mathcal{D}_{ev}$ or $\mathcal{D}_\infty$). The reason for this last choice of notation will be made clear later (see Section 2.2.7). The intersection with $\mathcal{D}_+^+$ of any subset $\mathcal{D}_*$ of $\mathcal{D}_+$ is denoted $\mathcal{D}_*^+$. Here we discuss the basic geometric and topological properties of these sets.

Proposition 2.2.2.

(a) $\mathcal{D}_+$ is a pointed polyhedral closed convex cone, whose extreme rays are those generated by the canonical dissimilarities $d^{\{i,j\}}$.

(b) $\mathcal{D}_+^+$ is just the interior of $\mathcal{D}_+$, being the positive orthant of $\mathcal{D}$ in the canonical basis. It is therefore open and convex and closed under positive scalar multiplication. It is not in general a cone, as $d_0 \notin \mathcal{D}_+^+$ unless $n = 1$.

(c) $\mathcal{D}_{ev}$ is a pointed convex cone. If $n = 1$ or 2, $\mathcal{D}_{ev} = \mathcal{D}_+$ and so, in particular, $\mathcal{D}_{ev}$ is closed. If $n \geq 3$, it is neither open nor closed. Its extreme rays include those generated by the dichotomies.

(d) $\mathcal{D}_\infty$ is a pointed polyhedral closed convex cone, whose extreme rays include those generated by the dichotomies.

(For this and further information on the extreme rays of $\mathcal{D}_\infty$, see Avis (1980)).

Proof. Immediate, except for the following. As $d_0 \in \mathcal{D}_{ev}$, the set $\mathcal{D}_{ev}$ is never open. For $n \geq 3$, it is not closed. To see this consider the following dissimilarity $d[\varepsilon]$ which is even for all $\varepsilon > 0$, but not for $\varepsilon = 0$:

$$d(i,j) = \varepsilon, d(i,k) = 2, d(j,k) = 1 \text{ and, if } n \geq 4, \text{ all other } d(l,m) = 1 \, (l \neq m).$$

Clearly, $\mathcal{D}_{di} \subseteq \mathcal{D}_{ev}$. Let d_J be any dichotomy. Suppose that

$$d_J = \lambda d + (1-\lambda) d' \text{ for some } 0 < \lambda < 1 \text{ and for some } d, d' \in \mathcal{D}_{ev}.$$

Now d_J vanishes on $J \times J$. So, therefore, do d and d'. Similarly, d and d' vanish on $J^c \times J^c$. But d is even. Thus, d is constant on $J \times J^c$. Similarly, so is d'. Thus d and d' are both proportional to d_J. Hence, d_J is an extreme ray of $\mathcal{D}_{ev}$. A fortiori, it is an extreme ray of $\mathcal{D}_\infty$, as $\mathcal{D}_\infty \subseteq \mathcal{D}_{ev}$. ∎

2.2.4. Stability under increasing transformations

A dissimilarity d' is called an increasing transformation of a dissimilarity d if both tied values and strict inequalities in d are preserved in d'. That is, if $\forall (i,j) \in I^2$, $\forall (k,l) \in I^2$,

$$\left[d(i,j) = d(k,l) \Rightarrow d'(i,j) = d'(k,l) \right] \text{ and } \left[d(i,j) < d(k,l) \Rightarrow d'(i,j) < d'(k,l) \right].$$

Equivalently, if $\forall (i,j) \in I^2$, $\forall (k,l) \in I^2$,

$$d(i,j) \leq d(k,l) \Leftrightarrow d'(i,j) \leq d'(k,l).$$

In this case, we write $d' \overset{\angle}{\sim} d$. Clearly, $\overset{\angle}{\sim}$ is an equivalence relation on $\mathcal{D}_+$. For a discussion of this relation in terms of preordonnances, and of other possible definitions of nondecreasing transformations between dissimilarities, see Shepard (1962), Kruskal (1964), Guttman (1968), Lerman (1970), Benzécri (1973), Fichet (1983), Fichet and Gaud (1987), and Critchley (1986a, Section 2), (noting that the term preordonnance had not yet been introduced in the first three of these papers).

A nonempty subset $\mathcal{D}_*$ of $\mathcal{D}_+$ is said to be stable under increasing transformations if

$$\left(d \in \mathcal{D}_* \text{ and } d' \overset{\angle}{\sim} d \right) \Rightarrow d' \in \mathcal{D}_*.$$

It is convenient to anticipate Section 2.3.5 slightly by introducing here the set $\mathcal{D}_u$ of all those dissimilarities which obey the ultrametric inequality:

$$d(i,j) \leq \max\{d(i,k), d(j,k)\}.$$

Note that on $\mathcal{D}_+$ this inequality is indeed stronger than the metric or triangle inequality, as the name suggests, so that $\mathcal{D}_u \subseteq \mathcal{D}_\infty$. For future reference, for any $d \in \mathcal{D}_+$ and for any $\alpha > 0$, let $d^{(\alpha)}$ be the dissimilarity defined by $d^{(\alpha)}(i,j) = (d(i,j))^\alpha$ and, for any subset $\mathcal{D}_*$ of $\mathcal{D}_+$, let $\mathcal{D}_*^{(\alpha)} = \{d^{(\alpha)} | d \in \mathcal{D}_*\}$. Where confusion will not arise we sometimes omit the parentheses around α.

Proposition 2.2.3.

(a) The sets $\mathcal{D}_+$, $\mathcal{D}_+^+$, $\mathcal{D}_{ev}$, $\mathcal{D}_{rs}$, $\mathcal{D}_{di}$ and $\mathcal{D}_u$ are stable under increasing transformations.

(b) If $n = 1$ or 2, $\mathcal{D}_\infty = \mathcal{D}_{ev} = \mathcal{D}_+$. Otherwise, $\mathcal{D}_\infty$ is not stable under increasing transformations.

Proof. The proofs of (a) and of the $n \leq 2$ part of (b) are obvious. Consider then $\mathcal{D}_\infty$ when $n > 2$. For $n = 3$, let $d(i,j) = d(j,k) = 1$ and $d(i,k) = 2$. Then $d^{(2)}$ is an increasing transformation of d. Essentially the same example can be used for any $n \geq 3$ by replicating one of the points as often as necessary. ∎

Proposition 2.2.4. Let $\mathcal{D}_*$ be any nonempty subset of $\mathcal{D}_\infty$. Then:

$$\mathcal{D}_* \text{ is stable under increasing transformations} \Rightarrow \mathcal{D}_* \subseteq \mathcal{D}_u.$$

Proof. It suffices to show that any $d \in \mathcal{D}_*$ satisfies the ultrametric inequality. Let $(i, j, k) \in I^3$. Relabelling as necessary, we may suppose that:

$$d(i,j) \leq d(i,k) \leq d(j,k)$$

and it suffices now to establish that $d(i,k) = d(j,k)$. If $d(j,k) = 0$, this is immediate. Suppose now that $d(j,k) > 0$. Now $\forall \alpha > 0$, $d^{(\alpha)} \in \mathcal{D}_*$ and hence $d^{(\alpha)} \in \mathcal{D}_\infty$. Thus:

$$\forall \alpha > 0, (d(j,k))^\alpha \leq (d(i,j))^\alpha + (d(i,k))^\alpha.$$

Considering arbitrarily large values of α, this is clearly impossible unless $d(i,k) = d(j,k)$. ∎

2.2.5. The quotient space of an even dissimilarity

Let (I, d) be a dissimilarity space in which d is even. Then the binary relation $\overset{d}{\sim}$, or simply $\sim$, on I defined by

$$i \overset{d}{\sim} j \text{ if } d(i,j) = 0$$

is clearly an equivalence relation. The *quotient space* of (I, d) by $\sim$ is the dissimilarity space $(\tilde{I}, \tilde{d})$ where $\tilde{I}$ denotes the set $I/\sim$ of all $\sim$ equivalence classes of I and the dissimilarity $\tilde{d}$ on $\tilde{I}$ is (well-)defined by

$$\forall (\tilde{i}, \tilde{j}) \in \tilde{I}^2, \tilde{d}(\tilde{i}, \tilde{j}) = d(i,j)$$

where $\tilde{i}$ denotes the $\sim$ equivalence class to which i belongs. The *quotient map* is the map $q: I \to \tilde{I}$ which sends i to $\tilde{i}$.

Note that:

(a) $\overset{d}{\sim}$ is the identity relation $\Leftrightarrow d$ is definite.

And that:

(b) $\tilde{d}$ is definite, by definition.

2.2.6. Embeddability in a metric space

A dissimilarity space (I, d) is said to be (*isometrically*) *embeddable* in a metric space (X, ρ) if $\exists$ a function $\phi : I \to X$ such that:
$$\forall (i, j) \in I^2, \rho(\phi(i), \phi(j)) = d(i, j).$$

We call ϕ an (*isometric*) *embedding* of I in X and write $(I, d) \stackrel{\phi}{\hookrightarrow} (X, \rho)$, or simply $(I, d) \hookrightarrow (X, \rho)$. Frequently, we denote $\phi(i)$ by x_i.

The following result is now immediate.

Proposition 2.2.5. *Let (I, d) be a dissimilarity space. Then:*

(a) $\exists$ *a metric space (X, ρ) in which (I, d) can be embedded*
 $\Leftrightarrow d$ *is a semi-distance*
 $\Leftrightarrow \tilde{d}$ *is a distance*
 $\Leftrightarrow (I, d) \stackrel{q}{\hookrightarrow} (\tilde{I}, \tilde{d})$.

(b) $(I, d) \stackrel{\phi}{\hookrightarrow} (X, \rho) \Rightarrow \forall i \in I, \phi^{\leftarrow}(\{\phi(i)\}) = \tilde{i}$.
 In words, the subset of I that is isometrically embedded at $\phi(i)$ is precisely the $\sim$ equivalence class of i.

(c) *If (I, d) is embeddable in a metric space (X, ρ), then so is each of its subspaces. Consequently,*
$$(I, d) \stackrel{\phi}{\hookrightarrow} (X, \rho) \Rightarrow \left(\tilde{I}, \tilde{d}\right) \stackrel{\tilde{\phi}}{\hookrightarrow} (X, \rho)$$
where $\tilde{\phi}$ is (well-)defined by $\tilde{\phi}(\tilde{i}) = \phi(i)$.

Most of the particular types of dissimilarity d with which we shall be concerned are either defined or, equivalently, characterised by the property that (I, d) can be embedded in a particular type of metric space (X, ρ). We call a dissimilarity d "of T type" if (I, d) can be embedded in a metric space of T type. For example, $\mathcal{D}_u$ defined above comprises all the dissimilarities of ultrametric type. By virtue of Proposition 2.2.5 (c), if d is of T type, so is $\tilde{d}$.

Reflecting the fact that we are particularly interested in embeddings which admit a visual form, there are two general prototypes for (X, ρ) in what follows:

(a) a finite dimensional real vector space endowed with a distance geometry induced by (say) a Minkowski p-norm, or a certain inner-product.

And

(b) the vertex set of a positively valued, undirected, connected graph (often, tree) in which the distance between two vertices is the length of the shortest chain (path) joining them.

Now a dissimilarity d defined in terms of embeddability in a metric space is in general only a semi-distance, being a distance if and only if any such embedding is injective.

That is, if and only if d is definite. This arises since different points of I are permitted to be embedded at the same point of X. There are good theoretical and practical reasons why we allow this. Theoretically, in order to optimally approximate a given data dissimilarity by another dissimilarity d of prescribed type, it may well be necessary for d not to be definite. This reflects the fact that extrema of functions are often attained at the boundary of sets. Equally, in practice, non-definite data dissimilarities can easily arise. For example, with reference to dissimilarity (2.1) of Section 2.1.1, this happens when i and j turn out to be members of the same zoological species. Moreover, in such a case, we would decidedly want $\phi(i)$ to equal $\phi(j)$!

At the risk of some possible slight confusion, it is often convenient to abbreviate "a dissimilarity of T type" to "a T dissimilarity", or even to just "a T"! For example, we often speak of an ultrametric dissimilarity, or even just an ultrametric. The possible confusion arises in that an ultrametric in this abbreviated sense is an ultrametric in the formal mathematical sense if and only if it is definite. Again, we spoke in the introduction of dissimilarities being represented by Euclidean distances or by tree distances, when we should formally have spoken of Euclidean or tree *semi*-distances. However, in any particular instance, the context will make it clear what precisely is meant.

2.2.7. The set of all semi-distances

In this section, we explain and justify the choice of notation $\mathcal{D}_\infty$ used above for the set of all semi-distances on I. For each $1 \leq p < \infty$, $\mathcal{D}_p$ denotes the set of all dissimilarities embeddable in a finite dimensional real vector space endowed with the Minkowski p-norm. In other words, a dissimilarity d belongs to $\mathcal{D}_p$ if and only if the following holds:

$\exists$ a positive integer N and, $\forall i \in I, \exists x_i \equiv (x_{i1}, \ldots, x_{ik}, \ldots, x_{iN}) \in \mathbb{R}^N$ such that:

$$\forall (i,j) \in I^2, d(i,j) = \left\{\sum_{k=1}^{N} |x_{ik} - x_{jk}|^p\right\}^{1/p}.$$

We say d is of L_p-type. Letting p tend to infinity, we obtain the set $\mathcal{D}_{(\infty)}$ of all dissimilarities embeddable in ℓ_∞^N for some N. That is $d \in \mathcal{D}_{(\infty)}$ if and only if:

$\exists$ a positive integer N and, $\forall i \in I, \exists x_i \equiv (x_{i1}, \ldots, x_{ik}, \ldots, x_{iN}) \in \mathbb{R}^N$ such that:

$$\forall (i,j) \in I^2, d(i,j) = \max_{1 \leq k \leq N} \left\{|x_{ik} - x_{jk}|\right\}.$$

We have the following remarkable result essentially due to Frechet (1910) as described by Schoenberg (1938):

Proposition 2.2.6. *Every semi-metric space (I, d) with $n > 1$ is embeddable in ℓ_∞^{n-1}.*

Corollary 2.2.1. *The set $\mathcal{D}_{(\infty)}$ defined above is precisely the set $\mathcal{D}_\infty$ of all semi-distances on I.*

Proof of Proposition 2.2.6. It suffices to take $N = n-1$ and $x_{ik} = d(i,k)$ for $i = 1,\ldots,n$ and $k = 1,\ldots,N$ noting that, as d is a semi-distance:

$$\forall (i,j) \in I^2, i \neq j, \max_{1 \leq k \leq N}\{|d(i,k) - d(j,k)|\}$$

is $d(i,j)$, attained when $k = i$ or j. ■

Proof of Corollary 2.2.1

Clearly, $\mathcal{D}_{(\infty)} \subseteq \mathcal{D}_\infty$. The reverse inclusion holds by Proposition 2.2.6. ■
Because of these results we drop the brackets on $\mathcal{D}_{(\infty)}$ and speak unambiguously of $\mathcal{D}_\infty$.

2.2.8. Stability under replication

A dissimilarity space (I', d') is said to be *obtained by replication* from a dissimilarity space (I, d) if there exists a surjection $\psi : I' \to I$ such that

$$d'(i', j') = d(\psi(i'), \psi(j')).$$

Clearly, $i' \stackrel{\psi}{=} j'$ if $\psi(i') = \psi(j')$ defines an equivalence relation on I' each of whose equivalence classes in the pre-image under ψ of a different singleton subset of I. We say $\psi^\leftarrow(\{i\})$ comprises the set of *replicates* of i. The following result is immediate.

Proposition 2.2.7. *Suppose d is even. Then, in the above notation:*

(a) $d'(i', j') = 0 \Leftrightarrow (q \circ \psi)(i') = (q \circ \psi)(j')$

where, we recall, $q : I \to \tilde{I}$ is the quotient map.

(b) *In particular, d' is even and the quotient space of (I', d') by $\stackrel{d'}{\sim}$ is naturally isomorphic to that of (I, d) by $\stackrel{d}{\sim}$ via:*

$$\tilde{i}' \leftrightarrow \tilde{i}$$

where $i = \psi(i')$.

(c) $(I, d) \stackrel{\phi}{\hookrightarrow} (X, \rho) \Rightarrow (I', d') \stackrel{\phi \circ \psi}{\hookrightarrow} (X, \rho).$

For any positive integer n_1, a sequence $\{\mathcal{D}_*(n)\}_{n=n_1}^\infty$ with each $\mathcal{D}_*(n) \subseteq \mathcal{D}_+(n)$ is said to be *stable under replication* if $\forall n \geq n_1$ and $\forall d \in \mathcal{D}_*(n)$:

for every (I', d') obtained by replication from (I, d), $d' \in \mathcal{D}_*(|I'|)$.

Clearly, any sequence in which $\mathcal{D}_*(n)$ corresponds to embeddability in a certain type of metric space is stable under replication. Such sequences will be used later in connection with the following result.

Proposition 2.2.8. *Let $\{\mathcal{D}_*(n)\}_{n=n_1}^\infty$ and $\{\mathcal{D}_{**}(n)\}_{n=n_1}^\infty$ be two sequences of sets of dissimilarities, the second of which is stable under replication. Suppose also that $\forall n \geq n_1, \mathcal{D}_*(n) \subseteq \mathcal{D}_{**}(n)$.*

Then, either (a) the two sequences are identical, or (b) $\exists n_0 \geq n_1$ such that:

$$\mathcal{D}_*(n) = \mathcal{D}_{**}(n) \text{ if } n_1 \leq n < n_0$$
$$\text{while } \mathcal{D}_*(n) \subset \mathcal{D}_{**}(n) \text{ if } n \geq n_0.$$

Proof. If (a) does not hold, $\exists n \geq n_1$ such that $\mathcal{D}_*(n) \subset \mathcal{D}_{**}(n)$. Let n_0 be the minimum of all such n. It suffices to note that if $d \in \mathcal{D}_{**}(n_0)\backslash\mathcal{D}_*(n_0)$ then, replicating as necessary, $d \in \mathcal{D}_{**}(n)\backslash\mathcal{D}_*(n)$ for all $n \geq n_0$. ∎

2.3. The general structures of dissimilarity data analysis and their geometrical and topological nature

We study here the main structures occurring in dissimilarity data analysis and some useful extensions. We begin with the structures of multidimensional scaling and continue with the structures of classification. Most of these structures correspond to cones of $\mathcal{D}$ and most are included in the cone $\mathcal{D}_\infty$ of semi-distances.

2.3.1. Euclidean semi-distances

Definition 2.3.1. A dissimilarity d on I is said to be Euclidean iff there exists some finite-dimensional Euclidean space E and an embedding $\{x_i, i \in I\}$ of I in E such that:

$$\forall (i,j) \in I^2, \|x_i - x_j\| = d(i,j). \qquad \square$$

It is obvious, choosing an orthonormal basis of E, that this definition coincides with that of a semi-distance of L_2-type given in Section 2.2.7.

Then we can give the following obvious remarks:

- a Euclidean dissimilarity is necessarily a semi-distance (we will speak about Euclidean semi-distances).

- The set, denoted by $\mathcal{D}_e$, of Euclidean dissimilarities, is a cone.

- The dimensionality of any Euclidean embedding is constant and at most $(n-1)$. It will be called the dimension of (I, d) or of d.

Thus $\mathcal{D}_e$ appears as the increasing union of the cones $\mathcal{D}_{e(k)}$ $(k = 0, \ldots, (n-1))$ of Euclidean semi-distances with dimensionality at most k.

In particular, for $k = 2$ we have the usual visual display in which, we note, the choice of origin and of axes are arbitrary. For example, the alternative embeddings of $I = \{i, j, k, l\}$ shown in Figures 2.3.1 and 2.3.2 viewed as Euclidean planes are clearly visual displays of the same dissimilarity.

For $k = 1$, the cone $\mathcal{D}_{e(1)}$ will be also denoted by $\mathcal{D}_{ch}$. Indeed, a 1-dimensional Euclidean semi-distance is equivalent to a complete elementary (simple) chain associated to a weighted tree having only two terminal vertices.

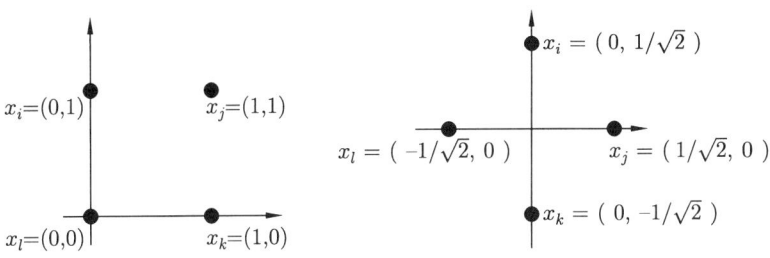

Figure 2.3.1 **Figure 2.3.2**
Alternative embeddings of $I = \{i, j, k, \ell\}$ in $\mathbb{R}^2$

The first general condition for embeddability in a Euclidean space or even in a Hilbert space was established by Schoenberg (1935, 1937) mentioning Menger (1931 a,b); see also Banach and Mazur in the book of Banach (1932) and Frechet (1935). In their footnotes these authors mention Gauss (1831) who noted the condition for a 3-dimensional space, Dirichlet (1850) who proved in greater detail the result, and Hermite (1850) and Minkowski (1891) for the extension to an n-dimensional space. The reader may find an interesting review of these developments in Lew (1978). Later other characterizations have been established, see Schoenberg (1938), Torgerson (1958). Here we give the first historical characterization in a similar form.

First we identify I with $\{1, \ldots, n\}$ and consider for every d of $\mathcal{D}$ the matrix: $W = \mathcal{T}(d)$ defined by:

$$\forall i, j = 1, \ldots, (n-1), W_{ij} = \tfrac{1}{2}[d(n,i) + d(n,j) - d(i,j)]$$

The mapping $\mathcal{T}$ from $\mathcal{D}$ onto the set $\mathcal{S}_{n-1}$ of symmetric matrices is a linear isomorphism since:

$\forall i = 1, \ldots, n, \quad d(n,i) = W_{ii},$
$\forall i, j = 1, \ldots, (n-1), \quad d(i,j) = W_{ii} + W_{jj} - 2W_{ij}.$

We have

Proposition 2.3.1. *For $d \in \mathcal{D}_+$, $d \in \mathcal{D}_e$ iff $\mathcal{T}(d^2)$ is p.s.d.*

When d is Euclidean, the dimension of (I, d) is the rank of $\mathcal{T}(d^2)$.

Proof. Let $W = \mathcal{T}(d^2)$. Suppose $d \in \mathcal{D}_e$. Then W is the scalar product matrix of vectors $(x_n - x_1), \ldots, (x_n - x_{n-1})$, so that W is p.s.d. Moreover, the rank of W is the rank of these vectors, i.e. the dimension of (I, d). Conversely, if W is p.s.d. let us consider an eigenvalue decomposition of W: $W = \sum_{k=1}^{h} \lambda_k y_k y_k^T$ ($h \leq n - 1$). Then, in $\mathbb{R}^h$ with an orthonormal basis, the vectors $x_i, i = 1, \ldots, (n-1)$ with coordinates $\sqrt{\lambda_k} y_k(i)$, $k = 1, \ldots, h$, and the vector x_n put at the origin, satisfy:

$\forall i = 1, \ldots, (n-1), \|x_i - x_n\|^2 = \|x_i\|^2 = W_{ii} = d^2(n,i)$
$\forall i, j = 1, \ldots, (n-1), \|x_i - x_j\|^2 = W_{ii} + W_{jj} - 2 <x_i, x_j> = W_{ii} + W_{jj} - 2W_{ij} = d^2(i,j).$ ∎

The previous result shows the interest in considering the cone $\mathcal{D}_e^2$ of squared Euclidean semi-distances. The isomorphism $\mathcal{T}$ maps $\mathcal{D}_e^2$ onto the cone $\mathcal{S}_{n-1}^+$ of p.s.d. matrices in $\mathcal{S}_{n-1}$. It is well known that $\mathcal{S}_{n-1}^+$ is a closed convex cone and has as interior the cone $\mathcal{S}_{n-1}^{++}$ of p.d. matrices. Thus, we have:

Proposition 2.3.2. $\mathcal{D}_e^2$ *is a closed convex cone, whose interior is the set of squared Euclidean (semi)-distances with maximum dimension.*

As an obvious consequence we have:

Corollary 2.3.1. $\mathcal{D}_e$ *is a closed cone.*

The interior of $\mathcal{D}_e$ has been shown to be the set of Euclidean (semi)-distances with maximum dimension by Fichet (1983) or Fichet and Gaud (1987).

Remark 2.3.1. For $n > 3$, $\mathcal{D}_e$ is never convex, hence strictly included in $\mathcal{D}_\infty$, as is shown in the following counter-example, (Critchley, 1980).

Let $I = \{i, j, k, l\}$. Let d' and d'' be the members of $\mathcal{D}_{e(1)} \equiv \mathcal{D}_{ch}$ corresponding to Figures 2.3.3 (a') and (a'') respectively, so that $d \equiv (d' + d'')$ is given by the following table:

	i	j	k	ℓ
i	0	1	2	5
j		0	3	4
k			0	5
ℓ				0

Note that j, k, l form a right-angled Euclidean triangle. The requirements that $d(i, j) = 1$ and $d(i, k) = 2$ place i at a point P in the interior of the line joining j and k, while $d(i, j) = 1$ and $d(i, l) = 5$ place i at a point Q on the extension of the line from l to j. See Figure 2.3.3 (b). This inconsistency shows that $d \notin \mathcal{D}_e$. Essentially the same example can be used for any $n \geq 4$ by replicating one of the points as often as necessary. □

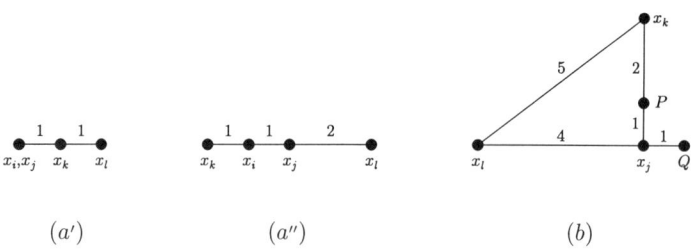

(a')   (a'')   (b)

Figure 2.3.3: **Configurations corresponding to** d', d'' **and** d

The previous counter-example shows that the cones $\mathcal{D}_{e(k)}$, $k = 1, \ldots, (n-1)$, and in particular $\mathcal{D}_{ch}$, are never convex for $n > 3$. For $\mathcal{D}_{ch}$, it is again true for $n = 3$, with an obvious counter-example. For other properties concerning $\mathcal{D}_{e(k)}$ we have:

Proposition 2.3.3. *For $k \leq (n-1)$, $\mathcal{D}_{e(k)}$ is a closed cone.*

Proof. Clearly $\mathcal{D}_{e(k)}$ is a cone. Now, let us consider a sequence $\{d_p\}$ of the cone $\mathcal{D}_{e(k)}$, converging to a limit d. Clearly, $\{d_p^2\} \to d^2$. By the characterization of $\mathcal{D}_e$, the matrices $\mathcal{T}(d_p^2)$ are p.s.d. with rank less than or equal to k. But the eigenvalues of a matrix depend continuously upon its elements (see, for example, Appendix D to Horn and Johnson, (1985)). Thus, $\mathcal{T}(d^2)$ satisfies the same property. ∎

2.3.2. Semi-distances of L_1-type

As a natural extension of Euclidean semi-distances, we may consider the set $\mathcal{D}_p$ of all semi-distances of L_p-type, i.e. such that (I, d) is embeddable in some space $\mathbb{R}^N$ endowed with the ℓ_p-norm, as defined in Section 2.2.7.

Conditions for embeddability of a metric space in an L_p-space have been established by several authors, such as Schoenberg (1938), Bretagnole, Dacunha-Castelle and Krivine (1966). Fichet (1986, 1988) has shown that the set of semi-distances of L_p-type, defined on a finite set I, forms a closed cone.

Now, we pay particular attention to semi-distances of L_1-type. Thus, we recall

Definition 2.3.2.

A dissimilarity d on I is said to be of L_1-type iff there exists an integer N and an embedding $\{x_i, i \in I\}$ of (I, d) in $\mathbb{R}^N$ such that

$$\forall (i,j) \in I^2, \|x_i - x_j\|_1 = d(i,j)$$

(where $\|\cdot\|_1$ stands for the ℓ_1-norm on $\mathbb{R}^N$). □

We will denote by $\mathcal{D}_1$ the set of dissimilarities of L_1-type defined on I.

Clearly, such dissimilarities are semi-distances. They have been advocated under a variety of names such as "(semi)-distances of L_1-type or L_1-embeddable" by Assouad and Deza (1982), "Hamming (semi)-distances" by Avis (1977 or 1981), "addressable (semi)-distances" by Blake and Gilchrist (1973), or more commonly "city-block or Manhattan (semi)-distances". When $N = 2$ we have an obvious visual display. Note that, unlike the Euclidean case, the axes are no longer arbitrary (although, of course, the origin remains so). For example when Figures 2.3.1 and 2.3.2 are viewed as L_1-planes, the two embeddings of I which they represent are no longer isometric.

(Semi-)Metric spaces embeddable in an L_1-space have been investigated in different areas by several authors, such as Schoenberg (1938), Bretagnolle, Dacunha-Castelle and Krivine (1966), Dor (1976), Assouad (1977, 1980(a),(b), 1984), Avis (1977, 1978, 1981), Fichet (1986, 1987), Le Calvé (1987) or Cailliez in unpublished papers.

Let us note that the cone of 1-dimensional semi-distances of L_1-type in nothing but $\mathcal{D}_{ch}$. Moreover, it is obvious that $\mathcal{D}_1 = \text{conv}(\mathcal{D}_{ch})$, the convex hull of $\mathcal{D}_{ch}$.

Now, we give a characterization of $\mathcal{D}_1$, showing the geometrical nature of this cone. This characterization has been observed by Avis (1977) and Cailliez. Here we use the proof of Fichet (1987).

Proposition 2.3.4. *$\mathcal{D}_1$ is a closed convex polyhedral cone. Its extreme rays are the rays spanned by the dichotomies.*

Proof. Clearly, every dichotomy is in $\mathcal{D}_1$. Conversely, since $\mathcal{D}_1 = \text{conv}(\mathcal{D}_{ch})$, we have only to prove that every element d of $\mathcal{D}_{ch}$ is a non-negative linear combination of dichotomies. Identify I with $\{i,\ldots,n\}$ according to an order defined by the chain. Then: $\forall i = 1,\ldots,(n-1)$, $d(i,i+1) = a_i$ (say). Considering the dichotomies $d_{J_i}, i = 1,\ldots,(n-1)$, where $J_i = \{1,\ldots,i\}$, we observe that:

$$\forall j,k = 1,\ldots,n, j < k, d(j,k) = a_j + \ldots + a_{k-1} = \sum_{i=1}^{n-1} a_i d_{J_i}(j,k).$$

Then the proof is complete since the rays generated by the dichotomies are extreme rays of $\mathcal{D}_\infty$, hence of $\mathcal{D}_1$. ∎

The cone $\mathcal{D}_1$ is called the Hamming cone by Avis (1981).

This proposition yields a numerical characterization of $\mathcal{D}_1$ in terms of linear programming techniques, as noted by Avis (1977). Denoting by $\{d_k, k \in K\}$ the family of dichotomies, d is in $\mathcal{D}_1$ iff there exists non-negative coefficients $\alpha_k, k \in K$ such that: $d = \sum_k \alpha_k d_k$. By projection on any basis of $\mathcal{D}$, we have to prove that the domain defined by $n(n-1)/2$ equations with $(2^{n-1}-1)$ nonnegative variables in nonempty. It is well known that by adding new parameters, this is transformed into a linear programming problem. But note the exponential nature of this programme. In (1987), Fichet solves the more general problem of the additive constant. Using the 2-dichotomy basis of $\mathcal{D}$ $(n > 4)$ (see Proposition 2.2.1) one has automatically a feasible solution for the corresponding linear programming problem.

Proposition 2.3.4 yields the following

Corollary 2.3.2. *A semi-distance d is of L_1-type iff there exists a positive integer K, nonnegative real numbers α_k, $k = 1,\ldots,K$ and a family of numbers $\{x_{ik}, i \in I, k = 1,\ldots,K\}$ equalling 0 or 1 such that:*

$$\forall (i,j) \in I^2, d(i,j) = \sum_k \alpha_k |x_{ik} - x_{jk}|$$

Proof. Sufficiency is obvious. Now, let $d \in \mathcal{D}_1$ and, according to Proposition 2.3.4, consider a dichotomy decomposition $d = \sum_k \alpha_k d_{J_k}$, $k = 1,\ldots,K$. Putting $x_{ik} = 1$ (resp. 0) iff $i \in J_k$ (resp. $i \notin J_k$), we get the required equality. ∎

Spaces of L_1-type are connected with spaces of symmetric difference type, studied by Kelly (1970,a). We give the

Definition 2.3.3. A dissimilarity d on I is said to be of symmetric difference type iff there exists a finite weighted set $Y = \{y_1, \ldots, y_K\}$ with nonnegative weights $w_1, \ldots, w_K$ and a family of subsets $\{B_i, i \in I\}$ of Y satisfying:

$$\forall (i,j) \in I^2, d(i,j) = \sum \{w_k | y_k \in B_i \Delta B_j\}.$$
□

A more general definition, valid in the infinite case, will be useful later. It appears through

Proposition 2.3.5. *A dissimilarity d on I is of symmetric difference type iff there exists a measure space $(X, \mathcal{F}, \mu)$, where μ is nonnegative and bounded, and a family of measurable subsets $\{A_i, i \in I\}$ such that:*

$$\forall (i,j) \in I^2, d(i,j) = \mu(A_i \Delta A_j).$$

Proof. Necessity is obvious. Conversely, define Y as the measurable partition of X formed by all nonempty subsets of the type $y = \cap_{i \in I} C_i$ where C_i is equal to A_i or A_i^c. For all $y \in Y$, let $w_y = \mu(y)$. For every $i \in I$ we define B_i to be $\{y \in Y | y \subseteq A_i\}$. Thus for every i and j in I, $y \in B_i \Delta B_j$ iff $y \subseteq A_i \Delta A_j$ and so $d(i,j) = \mu(\cup \{y | y \subseteq A_i \Delta A_j\}) = \sum \{w_y | y \in B_i \Delta B_j\}$. ∎

Spaces of symmetric difference type correspond exactly to spaces of L_1-type, as mentioned by Assouad (1980,a) and Avis (1981) and proved in Avis (1977).

Proposition 2.3.6. $d \in \mathcal{D}_1$ *iff d is of symmetric difference type.*

Proof. Let $d \in \mathcal{D}_1$. Using Corollary 2.3.2, we choose $Y = \{1, \ldots, K\}$ and for every $i \in I$ we define $B_i = \{k \in Y | x_{ik} = 1\}$. Then: $\forall (i,j) \in I^2, d(i,j) = \sum_k \{\alpha_k | k \in B_i \Delta B_j\}$. Conversely, let d be a dissimilarity of symmetric difference type. From Definition 2.3.3, for every $i \in I$ and $k = 1, \ldots, K$, put $x_{ik} = 1$ or 0 according as y_k is in B_i or is not in B_i. Then the equation of Corollary 2.3.2 holds with $\alpha_k = w_k$. ∎

2.3.3. Hypermetric semi-distances

The concepts behind hypermetric spaces were introduced by Deza (Tylkin) (1960) and rediscovered independently by Kelly (1970,a) who gave to these spaces their name. Since these dates, they have been investigated by several authors such as Deza (1962, 1969), Kelly (1970,a, 1975), Assouad (1980,a) and Avis (1981).

Here we define a hypermetric dissimilarity following the definition given by Kelly (1970,a) for a hypermetric space.

Definition 2.3.4. A dissimilarity d on I is said to be k-hypermetric (k a strictly positive integer) iff for every elements $i_1, \ldots, i_k, j_1, \ldots, j_{k+1}$ in I, we have:

$$\sum_{\ell=1}^{k} \sum_{m=1}^{(k+1)} d(i_\ell, j_m) \geq \sum_{1 \leq \ell < m \leq k} d(i_\ell, i_m) + \sum_{1 \leq \ell < m \leq k+1} d(j_\ell, j_m)$$

(with the convention that the first summation in the right part of the inequality is null when $k = 1$).

A dissimilarity that is k-hypermetric for all k is called hypermetric. □

The definition of k-hypermetricity is clearly equivalent to the following one, very often used by the above-mentioned authors:

Definition 2.3.5. A dissimilarity d on I is k-hypermetric iff for every elements $i_1, \ldots, i_{2k+1}$ in I and for every integers $\lambda_1, \ldots, \lambda_{2k+1}$ equalling -1 or 1 and such that $(\lambda_1 + \ldots + \lambda_{2k+1}) = 1$, we have:

$$\sum_{\ell=1}^{2k+1} \sum_{m=1}^{2k+1} \lambda_\ell \lambda_m d(i_\ell, i_m) \leq 0 \quad (k\text{-hypermetric inequality}).$$

□

The inequality occurring in this definition is called $(2k+1)$-polygonal by Deza and Assouad and $(2k+1)$-gonal by Avis. We will denote by $\mathcal{D}_{k-h}$ (resp. $\mathcal{D}_h$) the set of k-hypermetric (resp. hypermetric) dissimilarities. As proved by Kelly (1970,a), the sequence $\mathcal{D}_{k-h}$ is decreasing in k, i.e. $\mathcal{D}_{k-h} \supseteq \mathcal{D}_{(k+1)-h}$, so that $\mathcal{D}_h = \cap_k \mathcal{D}_{k-h}$. Indeed, let $i_1, \ldots i_{2k+1}$ and $\lambda_1, \ldots, \lambda_{2k+1}$ be as defined in the second definition. Let us choose any element i_{2k+2} (say) in I, and put $i_{2k+3} = i_{2k+2}$, $\lambda_{2k+2} = 1$, $\lambda_{2k+3} = -1$. Then, the $(k+1)$-hypermetric inequality yields the k-hypermetric inequality for the given points and the given coefficients. Kelly (1970,a) shows that the sequence is strictly decreasing while $2k+1 \leq n$. Clearly, the 1-hypermetric inequality corresponds to the triangular inequality, so that $\mathcal{D}_{1-h} = \mathcal{D}_\infty$. Thus, every (k)-hypermetric dissimilarity is a semi-distance, but note that (I, d) is an hypermetric space in the classical sense (Deza, 1960) if and only if d is definite.

The triangular inequality is the first of a series of more specific inequalities, the next, defined by the 2-hypermetric inequality, being the pentagon inequality, see Deza (1960). These specifications seem to have been the main motivation of Deza and Kelly in defining hypermetric spaces. Moreover a lot of metric structures are of hypermetric type. In addition to those which will be presented below, we would like to indicate two important results of Kelly (1970,a):

- 2-dimensional normed vector spaces are hypermetric.

- among all normed (and hence modular) lattices, the distributive ones are those which are 2-hypermetric.

Now, we may detail the geometrical and topological nature of the corresponding sets. We have the obvious:

Proposition 2.3.7.

- For every positive integer k, $\mathcal{D}_{k-h}$ is a closed convex polyhedral cone.

- $\mathcal{D}_h$ is a closed convex cone.

Note that the polyhedral nature of $\mathcal{D}_{k-h}$ is due to the finitude of I; the cone is the intersection of a finite number of half-spaces defined by all k-hypermetric inequalities. In papers to appear, Deza, Grishukhin and Laurent (1990 a,b) and Deza (1992) affirm that $\mathcal{D}_h$ is polyhedral too.

The following proposition gives an equivalent definition of hypermetricity, which is very often taken as the definition, as in Kelly (1972), Assouad (1980,a), Avis (1981) and Assouad and Deza (1982).

Proposition 2.3.8. *A dissimilarity d on I is hypermetric iff for every strictly positive integer k, for every elements $i_1, \ldots, i_k$ in I, for every integers $\lambda_1, \ldots, \lambda_k$ which sum to 1,*

$$\sum_{\ell=1}^{k} \sum_{m=1}^{k} \lambda_\ell \lambda_m d(i_\ell, i_m) \leq 0.$$

Proof. For every positive integer k, it is evident that every element of $\mathcal{D}_{k-h}$ satisfies the inequality of the proposition (for the integer $(2k+1)$).

Conversely, let d satisfy the inequality of the proposition for an integer k. For every ℓ, writing $\lambda_\ell (\lambda_\ell \neq 0)$ as the sum of $|\lambda_\ell|$ integers equalling $+1$ or -1 according to the sign of λ_ℓ, and duplicating $|\lambda_\ell|$ times the point i_ℓ, then d is seen to be in $\mathcal{D}_{k'-h}$ for some integer k'. ∎

In the finite case, we observe that the inequality in the proposition holds for every k, as soon as this inequality holds for every k less than or equal to n. This inequality generalizes for $k = (2k'+1)$ odd the $(2k'+1)$-gonal inequality (or k'-hypermetric inequality). For $k = 3$ we have an equivalence since, as will be shown below, every 3-point metric space is hypermetric. For $k = 5$, the equivalence still holds. Indeed Assouad (1980,a), mentioning Deza (1962) and Kelly (1968), has proved that every 5-point metric space satisfying the 5-gonal inequality, is of L_1-type, hence hypermetric, as will be seen below. Moreover we remark that the inequalities of the proposition for $k = 2$ and $d \in \mathcal{D}$ correspond to the constraints defining a dissimilarity.

Let us end this study on hypermetricity, by noting that to our knowledge, there does not exist any numerical procedure to establish whether or not a dissimilarity is hypermetric. In addition, there does not exist any associated visual representation of a hypermetric.

2.3.4. Quasi-hypermetric dissimilarities

The concept of quasi-hypermetric spaces has been defined by Kelly (1972). These spaces are those who satisfy the inequality of negative type, attributed by Assouad (1980,a), Bretagnolle, Dacunha-Castelle and Krivine (1966) to Schoenberg (1938). According to these notions, we give:

Definition 2.3.6. *A dissimilarity d on I is said to be quasi-hypermetric iff for every strictly positive integer k, for every elements $i_1, \ldots, i_k$ in I and for every real numbers $\lambda_1, \ldots, \lambda_k$ which sum to 0,*

$$\sum_{\ell=1}^{k} \sum_{m=1}^{k} \lambda_\ell \lambda_m d(i_\ell, i_m) \leq 0. \qquad \square$$

Let us note that (I, d) is a quasi-hypermetric space in the Kelly sense if and only if d is definite. In the finite case which is studied here, we remark that the inequality holds for every k whenever this inequality is true for every k less than or equal to n or, sufficiently, for $k = n$.

We will denote by $\mathcal{D}_{qh}$ the set of quasi-hypermetric dissimilarities. From the definition, it is clear that $\mathcal{D}_{qh}$ is a closed convex cone. This property also derives from the following proposition, based on Schoenberg's (1935) results and rediscovered by Kelly (1972). It shows the interest of quasi-hypermetric spaces in the field of data analysis.

Proposition 2.3.9. *The cone $\mathcal{D}_{qh}$ of quasi-hypermetric dissimilarities coincides with $\mathcal{D}_e^2$, the cone of squared Euclidean semi-distances.*

Proof. Identifying I with $\{1, \ldots, n\}$, let us consider the space $\mathbb{R}^n$ furnished with its canonical basis $\{e_1, \ldots, e_n\}$. The system $\{e_1 - e_n, \ldots, e_{n-1} - e_n\}$ forms a basis of the hyperplane H with equation $\lambda_1 + \ldots + \lambda_n = 0$. Let d be a dissimilarity on I. Define the symmetric bilinear form q on $\mathbb{R}^n$ by:

$$\forall (i,j) \in I^2, q(e_i, e_j) = \frac{1}{2} d(i,j).$$

Then we have: $\forall i, j = 1, \ldots, (n-1)$,

$$q(e_i - e_n, e_j - e_n) = \frac{1}{2}[d(i,j) - d(n,i) - d(n,j)].$$

Consequently, in the basis $\{e_1 - e_n, \ldots, e_{n-1} - e_n\}$, the matrix associated to the restriction of q to H, say q_H, is $-\mathcal{T}(d)$ (see Section 2.3.1). Using Proposition 2.3.1, $d \in \mathcal{D}_e^2$ iff $-q_H$ is p.s.d., that is: for every real numbers $\lambda_1, \ldots, \lambda_n$ which sum 0, $\sum_{i=1}^n \sum_{j=1}^n \lambda_i \lambda_j d(i,j) \leq 0$. But clearly, this condition is equivalent to the one given for quasi-hypermetricity. ∎

Remark 2.3.2. Using Proposition 2.3.9, it is easy to construct a dissimilarity in $\mathcal{D}_{qh}$ which is not a semi-distance whenever $n > 2$. However, every element d in $\mathcal{D}_{qh}$ is even. Indeed $\sqrt{d}$ is Euclidean and hence even, so that d is even by Proposition 2.2.3(a). Another easy proof can be obtained directly from Definition 2.3.6. □

Finally we note that, whatever the interest of the notion of quasi-hypermetricity, there does not exist any visual display associated to this structure.

2.3.5. Ultrametric semi-distances

It seems that ultrametric spaces appeared in number theory with M. Krasner (1944). They were amply developed in pure mathematics and some recent applications appeared in physics for modelling the absence of intervening states (spin glasses). Since the sixties they are widely used in taxonomy in connection with hierarchical classification.

Definition 2.3.7. *A dissimilarity d on I is said to be ultrametric iff:*

$$\forall (i,j,k) \in I^3, d(i,j) \leq \max[d(i,k), d(j,k)] \quad \text{(ultrametric inequality)}. \quad \square$$

Clearly an ultrametric dissimilarity d is a semi-distance, and (I, d) is an ultrametric space iff d is a distance. We will speak about ultrametric semi-distances, or more simply ultrametrics. We recall, see Proposition 2.2.3, that any increasing transformation of an ultrametric is also an ultrametric.

Ultrametricity yields a special geometry. In particular, among the three numbers $d(i,j)$, $d(i,k)$ and $d(j,k)$ two are equal and not less than the third. That is, every triangle is isoceles with a narrow base.

Ultrametrics play a fundamental role in clustering through what is called hierarchical classification. Two bijections have been established: a one-to-one correspondence between ultrametrics and dendrograms, Jardine and Sibson (1971), Johnson (1967), and a one-to-one correspondence between ultrametrics and indexed hierarchies, Benzecri (1973). For the link between these bijections and deep extensions, see Critchley and Van Cutsem (1994).

Dendrograms or indexed hierarchies lead to a well-known visual display, by means of a hierarchical tree. For example Figure 2.3.4 is a visual representation of the following dissimilarity:

	i	j	k	ℓ	m
i	0	1	3	3	3
j		0	3	3	3
k			0	2	2
ℓ				0	2
m					0

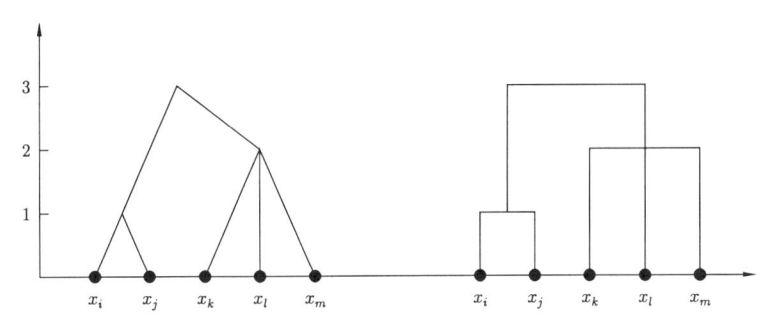

Figure 2.3.4: Two equivalent visual displays
of a simple ultrametric dissimilarity

The tree is oriented from a root at the highest level. The units of I are placed at the terminal vertices, with level 0. Each vertex of the tree corresponds to a class of the hierarchy, which contains the units which derive from it. In this diagram the (semi)-distance between two elements i and j of I is the level of the vertex (the smallest class containing i and j) from which i and j derive directly. Now we give the simple:

Proposition 2.3.10. *The set of ultrametrics, denoted by $\mathcal{D}_u$, is a closed cone.*

Proof. The conic nature is obvious, and if $\{d_p\}$, $p \in N$ is a sequence converging to d, the inequality $d(i,j) > \max[d(i,k), d(j,k)]$ would imply a similar inequality for d_p (p sufficiently large). ■

Remark 2.3.3. For $n > 2$, the cone $\mathcal{D}_u$ is non-convex. For a simple counter-example, define d and d' on $I = \{i, j, k\}$ by:

$$d(i,j) = d(i,k) = 2; \quad d(j,k) = 1$$
$$d'(i,j) = d'(j,k) = 2; \quad d'(i,k) = 1$$

Then $d, d' \in \mathcal{D}_u$; $(d + d') \notin \mathcal{D}_u$. □

2.3.6. Tree semi-distances

In this section, we deal with finite undirected connected graphs $G = (X, E)$. They are considered not to have any loops or multiple edges. Classically, a chain (path) of G is a finite sequence $\{x_1, \ldots, x_n\}$, $(n > 1)$ of X such that:

$$\forall j = 1, \ldots, (n-1), (x_j, x_{j+1}) \in E.$$

By hypothesis, $x_j \neq x_{j+1}$, $j = 1, \ldots, (n-1)$. A chain is elementary (simple) if $x_i = x_j (i \neq j)$ implies $\{i, j\} = \{1, n\}$.

Now, suppose that weights, i.e. strictly positive numbers, are assigned to each edge. We denote by $w(e)$, $e \in E$ these weights, and (G, w) the weighted graph. The length of any chain $C = \{x_1, \ldots, x_n\}$ is by definition additive, i.e.

$$L(C) = \sum \{w(x_j, x_{j+1}), j = 1, \ldots, (n-1)\}.$$

It is well known that the minimum length over all chains linking two different vertices defines a distance ρ on X (by putting $\rho(x,x) = 0, \forall x \in X$). If the weights were chosen to be non negative (this is not the case here), (X, ρ) would be a semi-metric space. Such metric spaces, defined by weighted connected graphs, play an important role in a lot of areas, such as operational research and computer science.

Now, let us consider a dissimilarity space (I, d) and a weighted connected graph (G, w), where $G = (X, E)$. If a mapping $\phi : I \to X$ realises an embedding of (I, d) in (X, ρ), see Section 2.2.6, necessarily d is a semi-distance. Conversely, every semi-metric space (I, d) is embeddable in the metric space defined by some weighted connected graph: choose the complete graph corresponding to the quotient space $(\tilde{I}, \tilde{d})$ with weights $w(\tilde{i}, \tilde{j}) = \tilde{d}(\tilde{i}, \tilde{j}), \tilde{i} \neq \tilde{j}$, and the mapping $q : I \to \tilde{I}$ (Section 2.2.5). For this reason, and for the graph aspect, only semi-metric spaces embeddable in particular graphs are of interest. For example, precise studies, in connection with the L_1-norm, have been explored for the bipartite graphs, Avis (1981), the gated graphs, Wilkeit (1990). The graphs defined by some particular polytopes, such as the cube, the octahedron, the icosahedron and the dodecahedron are hypermetric: see Kelly (1975). However the trees are by far those which have attracted most attention. They appeared

II. PARTIAL ORDER OF THE PRINCIPAL CLASSES OF DISSIMILARITY

for the classification of species and became of great interest for operational research and mathematical psychology, the latter following the work of Tversky.

Recall that a tree is a connected graph without elementary cycle (circuit), i.e. elementary chain $\{x_1, \ldots, x_n\}$ such that $x_1 = x_n$. In the field of data analysis, weighted trees are usually called additive trees ("additive" in connection with the definition of the induced distance) and the corresponding distance is called a tree-distance. For two different vertices, the distance is the length of the unique chain joining them. Now we give:

Definition 2.3.8. A dissimilarity d on I is said to be a tree dissimilarity (or a tree semi-distance) iff there exists a weighted tree (T, w), with $T = (X, E)$, and an embedding $\phi : I \to X$ such that: $\forall (i,j) \in I^2$, $d(i,j) = \rho(x_i, x_j)$ (where: $\forall i \in I, x_i = \phi(i)$ and ρ is the induced distance on (T, w)). □

From the definition, we get a useful visual display. For example Figure 2.3.5 represents the following dissimilarity:

	i	j	k	ℓ	m	n
i	0	2	3	6	7	7
j		0	3	6	7	7
k			0	3	4	4
ℓ				0	7	7
m					0	0
n						0

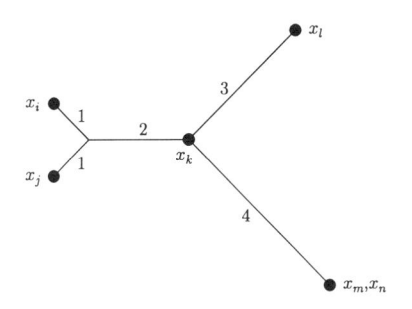

Figure 2.3.5: Visual display of a simple tree dissimilarity

An analytical and numerical characterization, which is very often chosen for defining tree semi-distances, is given by the four-point condition, stated as follows:

$$\forall (i,j,k,\ell) \in I^4, d(i,j) + d(k,\ell) \leq \max\left[d(i,k) + d(j,\ell), d(i,\ell) + d(j,k)\right] \quad (1)$$

Clearly the four-point condition is equivalent to: among the three numbers $d(i,j) + d(k,\ell)$, $d(i,k) + d(j,\ell)$, $d(i,\ell) + d(j,k)$, two numbers are equal and not less than the third.

Let us mention a very simple, but important remark. For any dissimilarity, the triangle inequality and the ultrametric inequality are obviously satisfied when two elements are identical, so that many proofs are simplified by considering distinct elements. This property is false for the four-point condition. Clearly, the inequality of (1) is equivalent to the triangle inequality when two elements are identical. Consequently the four-point condition is verified iff both conditions (a) and (b) below are satisfied:

(a) inequality (1) holds for distinct elements i, j, k, ℓ of I.

(b) triangle inequality.

As an immediate consequence we get: for $n \leq 3$, every semi-distance is a tree semi-distance. However, for $n \geq 4$, (a) does not imply (1), as proved by the following counter-example:

$$i \text{ fixed in } I; \quad \forall j \neq i, d(i,j) = 1; \quad \forall j \neq i, \forall k \neq i, \forall j \neq k, d(j,k) = 3.$$

In their pioneering works on tree-distances, the first authors discovered the four-point condition which is clearly verified by checking the possible configurations of elementary chains connecting four distinct vertices in a tree, as in Figure 2.3.6.

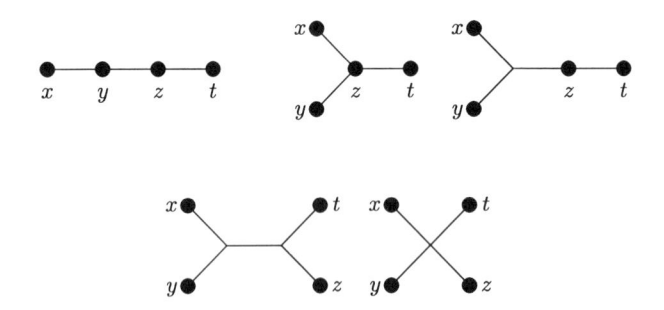

Figure 2.3.6: The five possible configurations of elementary chains connecting four distinct vertices of a tree

The converse statement, that is the existence of an embedding of a semi-metric space satisfying the four-point condition into a weighted tree, has been established by several authors. The first results seem due to Zaretskii (1965), Simoes-Pereira (1967) and Smolenskii (1969) for particular trees (with weights equal to 1). Generally, people mention Patrinos and Hakimi (1972), Buneman (1974) and Dobson (1974) for a complete and general proof. The proof of Patrinos and Hakimi is right, but a little bit long and complicated, while the proof of Dobson contains some errors. See Delclos (1987) for counter-examples associated to some of Dobson's assertions (Lemma 1, p.34), and an entirely modified version of the proof. The most elegant proof is the one of Buneman which is reproduced here. The first part is presented as a lemma, which is of some independent interest. The second part needs to be rather detailed, as Barthélemy and

Guenoche (1988) noted, in order to discuss the possible equalities between points. We give an alternative proof of the second part here.

Lemma 2.3.1. *Let d satisfy the four-point condition $(n > 2)$. Then there exist two different elements i and j in I such that:*

$$\forall k, \ell \in I\backslash\{i,j\}, \quad d(i,k) + d(j,\ell) = d(i,\ell) + d(j,k).$$

Proof. Select an unordered triple $\{i, j, k\}$ of distinct elements on I, such that $d(i,k) + d(j,k) - d(i,j)$ is maximal. Then:

$$\forall \ell \in I\backslash\{j\}, \quad d(k,\ell) + d(j,k) - d(j,\ell) \leq d(i,k) + d(j,k) - d(i,j)$$

i.e.: $d(k,\ell) + d(i,j) \leq d(i,k) + d(j,\ell)$. Similarly:

$$\forall \ell \in I\backslash\{i\}, d(k,\ell) + d(i,j) \leq d(i,\ell) + d(j,k).$$

Since d satisfies the four-point condition, we deduce:

$$\forall \ell \in I\backslash\{i,j\}, \quad d(i,k) + d(j,\ell) = d(i,\ell) + d(j,k).$$

Therefore: $\forall \ell, m \in I\backslash\{i,j\}, d(i,m) - d(j,m) = d(i,\ell) - d(j,\ell)$. ∎

Proposition 2.3.11. *d is a tree dissimilarity iff d satisfies the four-point condition.*

Proof. The necessary condition has already been given. The sufficient condition is proved inductively with respect to the cardinality n of I. If $n \leq 2$, or $n = 3$ and d non definite, the condition is obvious. If $n = 3$, $I = \{i, j, k\}$, d definite, then we use either the embedding shown in Figure 2.3.7 when $d(i,k) = d(i,j) + d(j,k)$ (up to a permutation) or the embedding shown in Figure 2.3.8 with,

$$a_i = \tfrac{1}{2}[d(i,j) + d(i,k) - d(j,k)] \tag{1}$$
$$a_j = \tfrac{1}{2}[d(j,i) + d(j,k) - d(i,k)]$$
$$a_k = \tfrac{1}{2}[d(k,i) + d(k,j) - d(i,j)]$$

when all triangle inequalities are strict. Hence we suppose $n > 3$. Let $\{i, j\}$ be as in the lemma. Let $k \in I\backslash\{i,j\}$ and a_i, a_j, a_k be as defined in (1). By the lemma: $\forall \ell \in I\backslash\{i,j\}$, $d(i,k) - d(j,k) = d(i,\ell) - d(j,\ell)$, so that a_i, and similarly a_j, does not depend on the choice of k in $I\backslash\{i,j\}$. Thus, we have defined a non-negative sequence $\{a_k, k \in I\}$. Let us add a new element u to I and extend d by putting:

$$\forall k \in I, d(u,k) = a_k \tag{2}$$

We have:

$$d(i,j) = a_i + a_j; \quad \forall k \in I\backslash\{i,j\}, \ d(i,k) = a_i + a_k; \quad d(j,k) = a_j + a_k. \tag{3}$$

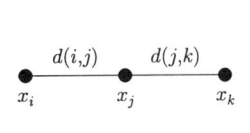

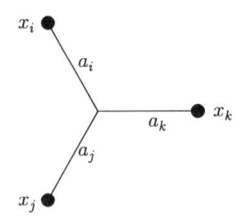

Figure 2.3.7 : Configuration for three metrically aligned points

Figure 2.3.8 : Configuration for three metrically non-aligned points

Writing $\bar{I} = I \cup \{u\}$, then d is a dissimilarity on $\bar{I}$. Now, we prove the triangle inequality, which has only to be established for 3 distinct elements of the type $\{u,i,j\}$, $\{u,i,k\}$ or $\{u,k,\ell\}$, $k,\ell \in I\backslash\{i,j\}$. The first two triples verify the triangle inequality by (2) and (3). Consider then the triple $\{u,k,\ell\}$. Now, $d(i,k) \leq d(i,\ell) + d(k,\ell)$ so that, by (2) and (3), $d(u,k) \leq d(u,\ell) + d(k,\ell)$. Similarly, $d(u,\ell) \leq d(u,k) + d(k,\ell)$. By the lemma and the four point condition,

$$d(i,j) + d(k,\ell) \leq d(i,k) + d(j,\ell)$$

whence, using (2) and (3), we find the other required inequality. Consequently d is a semi-distance on $\bar{I}$. Now, we prove the four-point condition on $\bar{I}$, which has only to be established for 4 distinct elements of the type $\{u,i,j,k\}$, $\{u,k,\ell,m\}$ or $\{u,i,k,\ell\}$, $k,\ell,m \in I\backslash\{i,j\}$. For the first quadruple, the result is immediate by (2) and (3). These same relations show that the four point condition on the second or third quadruple is equivalent to that on $\{i,k,\ell,m\}$ or $\{j,i,k,\ell\}$ respectively.

By the inductive hypothesis we have an embedding ϕ of $\bar{I}\backslash\{i,j\}$ into some weighted tree (X,w). Adding two new elements y_i and y_j to X and the new edges (y_i,u) and (y_j,u) with respective weights a_i and a_j to the tree, we obtain a new weighted tree. The embedding is also extended to $\bar{I}$ by putting $x_i = y_i$, $x_j = y_j$, and by (2) and (3), this embedding is isometric (if $a_i = 0$, or $a_j = 0$, or $a_i = a_j = 0$, we add only y_j, or y_i, or nothing, and we put $x_i = x_u$ or $x_j = x_u$ or $x_i = x_j = x_u$). ∎

Let us denote by $\mathcal{D}_t$, the set of tree-dissimilarities on I. Then we have:

Proposition 2.3.12. $\mathcal{D}_t$ *is a closed cone.*

Proof. The proof is similar to that of Proposition 2.3.10. ∎

Remark 2.3.4. For $n > 3$, the cone $\mathcal{D}_t$ is non-convex. For a simple counter-example, define d and d' on $I = \{i,j,k,\ell\}$ by:

$d(i,j) = d(k,\ell) = 1;\quad d(i,k) = d(i,\ell) = d(j,k) = d(j,\ell) = 2.$

$d'(i,k) = d'(j,\ell) = 1;\quad d'(i,j) = d'(k,\ell) = d'(i,\ell) = d'(j,k) = 2.$

Then $d,d' \in \mathcal{D}_t; (d+d') \notin \mathcal{D}_t$. □

Special trees induce particular tree semi-distances. In Section 2.3.1, we have introduced the cone $\mathcal{D}_{ch}$ of chain semi-distances. These semi-distances are associated to the weighted trees having only two terminal vertices. Recall that $\mathcal{D}_{ch}$ is clearly a closed cone which is nonconvex for $n > 2$.

The directed tree proposed for a visual display of dendrograms or indexed hierarchies may be regarded as a weighted tree. In this tree, the weight of an edge is one half the difference between the levels of corresponding vertices. There is a particular vertex, the root of the directed tree, at the same distance from each unit of I. Thus, an ultrametric is a particular tree semi-distance and Brossier (1986) has shown that the existence of a vertex equidistant from each unit is a necessary and sufficient condition for ultrametricity.

In Section 2.4, we will give a direct proof of the inclusion of $\mathcal{D}_u$ in $\mathcal{D}_t$.

Another particular case is given by star semi-distances in connection with star-graphs, as we now discuss.

2.3.7. Star semi-distances

Star-graphs are trees having only one nonterminal vertex. They seem to have been introduced in the field of data analysis by Carroll (1976), see also Le Calvé (1985). Now we give:

Definition 2.3.9. A dissimilarity d on I is said to be a star semi-distance iff there exist non negative numbers w_i, $i \in I$ such that:

$$\forall (i,j) \in I^2, \quad i \neq j \Rightarrow d(i,j) = w_i + w_j. \qquad \square$$

It is clear that (I, d) is isometrically embeddable in a weighted star graph. This yields an easy visual display. For example Figure 2.3.9 represents the following dissimilarity:

	i	j	k	ℓ	m
i	0	3	4	1	1
j		0	5	2	2
k			0	3	3
l				0	0
m					0

Note that in this example $w_l = w_m = 0$.

Thus any star semi-distance is a tree semi-distance.

Remark 2.3.5. The condition for embeddability in a star-graph is not sufficient. Only (tree) semi-distances embeddable in a weighted star-graph and having at most one unit placed at each terminal vertex are star semi-distances. We have the obvious necessary and sufficient condition: (I, d) is embeddable in a weighted star-graph iff the quotient distance $\tilde{d}$ is a star-distance. $\qquad \square$

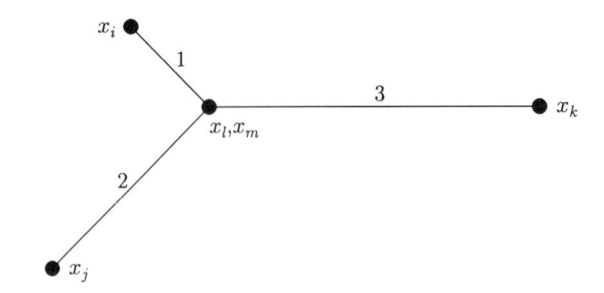

Figure 2.3.9: Visual display of a simple star semi-distance

Proposition 2.3.13. *The set $\mathcal{D}_s$ of star semi-distances is a closed convex polyhedral cone. For $n > 2$, its extreme rays are those spanned by the 1-dichotomies $d_{\{i\}}$, $i \in I$ and are linearly independent.*

Proof. For $n \leq 2, \mathcal{D}_s = \mathcal{D}_+$, and for $n > 2$, in the notation of Definition 2.3.9:

$$d \in \mathcal{D}_s \quad \text{iff} \quad d = \Sigma w_i d_{\{i\}}.$$

Now let: $d_0 = \Sigma a_i d_{\{i\}}$ and $a. = \Sigma a_i$:

$$\forall (i,j) \in I^2, \quad i \neq j, \quad a_i + a_j = 0.$$

Summing for fixed i over all $j \neq i$, gives: $(n-2)a_i + a. = 0$.

Summing now over all i, gives: $a. = 0$ and $a_i = 0$, $\forall i \in I$. ∎

Remark 2.3.6. For $n > 3$, $\mathcal{D}_s$ does not generate $\mathcal{D}$, and consequently the interior of $\mathcal{D}_s$ is empty. □

2.3.8. Robinsonian dissimilarities

Robinsonian dissimilarities (or similarities) appeared with Robinson (1951) for seriation problems in the field of archaeological sciences, see also Gelfand (1971), Hodson, Kendall, Tautu (1971). For comparing objects observed at different dates, Robinson stipulates a model in which the dissimilarity between objects is a nondecreasing function of the difference in their dates. Thus, on the set of objects we have an order defined by the time and a dissimilarity with the property that, when the objects are in their correct time order, the dissimilarity matrix $(d_{ij}) = (d(i,j))$ has elements which do not decrease when you move away from the main diagonal along any row or column. Such a matrix will be called Robinsonian.

The notion of order underlies ultrametricity via the dendrogram or the hierarchy. Only certain orders yield a visual display without crossings, specifically the orders for which the hierarchical clusters are intervals. These orders will be called compatible.

II. PARTIAL ORDER OF THE PRINCIPAL CLASSES OF DISSIMILARITY

After some pioneering works, either in connection with seriation, Hubert (1974), or with the choice of particular compatible orders to improve the visual display, the main result was obtained by Brossier (1980), see also Diday (1982, 1983) and Gaud (1983). The compatible orders are the orders for which the ultrametric dissimilarity matrix is Robinsonian.

Knowledge of a compatible order gives an easy procedure for constructing the hierarchy from an ultrametric, as we illustrate in Figure 2.3.10 for the following dissimilarity:

	i	j	k	ℓ	m	n
i	0	1	1	3	3	3
j		0	1	3	3	3
k			0	3	3	3
ℓ				0	3	3
m					0	2
n						0

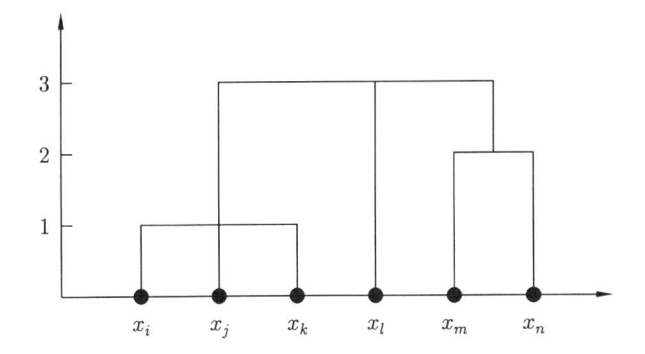

Figure 2.3.10: Construction of a hierarchy from a compatible order

Moreover, the same procedure permits us to extend the visual display to any, not necessarily ultrametric, Robinsonian matrix. Consider the following dissimilarity and its visual representation in Figure 2.3.11.

The graphical representation corresponds to a pseudo-hierarchy, called "pyramid" by Diday. It is associated with a graph, directed from a root at the highest level. This is a pure extension of the diagram proposed for a hierarchy. In particular, the clusters and the dissimilarity are observed in a similar way. However, note the existence of overlapping clusters.

Diday (1984) has proposed a one-to-one correspondence between (definite) Robinsonian matrices and "pyramids" (pseudo-hierarchies). However, we have only a weak index: two clusters, one being strictly included in the second, may be at the same level. Consequently, there does not exist any correct visual display. Strictly indexed pseudo-hierarchies will be discussed in the next section.

	i	j	k	ℓ	m	n
i	0	1	2	3	3	3
j		0	1	3	3	3
k			0	3	3	3
ℓ				0	1	2
m					0	1
n						0

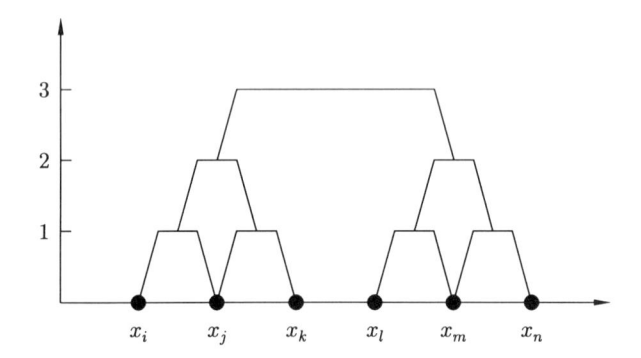

Figure 2.3.11: Construction of a pseudo-hierarchy from a compatible order

Now, given an order $\leq$ on I, we may identify I with $\{1,\ldots,n\}$ in such a manner that the order becomes the natural order between integers.

Definition 2.3.10. A dissimilarity d on I and an order $\leq$ on I are said to be right-compatible (resp. left-compatible) iff:

$$(i \leq j \leq k) \quad \text{implies} \quad d(i,j) \leq d(i,k) \quad (\text{resp.}\, d(j,k) \leq d(i,k)).$$

The dissimilarity and the order are compatible iff they are right and left-compatible. □

When the values taken by d are arranged as a matrix $(d(i,j))$, then "right" corresponds to "row" and "left" to "column". Clearly, an order is right (resp. left)-compatible iff its converse is left (resp. right)-compatible. Consequently, when an order $\leq$ is compatible (with d), its converse is too.

We have: $(i \leq j \leq k) \Rightarrow d(i,k) \geq \max[d(i,j), d(j,k)]$

Definition 2.3.11. A dissimilarity d on I is said to be Robinsonian iff it admits a compatible order. □

Proposition 2.3.14. *The set $\mathcal{D}_r$ of Robinsonian dissimilarities is a closed cone.*

Proof. The conic nature is obvious. Now, some easy limit considerations show that the set of dissimilarities compatible with a fixed order, is closed. But, $\mathcal{D}_r$ is the finite union, over all orders on I, of such closed sets. ∎

Remark 2.3.7.

- For $n > 2$, a Robinsonian dissimilarity is not necessarily even, hence not a semi-distance.

	i	j	k
i	0	0	2
j		0	1
k			0

- $\mathcal{D}_r$ is stable under increasing transformations.

- For $n > 3$, $\mathcal{D}_r$ is nonconvex. For a counter example, consider d and d' on $I = \{i, j, k, l\}$ as follows, Durand (1989):

$d:$

	i	j	k	ℓ
i	0	1	2	3
j		0	2	3
k			0	3
ℓ				0

$d':$

	i	ℓ	k	j
i	0	1	2	3
ℓ		0	2	3
k			0	3
j				0

$(d + d'):$

	j	i	k	ℓ
j	0	4	5	6
i		0	4	4
k			0	5
ℓ				0

□

2.3.9. Strongly-Robinsonian dissimilarities

In the previous section we mentioned the one-to-one correspondence proposed by Diday. This correspondence connects (definite) Robinsonian dissimilarities to weakly indexed (definite) pseudo-hierarchies, and may be extended to even Robinsonian dissimilarities. At the same time, Fichet (1984) established a bijection between (strictly) indexed pseudo-hierarchies and a subset of particular Robinsonian dissimilarities. He called them strongly-Robinsonian and gave a characterization. An exact visual display is therefore possible. The link between the two approaches is simple: it may be proved that the bijection of Fichet is the restriction to strongly-Robinsonian dissimilarities of the bijection of Diday. For this link and a unified presentation see Durand and Fichet (1988).

Definition 2.3.12. A dissimilarity d on I and order $\leq$ on I are said to be right strongly-compatible (resp. left strongly-compatible) iff:

(i) d and $\leq$ are right-compatible (resp. left-compatible).

(ii) $[i \leq j \leq k, d(i,k) = d(j,k)]$ implies: $\forall \ell \geq k, d(i,\ell) = d(j,\ell)$

(resp. $[i \leq j \leq k, d(i,k) = d(i,j)]$ implies: $\forall \ell \leq i, d(\ell,k) = d(\ell,j)$).

The dissimilarity and the order are strongly-compatible iff they are right and left strongly-compatible. □

Clearly an order is right (resp. left) strongly-compatible iff its converse is left (resp. right) strongly-compatible. Therefore, when an order is strongly-compatible (with d), its converse is too. Durand (1989) has shown that when d admits a strongly-compatible order, then every compatible order is strongly-compatible.

Definition 2.3.13. A dissimilarity d on I is said to be strongly-Robinsonian iff it admits a strongly-compatible order. □

Then we have the obvious:

Proposition 2.3.15. *The set $\mathcal{D}_{sr}$ of strongly-Robinsonian dissimilarities is a cone.*

Clearly, $\mathcal{D}_{sr}$ is stable under increasing transformations so that, for $n > 2$ a strongly-Robinsonian dissimilarity is not necessarily a semi-distance. However we have:

Proposition 2.3.16. *Every strongly-Robinsonian dissimilarity is even.*

Proof. Let $\leq$ be an order strongly-compatible with d, and suppose $d(i,j) = 0$. Without loss of generality we may suppose $i \leq j$. If $i \leq k \leq j$, then $d(i,k)$ and $d(j,k)$ are at most $d(i,j)$, hence null. If $i \leq j \leq k$, $d(i,j) = d(j,j) = 0$ implies $d(i,k) = d(j,k)$. If $k \leq i \leq j$, $d(i,i) = d(i,j) = 0$ implies $d(k,i) = d(k,j)$. ∎

Remark 2.3.8.

- Every chain is clearly strongly-Robinsonian.

- For $n > 2$, the cone $\mathcal{D}_{sr}$ is not closed and not open.

The proof is the same as the one given in the proof of Proposition 2.2.2(c), for the cone $\mathcal{D}_{ev}$ with $\varepsilon < 1$.

- For $n > 3$, the cone $\mathcal{D}_{sr}$ is nonconvex.

The counter-example given in Remark 2.3.7 is valid for strongly-Robinsonian dissimilarities. □

2.4. Inclusions

In the previous two sections we have introduced most of the usual structures occuring in the field of dissimilarity analysis. All these structures are associated with cones of the vector space $\mathcal{D}$. In this section we establish each of the inclusions between these cones shown in the summary Figure 2.4.1. Section 2.4.1 notes those inclusions which follow directly from earlier sections. The other inclusions are proved in Section 2.4.2. Section 2.7 will be devoted to counter-examples showing that no other inclusions between these cones are valid for all n. Thus, Figure 2.4.1 represents the complete partial order by inclusion of the subsets of $\mathcal{D}$ which it contains.

2.4.1. Some immediate inclusions

Section 2.2 emphasizes the cones corresponding to either general notions, $\mathcal{D}_+$ (dissimilarities), $\mathcal{D}_{ev}$ (even dissimilarities), $\mathcal{D}_\infty$ (semi-distances), or particular dissimilarities, $\mathcal{D}_{rs}$ (regular simplex distances) or $\mathcal{D}_{di}$ (rays generated by dichotomies). We have the following obvious inclusions: $\mathcal{D}_{rs} \subseteq \mathcal{D}_\infty \supseteq \mathcal{D}_{di}; \mathcal{D}_\infty \subseteq \mathcal{D}_{ev} \subseteq \mathcal{D}_+$. Recall that in Section 2.3 we introduced the cones $\mathcal{D}_e$ (Euclidean semi-distances), $\mathcal{D}_{ch}$ (chain semi-distances), $\mathcal{D}_1$ (semi-distances of L_1-type), $\mathcal{D}_h$ (hypermetric semi-distances), $\mathcal{D}_{qh}$ (quasi-hypermetric dissimilarities), $\mathcal{D}_u$ (ultrametric semi-distances), $\mathcal{D}_t$ (tree semi-distances), $\mathcal{D}_s$ (star semi-distances), $\mathcal{D}_r$ (Robinsonian dissimilarities) and $\mathcal{D}_{sr}$ (strongly-Robinsonian dissimilarities).

From the definitions, some inclusions are obvious $\mathcal{D}_{ch} \subseteq \mathcal{D}_t \supseteq \mathcal{D}_s; \mathcal{D}_{sr} \subseteq \mathcal{D}_r$ while others are immediate consequences $\mathcal{D}_e \supseteq \mathcal{D}_{ch} \subseteq \mathcal{D}_1; \mathcal{D}_u \subseteq \mathcal{D}_\infty$ and $\mathcal{D}_{ch} \subseteq \mathcal{D}_{sr}$. Moreover, some definitions are equivalent to a condition of embeddability in certain metric spaces. Consequently, the corresponding dissimilarities obey the triangle inequality. Thus, $\mathcal{D}_e, \mathcal{D}_1, \mathcal{D}_t$ (hence $\mathcal{D}_{ch}, \mathcal{D}_s$) are included in $\mathcal{D}_\infty$. Again, hypermetric spaces have been introduced in order to generalize the triangle inequality, so that $\mathcal{D}_h$ is included in $\mathcal{D}_\infty$. See Section 2.3.3. There are three types of dissimilarities introduced in Section 2.3 which are not even by virtue of being semi-distances. Of these two have been proved to be even by other means. From Section 2.3.4 and Proposition 2.3.16 we have: $\mathcal{D}_{qh} \subseteq \mathcal{D}_{ev} \supseteq \mathcal{D}_{sr}$. The third, the Robinsonian dissimilarities, are not necessarily even. See Remark 2.3.7. Finally, the particular cones $\mathcal{D}_{rs}$ and $\mathcal{D}_{di}$ verify some obvious inclusions. Both are included in $\mathcal{D}_u$ and $\mathcal{D}_{sr}$ (hence $\mathcal{D}_r$). $\mathcal{D}_{rs}$ is included in $\mathcal{D}_s$ (hence

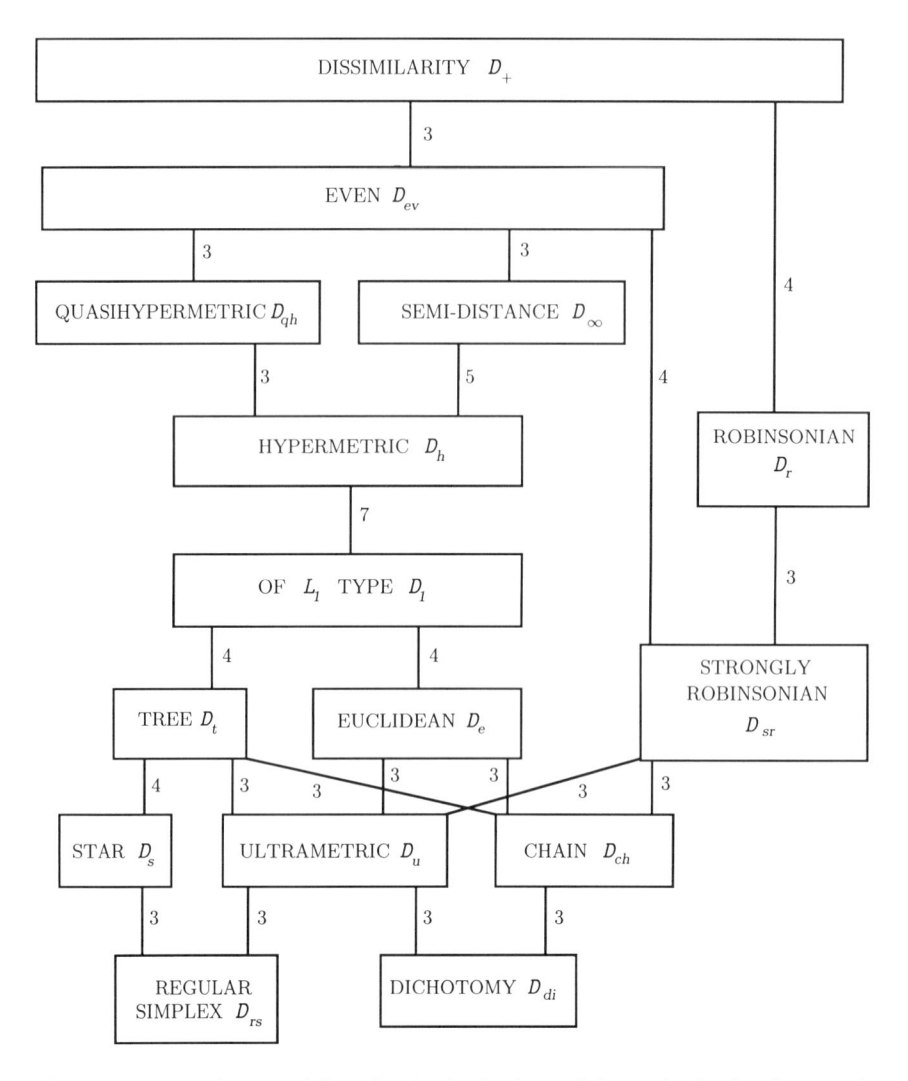

```
                    ┌─────────────────────────────┐
                    │   DISSIMILARITY  $D_+$      │
                    └─────────────────────────────┘
                                 │ 3
                    ┌─────────────────────────────┐
                    │        EVEN  $D_{ev}$        │
                    └─────────────────────────────┘
                       │ 3              │ 3            │ 4
        ┌──────────────────────┐  ┌──────────────────────┐
        │ QUASIHYPERMETRIC $D_{qh}$ │  │ SEMI-DISTANCE $D_\infty$ │
        └──────────────────────┘  └──────────────────────┘
                       │ 3              │ 5       │ 4   ┌──────────────┐
                       └────┬───────────┘         │     │ ROBINSONIAN  │
                    ┌─────────────────────────┐   │     │    $D_r$     │
                    │   HYPERMETRIC  $D_h$    │   │     └──────────────┘
                    └─────────────────────────┘   │            │ 3
                                 │ 7              │
                    ┌─────────────────────────┐   │
                    │    OF $L_1$ TYPE $D_l$  │   │
                    └─────────────────────────┘   │
                            │ 4        │ 4        │     ┌──────────────┐
                                                        │   STRONGLY   │
              ┌──────────┐    ┌────────────────┐        │ ROBINSONIAN  │
              │ TREE $D_t$│    │ EUCLIDEAN $D_e$│       │    $D_{sr}$  │
              └──────────┘    └────────────────┘        └──────────────┘
                 │ 4   │ 3    │ 3    │ 3         │ 3         │ 3
        ┌─────────┐  ┌──────────────────┐  ┌───────────────┐
        │STAR $D_s$│  │ULTRAMETRIC $D_u$ │  │ CHAIN $D_{ch}│
        └─────────┘  └──────────────────┘  └───────────────┘
              │ 3          │ 3        │ 3          │ 3
        ┌─────────────┐              ┌────────────────┐
        │   REGULAR   │              │ DICHOTOMY $D_{di}$│
        │SIMPLEX $D_{rs}$│           └────────────────┘
        └─────────────┘
```

Figure 2.4.1: The partial order by inclusion of the principal subsets of $\mathcal{D}_+$

Note. A line, with an integer n_0 beside it, joining the box containing $\mathcal{D}_*$ to a higher box containing $\mathcal{D}_{**}$ indicates that $\mathcal{D}_* \subseteq \mathcal{D}_{**}$ with strict inclusion if and only if $n \geq n_0$.

$\mathcal{D}_t$) and $\mathcal{D}_{di}$ is included in $\mathcal{D}_{ch}$ (hence $\mathcal{D}_t$). Both are included in $\mathcal{D}_e$ and $\mathcal{D}_1$. This is immediate for $\mathcal{D}_{di}$, and for $\mathcal{D}_{rs}$ the inclusions derive from simple arguments. Recall that d_1 is the dissimilarity defined by $d_1(i,j) = 1$ whenever $i \neq j$. Identifying I with $\{1,\ldots,n\}$, it suffices to consider the vectors x_i, $i \in I$ of $\mathbb{R}^n$, with coordinates x_{ik} equalling $1/\sqrt{2}$ (resp. $1/2$) if $k = i$ and 0 else, to observe that d_1 is Euclidean with maximum dimensionality (resp. of L_1-type).

2.4.2. Other inclusions

From a theorem of Bretagnolle, Dacunha-Castelle and Krivine (1966, Theorem 2, p.238), associated with a property established by Schoenberg (1938, Corollary 1, p.527 or Theorem 5, p.534) we deduce that every Hilbert space is embeddable in an L_1-space. Moreover, considering the extension of Schoenberg's property given by Bretagnolle, Dacunha-Castelle and Krivine (1966, Theorem 7, p.251) a more general result states: every L_p-space ($1 \leq p \leq 2$) is embeddable in an L_1-space. Note that this property is not valid for $p > 2$. See Dor (1976). Let us also remark that the necessary and sufficient conditions established by the above-mentioned authors are valid only for normed spaces. However, the necessary condition for embeddability of (finite) metric spaces clearly holds. Moreover, in the finite case, some easy limit considerations show that embeddability in an L_1-space is equivalent to embeddability in l_1^N for some N. See Le Calvé (1987). The mathematical tools used in the previous results are rather sophisticated. In the finite case which is studied here, we propose a new and simple proof of the fact that every Euclidean semi-distance is of L_1-type. It is based on simple results given in Section 2.3.2 and on a nice property due to Kelly (1970,b). We present this property and his elegant proof as the

Lemma 2.4.1. *Let S be a sphere in the Euclidean space $\mathbb{R}^N$, $N > 1$, with radius $r > 0$. Let $\{x_i, i \in I\}$ be a finite family of vectors in S. For every i and j in I, define $d(i,j)$ as the geodesic distance $\widehat{x_i x_j}$, i.e. the arclength given by a common great circle. Then d is of symmetric difference type.*

Proof. Let $(B,\mathcal{F},\mu)$ be the measure space, where B is the ball associated with S, $\mathcal{F}$ the set of Borel subsets of B and μ the Lebesgue measure, defining the volume. For every $i \in I$, denote by B_i the closed half-ball opposite to x_i. Here our proof differs somewhat from that of Kelly who considers the hemisphere opposed to x_i. Now, let x_i and x_j be two vectors and consider the plane P of a great circle containing them. Then $B_i \Delta B_j$ is the subset of points in B which project onto the shaded domain of P given in Figures 2.4.2 (a) and (b) according as the angle θ_{ij} between x_i and x_j is acute or obtuse.

The Lebesgue measure being invariant under rotation, we have:

$$\mu(B_i \Delta B_j) = (2\theta_{ij}/2\pi)V$$

where V stands for the volume of B. Since $d(i,j) = r\theta_{ij}$, $d(i,j)$ is proportional to $\mu(B_i \Delta B_j)$. The proof is complete by Proposition 2.3.5. ∎

We note in passing that Kelly's result shows that the arclength is a distance.

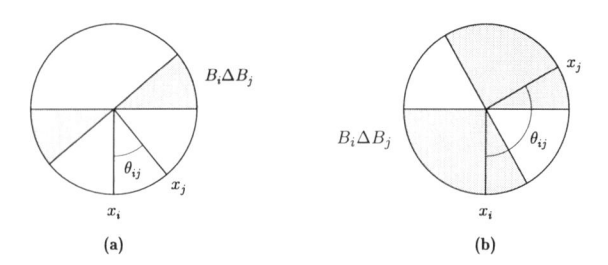

Figure 2.4.2: The plane P

Now we give

Proposition 2.4.1. *Every Euclidean semi-distance is of L_1-type, i.e. $\mathcal{D}_e \subseteq \mathcal{D}_1$.*

Proof. The result is obvious if $n \leq 2$. Hence, we suppose $n > 2$. Let $d \in \mathcal{D}_e$ and suppose that (I, d) has maximum dimensionality $(n-1)$. Consider a Euclidean embedding $\{x_i, i \in I\}$ in some Euclidean space E with dimension $(n-1) > 1$. Since the vectors form a simplex, they admit a circumscribed sphere with radius $r_0 > 0$ (say) and centre x_0 (say). We can always embed E in a Euclidean space E' with dimension n. Let D be the straight line perpendicular to E and passing through x_0. Then: $\forall x \in D$, $\forall i \in I$, $\|x - x_i\|^2 = \|x - x_0\|^2 + r_0^2$. Thus, for every $r \geq r_0$, there exists a sphere with radius r containing the $x_i, i \in I$. Therefore, for every $r \geq r_0$ we may define a geodesic distance d_r by the corresponding arclength distances. We have the classical relationship between arclength and chord distances: $\sin(d_r/2r) = d/2r \Leftrightarrow d_r/2r = \sin^{-1}[d/2r]$. Thus: $\lim_{r \to \infty} d_r = d$. By the previous lemma, for every $r \geq r_0$, d_r is of symmetric difference type, hence of L_1-type. See Proposition 2.3.6. Since $\mathcal{D}_1$ is closed, we deduce: $d \in \mathcal{D}_1$. Now, clearly every Euclidean semi-distance is the limit of some Euclidean semi-distances with maximum dimensionality. Thus the proof is complete since $\mathcal{D}_1$ is closed. ∎

Kelly (1970,a, 1972, 1975) proved that every metric space of symmetric difference type is embeddable in some hypermetric space. Recall that the equivalence between metric spaces of symmetric difference type and metric spaces of L_1-type was established later by Avis (1977). In Proposition 2.3.6, we have given a short proof of this equivalence in the finite case. Thus, metric spaces of L_1-type are hypermetric. Since Kelly's work, several authors such as Assouad (1980,a, 1982, 1984), Avis (1981) and Assouad and Deza (1982) obtained or mentioned a direct proof of this result. Again, we must recall the pioneering work of Deza (1960) who studied particular finite metric spaces of L_1-type: those which are embeddable on a hypercube. Deza showed that they obey the n-polygonal inequality (when n is odd).

Now, in the finite case, we give a short proof of

Proposition 2.4.2. *Every semi-distance of L_1-type is hypermetric, i.e. $\mathcal{D}_1 \subseteq \mathcal{D}_h$.*

Proof. Since $\mathcal{D}_h$ is a convex cone and $\mathcal{D}_1$ is the conic-convex hull of dichotomies, we have only to prove that every dichotomy is hypermetric. Let d_J ($\phi \subset J \subset I$) be any dichotomy. Using Proposition 2.3.8, we have to prove that for every positive integer k, for every collection of integers $\lambda_1, \ldots, \lambda_k$ which sum to 1, and for every collection of elements $i_1, \ldots, i_k$ in I:

$$2 \sum_{\ell \mid i_\ell \in J} \sum_{m \mid i_m \notin J} \lambda_\ell \lambda_m \leq 0.$$

But note that the left hand quantity is equal to:

$$1 - \left[\sum_{\ell \mid i_\ell \in J} \lambda_\ell \right]^2 - \left[\sum_{m \mid i_m \notin J} \lambda_m \right]^2,$$

which is an integer strictly less than 1. ∎

The following proposition is due to Kelly (1972). We reproduce his simple proof.

Proposition 2.4.3. *Every hypermetric semi-distance is quasi-hypermetric, i.e. $\mathcal{D}_h \subseteq \mathcal{D}_{qh}$.*

Proof. Let $d \in \mathcal{D}_h$. Let k be a positive integer and $i_1, \ldots, i_k$ be k elements of I. For every collection of integers $\lambda_1, \ldots, \lambda_k$ which sum to 0, we have $\sum_{\ell=1}^{k-1} \lambda_\ell + (\lambda_k + 1) = 1$. Since $d \in \mathcal{D}_h$, we deduce by Proposition 2.3.8:

$$\sum_{\ell=1}^{k} \sum_{m=1}^{k} \lambda_\ell \lambda_m d(i_\ell, i_m) + 2 \sum_{\ell=1}^{k-1} \lambda_\ell d(i_\ell, i_k) \leq 0. \tag{1}$$

Similarly: $\sum_{\ell=1}^{k-1} (-\lambda_\ell) + (-\lambda_k + 1) = 1$, so that:

$$\sum_{\ell=1}^{k} \sum_{m=1}^{k} \lambda_\ell \lambda_m d(i_\ell, i_m) - 2 \sum_{\ell=1}^{k-1} \lambda_\ell d(i_\ell, i_k) \leq 0. \tag{2}$$

Addition of (1) and (2) yields:

$$\sum_{\ell=1}^{k} \sum_{m=1}^{k} \lambda_\ell \lambda_m d(i_\ell, i_m) \leq 0$$

This last inequality holds for rational numbers by homogeneity, hence for real numbers by continuity. ∎

Remark 2.4.1. Proposition 2.4.2 and Proposition 2.4.3 yield: $\mathcal{D}_1 \subseteq \mathcal{D}_{qh}$. In other words, every semi-distance of L_1-type has a Euclidean square root. A direct and simpler proof is obtained by observing that any dichotomy is in $\mathcal{D}_{qh}$. By similar arguments, Fichet (1986, 1988) has shown that any semi-distance of L_1-type has a pth root of L_p-type ($p \geq 1$). Using Proposition 2.4.1 we also get $\mathcal{D}_e \subseteq \mathcal{D}_{qh}$. Every Euclidean semi-distance has a Euclidean square root. This property was established for the first time by Schoenberg (1937), extending a result of Szegö, valid only for Euclidean distances with dimensionality at most three. Moreover, the Euclidean square root is shown to have maximum dimensionality. The proof of Schoenberg is rather complicated, by using "transcendental means". In his paper the author writes: "although this theorem is algebraic in nature ... an algebraic proof would probably be difficult and complicated". More recently, a simple and algebraic proof has been obtained. See Joly and Le Calvé (1986). □

In Section 2.3.5 we recalled the bijection between ultrametrics and indexed hierarchies, and we gave the usual visual display by means of an oriented tree. This oriented tree with a level associated with each vertex (cluster) may be regarded as an undirected tree with a positive weight on each edge, defined by one half the difference between the two levels of the corresponding vertices. Then, clearly, the ultrametric distance given by the oriented tree is the tree distance defined by the weighted (undirected) tree. Thus, every ultrametric is a tree semi-distance. But to justify the result, in this way, we have used mathematical properties not established in this paper. There are a lot of simple direct proofs of the inclusion, the first being the one of Kelly (1970,a) before the four-point condition. Here we give a simple version of the proof of Dobson (1974).

Proposition 2.4.4. *Every ultrametric is a tree semi-distance, i.e.* $\mathcal{D}_u \subseteq \mathcal{D}_t$.

Proof. Let $d \in \mathcal{D}_u$. Since every ultrametric obeys the triangle inequality, we have only to prove the four-point condition for four distinct elements i, j, k, ℓ of I. See Section 2.3.6. Relabelling as necessary, we may suppose that $d(i,j) = \min\{d(u,v) | u \neq v, \ u, v \in \{i,j,k,\ell\}\}$ and $d(i,k) \leq d(i,\ell)$. Then by the triangle condition of ultrametricity, we have:
$$d(i,j) \leq d(i,k) = d(j,k) \leq d(i,\ell) = d(j,\ell) \ .$$
Consequently, $d(k,\ell) \leq d(i,\ell) = d(j,\ell)$, so that:
$$d(i,j) + d(k,\ell) \leq d(i,k) + d(j,\ell) = d(i,\ell) + d(j,k) \ .$$
■

Kelly (1975) showed that any tree semi-distance is hypermetric. A stronger property holds, which is intuitive from the definition: every tree semi-distance is of L_1-type. This has been proved by Fichet (1986, 1987, 1988) and Cailliez in unpublished papers. Following this property, a decomposition of semi-distances of L_1-type has been proposed by Bandelt and Dress (1990, 1992) in connection with classification.

Proposition 2.4.5. *Every tree semi-distance is of L_1-type, i.e. $\mathcal{D}_t \subseteq \mathcal{D}_1$.*

Proof. For a given element d in $\mathcal{D}_t$, let $\{x_i, i \in I\}$ be an embedding of (I, d) in some weighted tree (T, w), with $T = (X, E)$. Without loss of generality, we may suppose that every terminal vertex of T is one of the $x_i, i \in I$. If we suppress an edge $e \in E$ in the tree, we obtain two connected components for the new graph $(X, E \backslash \{e\})$, defining a bipartition (X_e, X_e^c) of X. Denote by (I_e, I_e^c) the induced bipartition of I: $i \in I_e$ iff $x_i \in X_e$. From the condition imposed on the tree, we have: $\phi \subset I_e \subset I$. Thus we may define a family of dichotomies $\{d_{I_e}, e \in E\}$, more simply denoted by $\{d_e, e \in E\}$. We show that: $d = \sum_{e \in E} w(e) d_e$. Let i and j be any two elements of I. If $x_i = x_j$, then clearly $d(i, j) = 0$ and: $\forall e \in E, d_e(i, j) = 0$. Now, suppose $x_i \neq x_j$ and consider the elementary chain $\{y_1 = x_i, y_2, \ldots, y_p = x_j\}$ connecting x_i and x_j. The edges of the chain $(x_j, x_{j+1}), j = 1, \ldots, (p-1)$ are denoted by $e_j, j = 1, \ldots, (p-1)$ respectively. Then $d(i, j) = \sum_{j=1}^{p-1} w(e_j)$. But clearly, x_i and x_j are connected in the graph $\{X, E \backslash \{e\}\}$ iff e does not belong to $\{e_j, j = 1, \ldots, (p-1)\}$ so that: $d_e(i, j) = 1$ iff $e \in \{e_j, j = 1, \ldots, (p-1)\}$. This completes the proof. ∎

Remark 2.4.2. In another proof, Fichet (1986, 1988) obtained the following stronger result: every tree semi-distance has a pth root of L_p-type ($p \geq 1$). In particular, for $p = 2$, we have: any $d \in \mathcal{D}_t$ has a Euclidean square root. In this paper, this is a consequence of Propositions 2.4.2, 2.4.3, 2.4.5 and 2.3.9. A more precise statement holds from a result established by Le Calvé (1985): if d is a tree distance, then its square root is Euclidean with maximum dimensionality. □

Kelly (1970,a) established that every ultrametric space is hypermetric. This property may be improved: every ultrametric d on I is Euclidean and has the maximum Euclidean dimensionality $(n-1)$ if and only if it is definite. The result is due to Holman (1972). Since then several proofs with or without the dimensionality property have been proposed. We indicate the proof of J. Gower, given in Escoufier (1975), which is iteratively constructive and which needs a minor improvement at one point. In the book of Cailliez and Pages (1976, p.260) the reader will find a simple proof based on a decomposition of the given ultrametric. The proof of Fichet (1983) is rather similar, by using the extreme rays of an ultrametric preordonnance, i.e. the cone of all ultrametrics obtained from a given ultrametric by a positive nondecreasing transformation. More recently, Aschbacher, Baldi, Baum and Wilson (1987) seem to have rediscovered the embedding. Their proof is original and uses determinants. Let us observe that the embedding is an immediate consequence of inclusions established above. Indeed, if d is ultrametric so is d^2. Then d^2 is in $\mathcal{D}_t \subseteq \ldots \subseteq \mathcal{D}_{qh}$. Thus, by Proposition 2.3.9: $d \in \mathcal{D}_e$. Again the property for the dimensionality derives from one of two results mentioned above, but not proved. Indeed, suppose $d \in \mathcal{D}_u$, with d definite. Then $d^2 \in \mathcal{D}_e$ and $d^2 \in \mathcal{D}_t$. We may use the results of Schoenberg (1937) and Le Calvé (1985) mentioned in the Remark 2.4.1 and the Remark 2.4.2.

We give here a direct and complete proof. It has been given by the second author in his courses for several years.

Proposition 2.4.6. *Let d be an ultrametric. Then d is Euclidean, i.e. $\mathcal{D}_u \subseteq \mathcal{D}_e$. The Euclidean dimension of (I,d) is $(|\tilde{I}|-1)$.*

Proof. We proceed by induction. The result is trivially true for $n \leq 3$. Now, suppose $|I|=n>3$. Let $d \in \mathcal{D}_u$. First, suppose d non definite. Since d is even, we may consider the quotient space $(\tilde{I}, \tilde{d})$ – see Section 2.2.5. Clearly, $\tilde{d}$ is ultrametric. By the inductive hypothesis, $\tilde{d}$ is Euclidean and has dimension $(|\tilde{\tilde{I}}|-1) = (|\tilde{I}|-1)$, and clearly d satisfies the same properties. Now suppose d definite. Let $a = \min\{d(i,j)|(i,j) \in I^2, i \neq j\} > 0$. Put $d' = d - ad_1$, where d_1 stands for the distance of the unit simplex. Then d' and $d'^{\frac{1}{2}}$ are trivially ultrametric and non definite, so that they are Euclidean. Recall that d_1 is Euclidean with maximum dimensionality. Now, we use the linear isomorphism $\mathcal{T}$ introduced in Section 2.3.1. We have $\mathcal{T}(d^2) = \mathcal{T}(d'^2) + 2a\mathcal{T}(d') + a^2\mathcal{T}(d_1^2)$. Using Proposition 2.3.1, we see that $\mathcal{T}(d'^2)$ and $\mathcal{T}(d')$ are p.s.d. and that $\mathcal{T}(d_1^2)$ is positive definite, so that $\mathcal{T}(d^2)$ is positive definite, and d is Euclidean of maximal dimensionality. ∎

In Section 2.3.9 we recalled the papers of Diday (1982, 1983) showing that every ultrametric admits a compatible order. Using this fact, Durand (1989) established that any order compatible with an ultrametric is strongly-compatible. We give here a direct proof. This is the one proposed by Fichet (1984) in introducing strongly-Robinsonian dissimilarities.

Proposition 2.4.7. *Every ultrametric is strongly-Robinsonian, i.e. $\mathcal{D}_u \subseteq \mathcal{D}_{sr}$.*

Proof. The result is obvious if $n \leq 2$ or $d = d_0$. Now we proceed by induction. Suppose $n > 2$ and let $d \in \mathcal{D}_u$ with $d \neq d_0$. Put $a = \max\{d(i,j)|(i,j) \in I^2\} > 0$. Then, from the ultrametric inequality, the relation $\sim$ on I defined by $i \sim j \Leftrightarrow d(i,j) < a$, is an equivalence. Denote by $\{I_1, \ldots, I_p\}$ the associated partition of I. Necessarily: $p > 1$. Clearly, for every $\ell = 1, \ldots, p$, the restriction d_{I_ℓ} of d to I_ℓ is ultrametric, so that by the inductive hypothesis, there exists an order $\leq_\ell$ on I_ℓ strongly-compatible with d_{I_ℓ}. Define the order $\leq$ on I by: $i \leq j$ iff $(i \in I_\ell, j \in I_m, \ell < m)$ or $(i \in I_\ell, j \in I_\ell, i \leq_\ell j)$. Note that the order used on the members of the partition is arbitrary. Then the corresponding block matrix is strongly-Robinsonian. ∎

2.5. The convex hulls

The mathematical structures introduced in Sections 2.2 and 2.3 are associated with cones of the vector space $\mathcal{D}$. Many cones have been shown to be convex: $\mathcal{D}_+$ (dissimilarities), $\mathcal{D}_{ev}$ (even dissimilarities), $\mathcal{D}_\infty$ (semi-distances), $\mathcal{D}_{rs}$ (regular simplex distances), $\mathcal{D}_s$ (star semi-distances), $\mathcal{D}_1$ (semi-distances of L_1-type), $\mathcal{D}_h$ (hypermetrics) and $\mathcal{D}_{qh}$ (quasi-hypermetrics). Moreover $\mathcal{D}_1$ is the convex hull of the cones $\mathcal{D}_{di}$ (dichotomies) and $\mathcal{D}_{ch}$ (chain semi-distances), i.e. $\mathcal{D}_1 = \text{conv}(\mathcal{D}_{di}) = \text{conv}(\mathcal{D}_{ch})$. See Section 2.3.2. The aim of this section is to characterize the convex hull of the other cones: $\mathcal{D}_e$ (Euclidean semi-distances), $\mathcal{D}_u$ (ultrametrics), $\mathcal{D}_t$ (tree semi-distances), $\mathcal{D}_r$ (Robinsonian dissimilarities) and $\mathcal{D}_{sr}$ (strongly-Robinsonian dissimilarities). The reader will observe that it does not necessitate introducing new structures. First we have the following easy result, due to Fichet (1986, 1988).

Proposition 2.5.1. *The cone of semi-distances of L_1-type is the convex hull of Euclidean semi-distances, ultrametrics and tree semi-distances, i.e. $\mathcal{D}_1 = \mathrm{conv}(\mathcal{D}_e) = \mathrm{conv}(\mathcal{D}_u) = \mathrm{conv}(\mathcal{D}_t)$.*

Proof. Clearly $\mathcal{D}_{di} \subseteq \mathcal{D}_e$ and $\mathcal{D}_{di} \subseteq \mathcal{D}_u$. Using Propositions 2.4.1, 2.4.4 and 2.4.5, we have:
$\mathcal{D}_{di} \subseteq \mathcal{D}_e \subseteq \mathcal{D}_1$; $\mathcal{D}_{di} \subseteq \mathcal{D}_u \subseteq \mathcal{D}_t \subseteq \mathcal{D}_1$. Taking convex hulls gives the result. ∎

We also have the following easy

Proposition 2.5.2. *The cone of dissimilarities is the convex hull of Robinsonian dissimilarities, i.e. $\mathcal{D}_+ = \mathrm{conv}(\mathcal{D}_r)$.*

Proof. The result is trivial if $n = 1$. Suppose then $n > 1$. Clearly, $\mathrm{conv}(\mathcal{D}_r) \subseteq \mathcal{D}_+$. Now, let $d \in \mathcal{D}_+$. Using the canonical basis introduced in Section 2.2.1: $d = \sum_{\{i,j\}} d(i,j) d^{\{i,j\}}$, the result follows, since any basic vector is trivially Robinsonian. ∎

We end this section with a result due to Durand (1989). The proof is rather simplified.

Proposition 2.5.3. *The cone of even dissimilarities is the convex hull of strongly-Robinsonian dissimilarities, i.e. $\mathcal{D}_{ev} = \mathrm{conv}(\mathcal{D}_{sr})$.*

Proof. Since $\mathcal{D}_{sr} \subseteq \mathcal{D}_{ev}$ (see Proposition 2.3.16), we have: $\mathrm{conv}(\mathcal{D}_{sr}) \subseteq \mathcal{D}_{ev}$. Since $\mathcal{D}_{ev}$ and $\mathcal{D}_{sr}$ are stable under replication (Proposition 2.6.1), it suffices to prove that every definite dissimilarity d is the sum of some definite strongly-Robinsonian dissimilarities. That is obvious for $n \leq 3$. Hence suppose $n > 3$. Put $\varepsilon = \min\{d(i,j)|(i,j) \in I^2, i \neq j\} > 0$. Let a, b, c obey $0 < a < b < c < b+c < \varepsilon$. For every pair $\{i,j\} \subset I$ define the definite dissimilarity $\underline{d}^{\{i,j\}}$ of $\mathcal{D}$ by: $\underline{d}^{\{i,j\}}(i,j) = d(i,j) - c$; $\underline{d}^{\{i,j\}}(i,k) = \underline{d}^{\{i,j\}}(j,k) = b$ for $k \notin \{i,j\}$; $\underline{d}^{\{i,j\}}(k,l) = a$ for $\{k,l\} \cap \{i,j\} = \phi$. Clearly any order having i and j as terminal elements is strongly-compatible with $\underline{d}^{\{i,j\}}$, so that all $\underline{d}^{\{i,j\}}$ are strongly-Robinsonian. We prove that we may choose a, b, c in such a way that $d = \sum_{\{i,j\}} \underline{d}^{\{i,j\}}$. For every pair $\{k,\ell\} \subset I$, we have:

$$\sum_{\{i,j\}} \underline{d}^{\{i,j\}}(k,\ell) = d(k,\ell) - c + 2(n-2)b + [n(n-1)/2 - (2n-3)]a$$

$$= d(k,\ell) - c + 2(n-2)b + [(n-3)(n-2)/2]a$$

Choosing $c = 2(n-2)b + [(n-3)(n-2)/2]a$, with for example $b = 2a$, a sufficiently small, we get the announced equality. ∎

Remark 2.5.1. In the proof of Proposition 2.5.3, we obtained, with the notation of Section 2.2.3: $\mathcal{D}_+^+ = \mathrm{conv}(\mathcal{D}_{sr}^+)$. □

2.6. When are the inclusions strict?

Consider again the inclusions between subsets of dissimilarities shown in Figure 2.4.1. In an obvious terminology, it will suffice to consider direct child-parent inclusions. Recall that each such inclusion $\mathcal{D}_* \subseteq \mathcal{D}_{**}$ holds for all n for which both subsets are defined. (Clearly, each subset $\mathcal{D}_*$ is defined for all $n \geq n_1$, with $n_1 = 2$ for $\mathcal{D}_{rs}$ and for $\mathcal{D}_{di}$, and with $n_1 = 1$ otherwise.) In this section we show that, for each such inclusion, there exists an n_0 such that $\mathcal{D}_* \subset \mathcal{D}_{**}$ if $n \geq n_0$ while $\mathcal{D}_* = \mathcal{D}_{**}$ if $n_1 \leq n < n_0$, where n_1 is the least n for which both subsets are defined. Our principal tool here is Proposition 2.2.8. We also establish what each n_0 is. This requires some nontrivial results and a number of counterexamples. For clarity, each n_0 is the number indicated by the line joining $\mathcal{D}_*$ to $\mathcal{D}_{**}$ in Figure 2.4.1.

Proposition 2.6.1. *Apart from the sequence $\{\mathcal{D}_s(n)\}_1^\infty$, each of the sequences $\{\mathcal{D}_*(n)\}_{n_1}^\infty$ with $\mathcal{D}_*$ shown in Figure 2.4.1 is stable under replication.*

Note that a dissimilarity space (I', d') obtained by replication from a dissimilarity space (I, d) where d is a star semi-distance, is embeddable in a weighted star-graph, although d' is not in general a star semi-distance. See Remark 2.3.5.

Proof. This is immediate for each sequence which corresponds to embeddability in a certain type of metric space and for the other sequences $\{\mathcal{D}_+\}_1^\infty$, $\{\mathcal{D}_{ev}\}_1^\infty$, $\{\mathcal{D}_{qh}\}_1^\infty$, $\{\mathcal{D}_r\}_1^\infty$ and $\{\mathcal{D}_{sr}\}_1^\infty$ by reference to the definition. ∎

The following proposition places lower bounds on n_0 for each of the inclusions in Figure 2.4.1. Parts (a) to (c) are obvious. Part (d) dates back to Deza (1962). We believe the proof of (d) given below to be original. See also Bandelt and Dress (1990, 1992). Part (e) is proved in Deza (1962, Theorems 3 and 4) and (f) in Avis and Mutt (1989). The first proof is rather long, and the second required the aid of a computer to perform an exhaustive (combinatorial) study. Both proofs are therefore omitted here.

Proposition 2.6.2.

(a) If $n = 1$, $\mathcal{D}_s = \mathcal{D}_u = \mathcal{D}_{ch} = \mathcal{D}_+$.

(b) If $n = 2$, $\mathcal{D}_{rs} = \mathcal{D}_{di} = \mathcal{D}_+$.

(c) If $n = 3$, $\mathcal{D}_s = \mathcal{D}_e = \mathcal{D}_\infty$ and $\mathcal{D}_{sr} = \mathcal{D}_{ev}$ and $\mathcal{D}_r = \mathcal{D}_+$.

(d) If $n = 4$, $\mathcal{D}_1 = \mathcal{D}_\infty$.

(e) If $n = 5$, $\mathcal{D}_1 = \mathcal{D}_h$.

(f) If $n = 6$, $\mathcal{D}_1 = \mathcal{D}_h$.

Proof of (d)

Let $d \in \mathcal{D}_\infty$. We show that $d \in \mathcal{D}_1$ in each of the three mutually exclusive and exhaustive cases:

(i) d is not definite.

(ii) d is definite and every triangle inequality is strict.

(iii) d is definite and, without loss of generality, $d(2,3) = d(1,2) + d(1,3)$.

We argue as follows:

(i) d not definite $\Rightarrow \tilde{d}$ is a tree semi-distance, by part (c).

$\Rightarrow d \in \mathcal{D}_1$, using $\mathcal{D}_t \subseteq \mathcal{D}_1$ and Proposition 2.6.1.

(ii) Let a be the largest real number such that $d' \equiv (d - ad_1) \in \mathcal{D}_\infty$. As $\mathcal{D}_1$ is a convex cone containing d_1, if $d' \in \mathcal{D}_1$ then $d \in \mathcal{D}_1$. But for d' a triangle inequality is an equality. It therefore suffices to prove (iii).

(iii) By Proposition 2.3.4, it suffices to note that

$$d = \alpha_2 d_{\{2\}} + \alpha_3 d_{\{3\}} + \alpha_4 d_{\{4\}} + \beta_2 d_{\{1,2\}} + \beta_3 d_{\{1,3\}}$$

in which each of the coefficients:

$$\beta_2 = \frac{1}{2}[d(1,3) + d(1,4) - d(3,4)], \quad \beta_3 = \frac{1}{2}[d(1,2) + d(1,4) - d(2,4)],$$

$$\alpha_2 = d(1,2) - \beta_3, \quad \alpha_3 = d(1,3) - \beta_2,$$

and

$$\alpha_4 = d(1,4) - (\beta_2 + \beta_3) = \frac{1}{2}[d(2,4) + d(3,4) - d(2,3)]$$

is nonnegative. ∎

Proposition 2.6.3. *Let $\mathcal{D}_* \subseteq \mathcal{D}_{**}$ denote any of the direct child-parent inclusions shown in Figure 2.4.1 and let both subsets be defined if and only if $n \geq n_1$. Then $\exists n_0 \geq n_1$ such that $\mathcal{D}_* = \mathcal{D}_{**}$ if $n_1 \leq n < n_0$ and $\mathcal{D}_* \subset \mathcal{D}_{**}$ if $n \geq n_0$. Moreover, each n_0 is as specified in Table 2.6.1. For any other inclusion in Figure 2.4.1, the existence of such an n_0, and its value, can then be deduced at once in the obvious way.*

Proof. For each inclusion, the penultimate column of Table 2.6.1 specifies which part(s) of Proposition 2.6.2 establish that $\mathcal{D}_* = \mathcal{D}_{**}$ if $n < n_0$. The final column of the same Table specifies one of the numbered counterexamples below as a dissimilarity d that belongs to $\mathcal{D}_{**}(n_0)$ but not to $\mathcal{D}_*(n_0)$. Propositions 2.2.8 and 2.6.1 complete the proof (the inclusion $\mathcal{D}_{rs} \subset \mathcal{D}_s$ for every $n \geq 3$ being obvious).

$\mathcal{D}_*$	$\mathcal{D}_{**}$	n_0	Part(s) of Proposition 2.6.2	Counterexample
$\mathcal{D}_{rs}$	$\mathcal{D}_s$	3	(b)	1
$\mathcal{D}_{rs}$	$\mathcal{D}_u$	3	(b)	2
$\mathcal{D}_{di}$	$\mathcal{D}_u$	3	(b)	2
$\mathcal{D}_{di}$	$\mathcal{D}_{ch}$	3	(b)	3
$\mathcal{D}_s$	$\mathcal{D}_t$	4	(a),(b),(c)	4
$\mathcal{D}_u$	$\mathcal{D}_t$	3	(a),(b)	1
$\mathcal{D}_u$	$\mathcal{D}_e$	3	(a),(b)	3
$\mathcal{D}_u$	$\mathcal{D}_{sr}$	3	(a),(b)	3
$\mathcal{D}_{ch}$	$\mathcal{D}_t$	3	(a),(b)	1
$\mathcal{D}_{ch}$	$\mathcal{D}_e$	3	(a),(b)	1
$\mathcal{D}_{ch}$	$\mathcal{D}_{sr}$	3	(a),(b)	1
$\mathcal{D}_t$	$\mathcal{D}_1$	4	(a),(b),(c)	5
$\mathcal{D}_e$	$\mathcal{D}_1$	4	(a),(b),(c)	5
$\mathcal{D}_1$	$\mathcal{D}_h$	7	(a) to (f)	6
$\mathcal{D}_h$	$\mathcal{D}_{qh}$	3	(a),(b)	7
$\mathcal{D}_h$	$\mathcal{D}_\infty$	5	(a),(b),(c),(d)	8
$\mathcal{D}_{qh}$	$\mathcal{D}_{ev}$	3	(a),(b)	9
$\mathcal{D}_\infty$	$\mathcal{D}_{ev}$	3	(a),(b)	7
$\mathcal{D}_{sr}$	$\mathcal{D}_{ev}$	4	(a),(b),(c)	5
$\mathcal{D}_{sr}$	$\mathcal{D}_r$	3	(a),(b)	10
$\mathcal{D}_{ev}$	$\mathcal{D}_+$	3	(a),(b)	10
$\mathcal{D}_r$	$\mathcal{D}_+$	4	(a),(b),(c)	5

Table 2.6.1: The values of n_0 for each child-parent inclusion in Figure 2.4.1

Each counterexample is given in the form of a symmetric table in which $d(i,j)$ occurs in the row labelled i and the column labelled j.

Ex.1	i	j	k	Ex.2	i	j	k	Ex.3	i	j	k
i	0	3	5	i	0	1	2	i	0	1	2
j		0	4	j		0	2	j		0	1
k			0	k			0	k			0

II. PARTIAL ORDER OF THE PRINCIPAL CLASSES OF DISSIMILARITY

Ex.4	i	j	k	ℓ
i	0	2	3	3
j		0	3	3
k			0	2
ℓ				0

Ex.5	i	j	k	ℓ
i	0	1	2	1
j		0	1	2
k			0	1
ℓ				0

Example 6 (Assouad, 1977, Proposition 2b; Avis, 1977, 1978, 1981 Theorem 2.3) give examples of dissimilarities $d \in \mathcal{D}_h \backslash \mathcal{D}_1$ when $n = 7$. Avis' example is easier to describe. It uses the graph G obtained from the complete graph K_7 by deleting the edge joining vertices 1 and 2 and that joining vertices 1 and 3. Each edge of G is then given unit weight and $d(i,j)$ is the usual shortest path length metric. Another example that is easy to describe is given by Assouad (1980,a). It uses the complete graph K_7 with a 5-cycle deleted.

Ex.7	i	j	k
i	0	1	4
j		0	1
k			0

Example 8 (Assouad, 1977, Proposition (2a)) gives the following example of a member of $\mathcal{D}_\infty$, (indeed of $\mathcal{D}_\infty \cap \mathcal{D}_{qh}$), which does not belong to $\mathcal{D}_h$. It corresponds to the squared distances between the 5 points in three dimensional Euclidean space whose coordinates are given in table Ex. 8* below.

Ex.8	i	j	k	ℓ	m
i	0	2	2	5	5
j		0	4	3	3
k			0	3	3
ℓ				0	4
m					0

Ex.8*	1	2	3
i	-1	0	0
j	0	1	0
k	0	-1	0
ℓ	1	0	1
m	1	0	-1

Ex.9	i	j	k
i	0	1	9
j		0	1
k			0

Ex.10	i	j	k
i	0	0	2
j		0	1
k			0

Note that this last example was given in Remark 2.3.7. ∎

2.7. The inclusions shown are exhaustive

In this section we establish that the inclusions shown in Figure 2.4.1 are exhaustive. That is, among the subsets of $\mathcal{D}_+$ shown there, no other inclusions are valid for all n for which the sets are defined.

Proposition 2.7.1. *Figure 2.4.1 presents the complete partial order by inclusion of the subsets of $\mathcal{D}_+$ which it contains.*

Proof. If $A \subseteq \ldots \subseteq Z$ and $A' \subseteq \ldots \subseteq Z'$ are two chains by inclusion of subsets of $\mathcal{D}_+$, then clearly:

$\exists$ no other inclusions among the sets $A, \ldots, Z, A', \ldots, Z'$

$\Leftrightarrow$ the sets $A \backslash Z'$ and $A' \backslash Z$ are nonempty.

In this case we say that the chains are parallel. Note that $A = Z$ and/or $A' = Z'$ is permitted. Consideration of Figure 2.4.1 shows that it is sufficient to establish that the following 9 pairs of chains are parallel:

A	Z	A'	Z'
$\mathcal{D}_{rs}$	$\mathcal{D}_s$	$\mathcal{D}_{di}$	$\mathcal{D}_{ch}$
$\mathcal{D}_u$	$\mathcal{D}_u$	$\mathcal{D}_{ch}$	$\mathcal{D}_{ch}$
$\mathcal{D}_s$	$\mathcal{D}_s$	$\mathcal{D}_u$	$\mathcal{D}_e$
$\mathcal{D}_s$	$\mathcal{D}_s$	$\mathcal{D}_{ch}$	$\mathcal{D}_r$
$\mathcal{D}_t$	$\mathcal{D}_t$	$\mathcal{D}_e$	$\mathcal{D}_e$
$\mathcal{D}_t$	$\mathcal{D}_{qh}$	$\mathcal{D}_{sr}$	$\mathcal{D}_r$
$\mathcal{D}_e$	$\mathcal{D}_\infty$	$\mathcal{D}_{sr}$	$\mathcal{D}_r$
$\mathcal{D}_{qh}$	$\mathcal{D}_{qh}$	$\mathcal{D}_\infty$	$\mathcal{D}_\infty$
$\mathcal{D}_{ev}$	$\mathcal{D}_{ev}$	$\mathcal{D}_r$	$\mathcal{D}_r$

This happens if and only if the 9 pairs of sets $A \backslash Z'$ and $A' \backslash Z$ are nonempty. In fact six of these conditions are redundant, given the other inclusion information in Figure 2.4.1. For example, $\mathcal{D}_{rs} \backslash \mathcal{D}_{ch} \neq \emptyset \Rightarrow \mathcal{D}_u \backslash \mathcal{D}_{ch} \neq \emptyset$. It is straightforward to show that it is sufficient that each of the twelve sets in Table 2.7.1 be nonempty.

The rightmost column of this table specifies a numbered example establishing that the corresponding set is nonempty. We use and extend the list of examples given in the proof of Proposition 2.6.3. In each case we give an example with the smallest possible value n_* of n. For $n < n_*$, the set in question is empty. The additional examples are:

Ex. 11: d_1 with $n = 3$.

Ex. 12: any star distance with $n = 4$ and $a_i = 0$ and $\min\{a_j, a_k, a_l\} > 0$.

Ex. 13: the star distance with $n = 4$, $a_i = 1$, $a_j = 2$, $a_k = 3$ and $a_l = 4$.

Ex. 14: the Euclidean distance generated by the 4 corners of the unit square.

II. PARTIAL ORDER OF THE PRINCIPAL CLASSES OF DISSIMILARITY   53

Set	n_*	Example	Set	n_*	Example
$\mathcal{D}_{rs}\backslash\mathcal{D}_{ch}$	3	11	$\mathcal{D}_{sr}\backslash\mathcal{D}_{qh}$	3	9
$\mathcal{D}_s\backslash\mathcal{D}_e$	4	12	$\mathcal{D}_{sr}\backslash\mathcal{D}_\infty$	3	7
$\mathcal{D}_s\backslash\mathcal{D}_r$	4	13	$\mathcal{D}_{ch}\backslash\mathcal{D}_u$	3	3
$\mathcal{D}_e\backslash\mathcal{D}_r$	4	14	$\mathcal{D}_e\backslash\mathcal{D}_t$	4	14
$\mathcal{D}_{qh}\backslash\mathcal{D}_\infty$	3	7	$\mathcal{D}_\infty\backslash\mathcal{D}_{qh}$	5	16
$\mathcal{D}_{di}\backslash\mathcal{D}_s$	4	15	$\mathcal{D}_r\backslash\mathcal{D}_{ev}$	3	10

TABLE 2.7.1: Examples establishing that the inclusions of Figure 2.4.1 are exhaustive

(Note: See the text for the Examples and for the meaning of n_*)

Ex. 15:

	i	j	k	ℓ
i	0	0	1	1
j		0	1	1
k			0	0
ℓ				0

Ex. 16: (unpublished work by Fichet and Le Calvé, cited in Joly and Le Calvé, (1986))

	i	j	k	ℓ	m
i	0	1	2	1	1
j		0	1	2	1
k			0	1	1
ℓ				0	2
m					0

This completes the proof. ∎

2.8. Discussion

The field of dissimilarity analysis as described above has existing or potential extensions in a number of directions. Some of these are briefly indicated in this final section. We have, somewhat arbitrarily, grouped our remarks under the following headings: (1) further mathematical study, (2) extensions to other types of data, (3) connections with neighbouring disciplines and (4) the future of dissimilarity analysis.

2.8.1. Further mathematical study

Among the many possibilities, we note here the following topics that merit further study:

(1) "Continuous" inclusions and transformations: For example, it is an immediate consequence of Bretagnolle, Dacunha-Castelle and Krivine (1966) that $\mathcal{D}_p$ is decreasing over $1 \leq p \leq 2$. Also, it can be shown that for all $\alpha > 0$ the set $\mathcal{D}_e^{(\alpha)}$ is increasing with α (Critchley, 1986,b) while clearly $\mathcal{D}_{rs}^{(\alpha)} = \mathcal{D}_{rs}$ and $\mathcal{D}_u^{(\alpha)} = \mathcal{D}_u$.

(2) Intersections of nonnested subsets of $\mathcal{D}_+$: It is of interest to study the intersection with $\mathcal{D}_+^+$ of each of the structures $\mathcal{D}_*$ studied above and, in particular, their convex hulls. Note for example that although $\text{conv}(\mathcal{D}_u) = \mathcal{D}_1$, $\text{conv}(\mathcal{D}_u^+)$ is strictly contained in $\mathcal{D}_1^+$. Assouad (1977, Proposition 2a), recalled as Example 8 in Section 2.6, has shown that $\mathcal{D}_{qh} \cap \mathcal{D}_\infty$ strictly contains $\mathcal{D}_h$ when $n = 5$. Recall that $\mathcal{D}_r \not\subseteq \mathcal{D}_{ev}$, although $\mathcal{D}_{sr} \subseteq \mathcal{D}_{ev}$, and that $\mathcal{D}_{sr} \not\subseteq \mathcal{D}_\infty$. It is therefore of interest to study $\mathcal{D}_r \cap \mathcal{D}_{ev}$, $\mathcal{D}_r \cap \mathcal{D}_\infty$ and $\mathcal{D}_{sr} \cap \mathcal{D}_\infty$. In Sections 2.3.8 and 2.3.9 we noted that it was necessary to restrict attention from $\mathcal{D}_r$ to $\mathcal{D}_r \cap \mathcal{D}_{ev}$ in order to arrive at a bijection with weakly-indexed pseudo-hierarchies (pyramids). In connection with $\mathcal{D}_\infty$, Durand (1989) gave a counterexample establishing that $\mathcal{D}_r \cap \mathcal{D}_\infty \not\subseteq \mathcal{D}_1$ when $n = 5$. This can be extended to $\mathcal{D}_{sr} \cap \mathcal{D}_\infty \not\subseteq \mathcal{D}_1$ by continuity arguments, noting that $\bar{\mathcal{D}}_{sr} = \mathcal{D}_r$. Recall also that $\mathcal{D}_{sr}$ and $\mathcal{D}_t$ are both extensions of $\mathcal{D}_u$. It is therefore of interest to study $\mathcal{D}_{sr} \cap \mathcal{D}_t$. As was effectively shown in Durand (1989), we have that $\mathcal{D}_{sr} \cap \mathcal{D}_t = \mathcal{D}_r \cap \mathcal{D}_t$. This set has been studied further by Batbedat (1989) and by Critchley (1994).

(3) Dimensionality theorems: For a review and extension of dimensionality theorems in multidimensional scaling and hierarchical cluster analysis, see Critchley (1986,a). In particular, Gower's conjecture that squared-Euclidean distance scaling never takes more dimensions than classical scaling is proved there.

(4) Approximations: Many interesting mathematical problems arise in connection with the general task of approximating a data dissimilarity by a representation dissimilarity as outlined in Section 2.1. There is a vast literature on this topic which we have not attempted to discuss in this chapter. Rather, we content ourselves here with noting the following in connection with least squares approximation. Suppose that $\mathcal{D}$ is endowed with an inner-product $< \cdot, \cdot >$ and let $\| \cdot \|$ denote the induced norm. Then, relative to this inner-product, the $(\mathcal{D}-)$ polar of a subset $\mathcal{D}_*$ of $\mathcal{D}$ is defined to be

$$\{d \in \mathcal{D} | \forall d_* \in \mathcal{D}_*, < d, d_* > \leq 0\}$$

and is denoted by $\mathcal{D}_*^*$. By definition, $\mathcal{D}_*^*$ is always a closed convex cone. Moreover, $\mathcal{D}_*$ is a closed convex cone $\Leftrightarrow \mathcal{D}_*^{**} = \mathcal{D}_*$. The importance of polar cones is illustrated in the following projection theorem that lies behind many practical methods of approximating an observed dissimilarity by a representation of a given type. (For these last two results, consult Rockafellar (1970).) Let $d \in \mathcal{D}$ and a nonempty closed convex cone $\mathcal{D}_* \subseteq \mathcal{D}$ be given. Then the problem:

$$\text{minimise } \|d - d_*\| \text{ over } d_* \in \mathcal{D}_*$$

has a unique solution $\hat{d}_*$ that, moreover, is characterised by:

(i) $\hat{d}_* \in \mathcal{D}_*$, (ii) $<\hat{d}_*, d - \hat{d}_*> = 0$ and (iii) $(d - \hat{d}_*) \in \mathcal{D}_*^*$.

One particularly useful example (Critchley, 1986,a) is that, with respect to the canonical Euclidean inner-product:

$$<d, d'> = \sum_i \sum_j d(i,j) d'(i,j),$$

one has:

$$\mathcal{D}_{qh}^* = \{d \in \mathcal{D} | B(d) \text{ is positive semi-definite}\}$$

where $B(d)$ is the matrix with $(i,j)^{\text{th}}$ element:

$$d(i,j) \text{ when } i \neq j, \text{ and } -\sum_{j \neq i} d(i,j) \text{ for } i = j$$

so that, in particular, $B(d)$ has zero row and column sums.

(5) Dissimilarities on sets of infinite cardinality: In this paper, we have restricted attention to dissimilarities on a finite set I. A natural extension is to sets I of infinite cardinality. Many instances of this extension have been mentioned in the paper in connection with embeddings in L_p-spaces. Another example arises when we wish to discuss (geodesic) distances between members of a family of probability distributions indexed by a continuously varying parameter. There are strong links between the finite and infinite cardinality cases. However, care is needed as some differences do arise. For example, Witsenhausen (1973, Corollary 1.1) has shown that a real normed space is hypermetric if and only if it is quasihypermetric. This contrasts with the present finite case where, as we have seen, hypermetric implies quasihypermetric but not conversely. An excellent general reference is the monograph by Wells and Williams (1975).

(6) Dissimilarities taking values in an ordered set: These have been studied, inter alia, by Janowitz (1978) and Barthélemy, Leclerc and Monjardet (1984). Recently it has been shown that a single, simple, order-theoretic result unifies and generalises seven familiar and fundamental bijections in mathematical classification. See Critchley and Van Cutsem (1994) who also extend these bijections to the infinite cardinality case. Moreover, dissimilarities taking values in an ordered set are connected in a natural way with preordonnances, cited in Section 2.2.4. These latter often arise in proximity analysis.

2.8.2. Extensions to other types of data

In this paper, we have concentrated on the case where the data take the form of a single dissimilarity in $\mathcal{D}_+$. This can be profitably extended in a number of ways, including the following:

(1) Rectangular not square data: Typically, $d(i,j)$ now measures the difference between objects i and j lieing in *distinct* sets I and J. For example, preferences of judges I between candidates J. See, for example, De Soete, Desarbo, Furnas and Carroll (1984) and Brossier (1987).

(2) The analysis of asymmetric data: There are many situations where symmetry does not necessarily hold. For example, flight times between given towns in Europe and North America are typically longer than those in the opposite direction because of the prevailing wind. For one recent review of models and methods in this area, together with some new proposals, the thesis by Bove (1989) may be consulted.

(3) Hybrid models in which the class of representations is some combination of "pure" classes. See, for example, the work of Carroll (1976) who establishes that

$$\mathcal{D}_u + \mathcal{D}_s \equiv \{d | d = d_u + d_s, d_u \in \mathcal{D}_u, d_s \in \mathcal{D}_s\} \subseteq \mathcal{D}_t.$$

(4) Problems with constraints on the representation: For example, all the points must lie on a Euclidean sphere.

(5) Three and higher way data: This area covers a number of logically distinct cases calling for separate consideration. It could, for example, include dissimilarities that are:

(a) independent replicates of the same experiment,

(b) observed at given values of relevant covariates,

or

(c) located in space, in time, or both.

For a recent review see, for example, Coppi and Bolasco (1989). In this context problems of finding a consensus often arise. See, for example, Barthélemy, Leclerc and Monjardet (1984, 1986).

Again, in particular situations it may be more convenient to define a dissimilarity measure between three (or more) objects. This approach may offer a new direction for theoretical and practical investigations. See Joly (1989).

2.8.3. Connections with neighbouring disciplines

The analysis of dissimilarity has close connections with a number of neighbouring disciplines. Exploring these provides useful insights and opens up the possibility of cross-fertilisation. They include links with spatial statistics, differential geometry and classical multivariate analysis. This last discipline is primarily concerned with data which, viewed geometrically, are inner-products rather than (squared) distances. Covariances and correlations, (more generally, association measures: see, for example, Zegers and Ten Berge (1985)), are examples of such data. Principal component and canonical correlation analysis are among the well-known corresponding methods of analysis. There are, of course, close links between the two deriving from the familiar expression generalising the Cosine Law. If ϕ embeds I in an inner-product space X then, in the usual notation,

$$\|\phi(i) - \phi(j)\|^2 = <\phi(i), \phi(i)> + <\phi(j), \phi(j)> -2<\phi(i), \phi(j)>.$$

These links are explored in depth in Critchley (1988) who obtains the spectral decomposition of the invertible linear map sending matrices of centred inner-products to their induced matrices of squared distances. Notwithstanding these links, certain fundamental differences remain. Notably, distances are, on the whole, more difficult to analyse mathematically. The principal reason for this is that, unlike an inner-product space, a distance space does not have a natural dual.

2.8.4. The future of dissimilarity analysis

We close with two observations on future prospects for dissimilarity analysis.

First we note that, despite recent advances, the subject is still for the most part Euclidean, static and second-order. That is, it is usually the case that Euclidean geometry is used to produce a fixed visual display which is of second order in the following senses: points are considered two-at-a-time, key quantities analysed are of second degree in the configuration coordinates (e.g. inner-products or squared distances) and the data consists of a two-way array. All of these constraints can productively be relaxed and, we believe, will be in future research. There are many other (differential) geometries apart from the Euclidean one, some of which we have reviewed above. Static visual displays are slowly giving way to dynamic ones. The potential of real-time interactive graphics is enormous. And higher order methods, in all three senses, can be explored.

Secondly, ideas from (statistical) modelling have much to offer exploratory data analysis (EDA, for short) and not just the other way round. Traditionally, dissimilarity analysis is seen as an EDA activity which can help (statistical) modelling, for example by identifying an outlying cluster of observations. Whether or not probability models can or should be introduced into dissimilarity analysis per se is a thorny question, not least because most plausible probability models for dissimilarity data are not tractable, and those that are tractable are often not very realistic. However, even while maintaining dissimilarity analysis as a purely EDA activity, we believe that it will prove helpful to transfer across to it ideas, if not entire methodologies, from the modelling world. The central questions "Which method for my data?" and "Does the chosen representation give a good fit?" remain. This suggests the widespread introduction of diagnostic checks, influence analysis, and robustness studies. And, perhaps most importantly of all, basic modelling considerations suggest replacing the essentially *point* estimate approach of much of dissimilarity analysis in which $\hat{d}_* \in \mathcal{D}_*$ represents d by a *set* or *interval* estimate approach in which a subset $\hat{\mathcal{D}}_*$ of $\mathcal{D}_*$ represents d.

Whatever the future holds for dissimilarity analysis, it will be best advanced by a healthy interplay between theory and practice, conducted internationally in a clear methodological framework, and exploiting the latest technology. We hope that this volume will make some small contribution in this direction.

Acknowledgements

The authors gratefully acknowledge the financial support provided by:

(a) an ESRC/CNRS joint award for Anglo-French collaboration on multivariate analysis.

(b) awards from the Ministère de la Recherche et de la Technologie and Warwick University Research and Innovations Fund for a European Network on Metric Structures for Dissimilarity Analysis.

References

The number(s) following a reference indicates, apart from the leading 2, the section(s) of this chapter where the reference is cited.

Aschbacher, M., Baldi, P., Baum, E.B., Wilson, R.M. (1987), Embeddings of ultrametric spaces in finite dimensional structures, *SIAM J. Alg. Disc. Meth.*, 8, pp. 564–577. [4.2].

Assouad, P. (1977), Un espace hypermétrique non plongeable dans un espace L^1, *C.R. Acad. Sci. Paris*, t. 285, Série A, pp. 361–363. [3.2, 6, 8.1].

Assouad, P. (1980a), Plongements isométriques dans L^1: aspect analytique, Séminaire d'Initiation à l'analyse, n° 14, Université de Paris-Sud, France, pp. 1–23. [3.2, 3.3, 3.4, 4.2, 6].

Assouad, P. (1980b), Caractérisations de sous-espaces normés de L^1 de dimension finie, Séminaire d'Analyse Fonctionnelle, Ecole Polytechnique, Palaiseau, France. [3.2].

Assouad, P. (1982), Sous-espaces de L^1 et inégalités hypermétriques, *C.R. Acad. Sci. Paris*, t. 294, Série 1, pp. 439–442. [4.2].

Assouad, P. (1984), Sur les inégalités valides dans L^1, *European J. Combin.*, 5, pp. 99–112. [3.2, 4.2].

Assouad, P., Deza, M. (1982), Metric subspaces of L^1, Publications Mathématiques d'Orsay, Dept. Mathématiques, Université de Paris-Sud, France. [3.2, 3.3, 4.2].

Avis, D. (1977), On the Hamming Cone, Technical Report 77-5. Dept. of Operations Research, Stanford University, USA. [3.2, 4.2, 6].

Avis, D. (1978), Hamming metrics and facets of the Hamming cone, Technical Report SOCS-78.4, School of Computer Science, McGill University, Montreal, Canada. [3.2, 6].

Avis, D. (1980), On the extreme rays of the metric cone, Canad. J. Math, 32, pp. 126–144. [2.3].

Avis, D. (1981), Hypermetric spaces and the Hamming Cone, Canad. J. Math, 33, pp. 795–802. [3.2, 3.3, 3.6, 4.2, 6].

Avis, D., Mutt (1989), All the facets of the six-point Hamming cone, European. J. Combin., 10, pp. 309–312. [6].

Banach, S. (1932), Théorie des opérations linéaires, Warsaw. [3.1].

Bandelt, H.J., Dress, A.W.M. (1990), A canonical decomposition theory for metrics on a finite set, Preprint 90-032, Diskrete Strukturen in der Mathematik, Universität Bielefeld 1, FRG,

Bandelt, H.J., Dress, A.W.M. (1992), A canonical decomposition theory for metrics on a finite set, Adv. Math., 1, pp. 47–104. [4.2, 6].

Barthélemy, J.P., Guénoche, A. (1988), Les arbres et les représentations des proximités, Masson, Paris. [3.6].

Barthélemy, J.P., Leclerc, B., Monjardet, B. (1984), Quelques aspects du consensus en classification, In Diday, E., et al., eds, Data Analysis and Informatics. North-Holland, Amsterdam, pp. 307–316. [8.1, 8.2].

Barthélemy, J.P., Leclerc, B., Monjardet, B. (1986), On the use of ordered sets in problems of comparison and consensus of classification, J. Classification, 3, pp. 187–224. [8.2].

Batbedat, A. (1989), Les dissimilarités Medas ou Arbas, Statist. Anal. Données, 14, pp. 1–18. [8.1].

Benzécri, J.P., et al. (1973), L'Analyse des Données. 1. La Taxinomie, Dunod, Paris. [2.4, 3.5].

Blake, I., Gilchrist, J. (1973), Addresses for graphs, IEEE. Trans. Inform. Theory, 19, pp. 683–688. [3.2].

Bove, G. (1989), New methods of representation of proximity data, Doctoral thesis, Università La Sapienza, Rome, Italy, (in Italian). [8.2].

Bretagnolle, J., Dacunha-Castelle, D., Krivine, J.L. (1966), Lois stables et espaces L^p, Ann. Inst. H. Poincaré Sect. B, II, n° 3, pp. 231–259. [3.2, 3.4, 4.2, 8.1].

Brossier, G. (1980), Représentation ordonnée des classifications hiérarchiques, Statist. Anal. Données, 5, pp. 31–44. [3.8].

Brossier, G. (1986), Problèmes de représentation des données par des arbres, Thèse d'Etat, Université de Haute Bretagne, Rennes 2, France. [3.6].

Brossier, G. (1987), Etude des matrices de proximité rectangulaires en vue de la classification, *Rev. Statist. Appl.*, 35, pp. 43–68. [8.2].

Buneman, P. (1974), A note on metric properties of trees, *J. Combin. Theory Ser. B*, 17, pp. 48–50. [3.6].

Cailliez, F., Pagès, J.P. (1976), *Introduction à l'Analyse des Données*, S.M.A.S.H., Paris. [4.2].

Carroll, J.D. (1976), Spatial, non-spatial and hybrid models for scaling, *Psychometrika*, 41, pp. 439–463. [3.7, 8.2].

Coppi, R., Bolasco, S. (1989), *Multiway Data Analysis*, North-Holland, Amsterdam. [8.2].

Crippen, G., Havel, T. (1988), *Distance Geometry and Molecular Conformation*, Wiley, New York. [1.1].

Critchley, F. (1980), Optimal norm characterisations of multidimensional scaling methods and some related data analysis problems, In Diday, E., et al., eds., *Data Analysis and Informatics*, North-Holland, Amsterdam, pp. 209–229. [1.2, 3.1].

Critchley, F. (1986a), Dimensionality theorems in hierarchical cluster analysis and multidimensional scaling, In Diday, E., et al., eds., *Data Analysis and Informatics 4*, North-Holland, Amsterdam, pp. 45–70. [2.4, 8.1].

Critchley, F. (1986b), Some observations on distance matrices, In de Leeuw, J., et al., eds., *Multidimensional Data Analysis*, DSWO Press, Leiden, pp. 53–60. [8.1].

Critchley, F. (1988), On certain linear mappings between inner-product and squared-distance matrices, *Linear Algebra Appl.*, 105, pp. 91–107. [8.3].

Critchley, F. (1994), On exchangeability-based equivalence relations induced by strongly Robinson and, in particular, by quadripolar Robinson dissimilarity matrices, In Van Cutsem, B., ed., *Classification and Dissimilarity Analysis*, Ch. 7, Lecture Notes in Statistics, Springer-Verlag, New York. [8.1].

Critchley, F., Fichet, B. (1993), The partial order by inclusion of the principal classes of dissimilarity on a finite set, and some of their basic properties, Joint Research Report, University of Warwick, U.K. and Université d'Aix-Marseille II, France. [1.2].

Critchley, F., Marriott, P.K., Salmon, M.H. (1992), Distances in statistics, In *Proceedings of the 36th meeting of the Italian Statistical Society*, CISU, Rome, pp. 36–60. [1.1].

Critchley, F., Van Cutsem, B. (1994), An order-theoretic unification and generalisation of certain fundamental bijections in mathematical classification, In Van Cutsem, B., ed., *Classification and Dissimilarity Analysis*, Ch. 4 and 5, Lecture Notes in Statistics, Springer-Verlag, New York. [3.5, 8.1].

Delclos, Th. (1987), Sur la représentation arborée en analyse des données, Mémoire D.E.A. Math. Appl. Université de Provence, Marseille, France. [3.6].

de Leeuw, J., Heiser, W., Meulman, J., Critchley, F. (1986), *Multidimensional Data Analysis*, DSWO Press, Leiden. [1.1].

De Soete, G., Desarbo, W.S., Furnas, G.W., Carroll, J.D. (1984), Tree representation of rectangular proximity matrices, In Degreef,E., Van Buggenhaut, J., eds., *Trends in Mathematical Psychology*, Elsevier Science Publishers, North-Holland, Amsterdam, pp. 377–392. [8.2].

Deza, M. (Tylkin) (1960), On Hamming geometry of unitary cubes, *Dokl. Acad. Nauk SSSR*, 134, pp. 1037–1040. [3.3, 4.2].

Deza, M. (Tylkin) (1962), On the realizibility of distance matrices in unit cubes, *Problemy Kibernetiki*, 7, pp. 31–45. [3.3, 6].

Deza, M. (Tylkin) (1969), Linear metric properties of unitary cubes (in Russian), *Proc. of the 4th Soviet Union Conference on Coding theory and transmission of information*, Moscow-Tashkent, pp. 77–85. [3.3].

Deza, M. (1992), Hypermetrics, ℓ_1-metrics and Delauney polytopes, *Proc. of Distancia 92. International Meeting on distance analysis*, Rennes. pp. 7–10. [3.3].

Deza, M., Grishukhin, V.P., Laurent, M. (1990a), Extreme hypermetrics and L-polytopes, Report No. 90668-OR. Institut für Okonometrie und Operations Research, Universität Bonn, FRG. [3.3].

Deza, M., Grishukhin, V.P., Laurent, M. (1990b), Hypermetric cone is polyhedral, *Combinatorica*, to appear. [3.3].

Diday, E. (1982), Croisements, ordres et ultramétriques: application à la recherche de consensus en classification automatique, Rapport de recherche, n° 144, I.N.R.I.A., Rocquencourt, France. [3.8, 4.2].

Diday, E. (1983), Croisements, ordres et ultramétriques, *Math. Sci. Hum.*, 2, pp. 31–54. [3.8, 4.2].

Diday, E. (1984), Une représentation visuelle des classes empiétantes : les pyramides, Rapport de recherche, n° 291, I.N.R.I.A., Rocquencourt, France. [3.8].

Dirichlet, G.L. (1850), Über die Reduction der positiven quadratischen formen mit drei unbestimmten ganzen zahlen, *J. Reine Angew. Math.*, 50, pp. 209–227. [3.1].

Dobson, A.J. (1974), Unrooted trees for numerical taxonomy, *J. Appl. Probab.*, 11, pp. 32–42. [3.6, 4.2].

Dor, L.E. (1976), Potentials and isometric embeddings in L_1, Israel J. Math., 24, pp. 260–268. [3.2, 4.2].

Durand, C. (1989), Ordres et graphes pseudo-hiérarchiques: théorie et optimisation algorithmique, Thèse de l'Université de Provence, Marseille, France. [3.8, 3.9 , 4.2, 5, 8.1].

Durand, C., Fichet, B. (1988), One-to-one correspondences in pyramidal representation: a unified approach, In Bock, H.H., ed., Classification and Related Methods of Data Analysis, North-Holland, Amsterdam, pp. 85–90. [3.9].

Escoufier, Y. (1975), Le positionnement multidimensionnel, Rev. Statist. Appl., 23, pp. 5–14. [4.2].

Fichet, B. (1983), Analyse factorielle sur tableaux de dissimilarité. Application aux données sur signes de présence-absence en médecine, Thèse de Biologie Humaine, Université d'Aix-Marseille II, France. [2.4, 3.1, 4.2].

Fichet, B. (1984), Sur une extension des hiérarchies et son équivalence avec certaines matrices de Robinson, Journées de Statistique, Montpellier, France. [3.9, 4.2].

Fichet, B. (1986), Data analysis: geometric and algebraic structures, First World Congress of Bernoulli Society, Tashkent, URSS, T1, pp. 75–77. [1.2, 3.2, 4.2, 5].

Fichet, B. (1987), The role played by L_1 in data analysis, In Dodge, Y., ed., Statistical Data Analysis based on the L_1-norm and Related Methods, North-Holland, Amsterdam, pp. 185–193. [2.2, 3.2, 4.2].

Fichet, B. (1988), L_p-spaces in data analysis, In Bock, H.H., ed., Classification and Related Methods of Data Analysis, North-Holland, Amsterdam, pp. 439–444. [3.2, 4.2, 5].

Fichet, B., Gaud, E. (1987), On Euclidean images of a set endowed with a preordonnance, J. Math. Psych., 31, pp. 24–43. [2.4, 3.1].

Fichet, B., Le Calvé, G. (1984), Structure géométrique des principaux indices de dissimilarité sur signes de présence-absence, Statist. Anal. Données, 9, pp. 11–44. [1.1].

Fraser, D.A.S., Massam, H. (1989), Mixed primal-dual bases algorithm for regression under inequality constraints: application to concave regression, Scand. J. Statist., 16, pp. 65–74. [1.2].

Fréchet, M. (1910), Les dimensions d'un ensemble abstrait, Math. Ann., 68, pp. 145–168. [2.7].

Fréchet, M. (1935), Sur la définition axiomatique d'une classe d'espaces vectoriels distanciés applicables vectoriellement sur l'espace de Hilbert, Ann. of Math., 36, pp. 705–718. [3.1].

Gaud, E. (1983), Représentation d'une préordonnance: étude de ses images euclidiennes, problèmes de graphes dans sa représentation hiérarchique, Thèse de 3ème cycle, Math. Appl., Université de Provence, Marseille, France. [3.8].

Gauss, K.F. (1831), Göttingsche gelehrte anzeigen, 2, 1075. [3.1].

Gelfand, A.E. (1971), Rapid seriation methods with archaeological applications, In Hodson, R.F., et al., eds., *Mathematics in the Archaeological and Historical Sciences*, Edinburgh University Press. [3.8].

Gower, J.C., Legendre, P. (1986), Metric and Euclidean properties of dissimilarity coefficients, *J. Classification*, 3, pp. 5–48. [1.1].

Guttman, L. (1968), A general nonmetric technique for finding the smallest coordinate space for a configuration of points, *Psychometrika*, 33, pp. 469–506. [2.4].

Hermite, C. (1850), Sur différents objets de la théorie des nombres, *J. Reine Angew. Math.*, 40, pp. 261–315. [3.1].

Hodson, R.F., Kendall, D.G., Tautu, P. (1971), *Mathematics in the Archaeological and Historical Sciences*, Edinburgh University Press. [3.8].

Holman, W. (1972), The relation between hierarchical and Euclidean models for psychological distances, *Psychometrika*, 37, pp. 417–423. [4.2].

Horn, R.A., Johnson, C.A. (1985), *Matrix Analysis*, Cambridge University Press. [3.1].

Hubert, L. (1974), Some applications of graph theory and related nonmetric techniques to problems of approximate seriation: the case of symmetric proximity measures, *British J. Math. Statist. Psych.*, 7, pp. 133–153. [3.8].

Janowitz, M.F. (1978), An order theoretic model for cluster analysis, *SIAM J. Appl. Math.*, 34, pp. 55–72. [8.1].

Jardine, N., Sibson, R. (1971), *Mathematical Taxonomy*, Wiley, London. [3.5].

Johnson, S.C. (1967), Hierarchical clustering schemes, *Psychometrika*, 32, pp. 241–254. [3.5].

Joly, S. (1989), On ternary distances $D(i,j,k)$, Second Conference of the International Federation of Classification Societies, Charlotteville, Virginia, USA. [8.2].

Joly, S., Le Calvé, G. (1986), Etude des puissances d'une distance, *Statist. Anal. Données*, 11, pp. 30–50. [4.2, 7].

Kelly, J.B. (1968), Products of zero-one matrices, *Canad. J. Math.* 20. pp. 298–329. [3.3].

Kelly, J.B. (1970a), Metric inequalities and symmetric differences, In Shisha, O., ed, *Inequalities II*, Academic Press, New York. pp. 193–212. [3.2, 3.3, 4.2].

Kelly, J.B. (1970b), Combinatorial inequalities, In Guy, R., Hanani, H., Saver, N., Schoenberg, J., ed., *Combinatorial Structures and their applications*, Gordon and Breach, New York, pp. 201–207. [4.2].

Kelly, J.B. (1972), Hypermetric spaces and metric transforms, In Shisha, O., ed, *Inequalities III*, Academic Press, New York, pp. 149–158. [3.3, 3.4, 4.2].

Kelly, J.B. (1975), Hypermetric spaces, In Kelly, L.M., ed, *The geometry of metric and linear spaces*, Springer-Verlag, Berlin, pp. 17–31. [3.3, 3.6, 4.2].

Krasner, M. (1944), Nombres semi-réels et espaces ultramétriques, *C.R. Acad. Sci. Paris*, 219, pp. 433–435. [3.5].

Kruskal, J.B. (1964), Multidimensional scaling by optimizing goodness of fit to a nonmetric hypothesis, *Psychometrika*, 29, pp. 1–27. [2.4].

Le Calvé, G. (1985), Distances à centre, *Statist. Anal. Données*, 10, pp. 29–44. [3.7, 4.2].

Le Calvé, G. (1987), L_1-embeddings of a data structure (I, D), In Dodge, Y., ed., *Statistical Data Analysis based on the L_1-norm and Related Methods*, North-Holland, Amsterdam, pp. 195–202. [3.2].

Lerman, I.C. (1970), *Les bases de la classification automatique*, Gauthier-Villars, Paris. [2.4].

Lew, J.S. (1978), Some counterexamples in multidimensional scaling, *J. Math. Psych.*, 17, pp. 247–254. [3.1].

Menger, K. (1931a), Bericht über metrische Geometrie, *Jahresber der deutschen Math.-Ver.*, 40, pp. 201–219. [3.1].

Menger, K. (1931b), New foundations of euclidean geometry, *Amer. J. Math.*, 53, pp. 721–745. [3.1].

Minkowski, H. (1891), Über die positiven quadratischen Formen und über kettenbruchahnliche Algorithm, *J. Reine Angew. Math.*, 107, pp. 278–297. [3.1].

Oxender, D., Fox, C. (1987), *Protein Engineering*, Alan Liss Inc., New York. [1.1].

Patrinos, A.N., Hakimi, S.L. (1972), The distance matrix of a graph and its tree realization, *Quart. Appl. Math.*, 30, pp. 255–269. [3.6].

Robinson, W.S. (1951), A method for chronological ordering of archaeological deposits, *American Antiquity*, 16, pp. 293–301. [3.8].

Rockafellar, R.T. (1970), *Convex Analysis*, Princeton University Press. [2.1, 8.1].

Schoenberg, I.J. (1935), Remarks to Maurice Fréchet's article "sur la definition axiomatique d'une classe d'espace distanciés vectoriellement applicable sur l'espace de Hilbert", *Ann. of Math.*, 36, pp. 724–732. [3.1, 3.4].

Schoenberg, I.J. (1937), On certain metric spaces arising from Euclidean spaces by a change of metric and their imbedding in Hilbert space, *Ann. of Math.*, 38, pp. 787–793. [3.1, 4.2].

Schoenberg, I.J. (1938), Metric spaces and positive definite functions, *Trans. Amer. Math. Soc.*, 44, pp. 522–536. [2.7, 3.1, 3.2, 3.4, 4.2].

Shepard, R.N. (1962), The analysis of proximities: multidimensional scaling with an unknown distance function I, II, *Psychometrika*, 27, pp. 125–140, pp. 219–246. [2.4].

Simoes-Pereira, J.M.S. (1967), A note on tree realizability of a distance matrix, *J. Combin. Theory Ser. B*, 6, pp. 303–310. [3.6].

Smolenskii, Y.A. (1969), A method for linear recording of graphs, *U.S.S.R. Comput. Math. and Math. Phys.*, 2, pp. 396–397. [3.6].

Torgerson, W.S. (1958), *Theory and methods of scaling*, Wiley, New York. [3.1].

Wells, J.H., Williams, L.R. (1975), *Embeddings and extensions in analysis*, Springer-Verlag, Berlin. [8.1].

Wilkeit, E. (1990), Isometric embeddings in Hamming graphs, *J. Combin Theory Ser. B*, 50, pp. 179–197. [3.6].

Witsenhausen, H.S. (1973), Metric inequalities and the zonoid problem, *Proc. Amer. Math. Soc.*, 40, pp. 517–520. [8.1].

Zaretskii, K. (1965), Constructing a tree on the basis of a set of distances between the hanging vertices, (in Russian), *Uspekhi. Mat. Nauk*, 20, pp. 90–92. [3.6].

Zegers, F.E., Ten Berge, J.M.F. (1985), A family of association coefficients for metric scales, *Psychometrika*, 50, pp. 17–24. [8.3].

Chapter 3.
Similarity functions[*]

Serge Joly[†]

Georges Le Calvé[†]

3.1. Introduction

The basic tool, in Statistics like in many branches of experimental sciences concerned with the study of information expressed in observations, is comparison analysis: in the field of statistical modelling, comparison to a theoretical model, in exploratory data analysis (EDA), comparison between data.

In EDA, these comparisons between data fall into two broad categories: analysis of similarities, when we measure how similar two objects look, analysis of dissimilarities, when we measure how different they are. These approaches are not contradictory, though each statistical technique is usually more specifically related to one or the other. For instance Principal Component Analysis (PCA) is related to the analysis of similarities (by means of covariances), Hierarchical cluster analysis (HCA) and additive trees to the analysis of dissimilarities (by means of distances) ; however, in both cases, we can associate with the index commonly used an index of the other category, in a natural way: with the covariance we can associate the Euclidean distance, while by taking the opposite of the ultrametric distance and adding a well chosen positive constant , we get an index of similarity.

Both examples show that these associations are basically worked out by means of decreasing functions ; it also appears that different models of functions are available: quadratic function (PCA), linear function (HCA).

We could of course use a linear link for the Euclidean geometry, or a quadratic link for HCA. Quite obviously these new indices would not be well adapted to the corresponding representation, and they would provide little if any lisibility.

This article is devoted to a number of functions that link indices of similarity and indices of dissimilarity. We call these links "similarity functions" (SF). We shall not make an analytical study of SF. We shall concern ourselves with their properties with respect to methods of representation.

---

[*] *In* Van Cutsem, B. (Ed.), (1994) *Classification and Dissimilarity Analysis*, Lecture Notes in Statistics, Springer-Verlag, New York.

[†] Université de Haute Bretagne, Rennes, France.

3.2. Definitions. Examples

3.2.1. Definitions

Let I be a finite set, $I = \{1, 2, ..., n\}$, and D and S be two $n \times n$-matrices.

Definition 3.2.1.

A $n \times n$-matrix D is a *dissimilarity on I* if and only if D is symmetric, with a null diagonal, and all the other terms are non-negative.

A $n \times n$-matrix D is a *semi-distance* if D is a dissimilarity and if the triangle inequality holds:
$$\forall (i, j, k) \in I^3, \ D_{ij} \leq D_{jk} + D_{kj}.$$

A $n \times n$-matrix D is *definite* (in the French literature "*propre*") if and only if
$$\forall (i, j) \in I^2, \ D_{ij} = 0 \iff i = j.$$

A $n \times n$-matrix D is *semi-definite* if and only if
$$D_{ij} = 0 \implies \left(\forall k \in I, \ D_{ik} = D_{jk} \right).$$

A semi-distance which is definite will be called a *distance*. □

Definition 3.2.2. A $n \times n$-matrix S is a *similarity on I* if and only if S is symmetric and
$$\forall (i, j) \in I^2, \ S_{ii} \geq S_{ij}$$

A similarity S is said to be *proper* if
$$\forall (i, j) \in I^2, \ S_{ii} > S_{ij}$$

A similarity S is said to be *normed* if
$$\forall i \in I, \ S_{ii} = 1.$$
□

Theorem 3.2.1. Given f a decreasing real function with $f(0) = 1$, and D a dissimilarity, let us define S such that
$$\forall (i, j) \in I^2, \ S_{ij} = f(D_{ij}).$$

Then S is a normed similarity. If f is strictly decreasing and D is definite, then S is proper.

Conversely, let g be a decreasing real function with $g(1) = 0$ and let S be a normed similarity. Then $D = g(S)$ is a dissimilarity.

If S is proper and g is strictly decreasing, then D is definite.

The proof is obvious. We will emphasize two points:

1. It is natural to consider using $g = f^{-1}$ and hence to take invertible functions. In that case any decreasing function is strictly decreasing.

2. Let g be a strictly decreasing function. In order for $g(S)$ to be a dissimilarity, S should have a constant diagonal. Then there is no point in looking for a dissimilarity $g(S)$ when S is not normed (to a multiplicative factor), and we restrict our choice of g's to invertible functions.

From now on, unless explicitly mentioned otherwise, we restrict ourselves to normed similarities and invertible functions.

3.2.2. Examples

We review here the main examples of application of Theorem 3.2.1.

3.2.2.1. Linear function

An example is $S = 1 - D$. Hence

$$\forall (i,j) \in I^2, \ s_{ij} = 1 - d_{ij} \quad \text{and} \quad d_{ij} = 1 - s_{ij}.$$

This is the most frequently used SF for qualitative "presence-absence" variables. (The reader will find a list of similarity indices for categorical variables in Appendix).

That kind of SF is well fitted to hierarchical representations (classifications, pyramids).

3.2.2.2. Homographic function

An example is

$$S = \frac{2}{1+D} - 1 \quad \Longleftrightarrow \quad S = \frac{1-D}{1+D}$$

$$D = \frac{2}{1+S} - 1 \quad \Longleftrightarrow \quad D = \frac{1-S}{1+S}$$

This SF is of interest mainly because of its analytical properties.

Many indices used for presence-absence variables are derived homographically from others. For example, Jaccard's distance is thus associated to the Sokal-Sneath-Anderberg similarity, and the Czenakowski-Dice distance to Jaccard's similarity. (See Appendix for definitions)

3.2.2.3. Quadratic function

An example is

$$S = 1 - \frac{1}{2}D^2 \quad \Longleftrightarrow \quad D = \sqrt{2(1-S)}$$

This formula is well fitted to the Euclidean representation, and specially to variables which are representable on a sphere centered at the origin O.

In that case s_{ij} is equal to the inner product $\overrightarrow{Oi} \cdot \overrightarrow{Oj}$ (the multiplicative coefficient $\frac{1}{2}$ for D^2 is necessary in order to represent s_{ij} as the inner product $\overrightarrow{Oi} \cdot \overrightarrow{Oj}$).

3.2.2.4. Exponential function

An example is

$$\forall\, (i,j) \in I^2, \ s_{ij} = e^{-d_{ij}} \quad \Longleftrightarrow \quad d_{ij} = -\ln s_{ij}$$

or, more generally

$$s_{ij} = e^{-d_{ij}^p}$$

This kind of SF is seldom used in data analysis. It is well adapted to multiplicatively transformed variables (economic growth rate, for instance).

The exponential $e^{-d_{ij}^p}$ is well fitted to representation in L^p-spaces.

Given $p = 2$ and the Euclidean distance

$$D^2 = \sum_i (x_i - y_i)^2$$

we once again observe the strong link between the Euclidean geometry and the normal distribution.

3.2.2.5. Circular function

An example is

$$\forall\, (i,j) \in I^2, \ s_{ij} = \cos d_{ij} \quad \Longleftrightarrow \quad d_{ij} = \operatorname{Arccos} s_{ij}$$

In this case, d_{ij} is set as an angle.

Angular distances are well adapted to such notions as "apparent distances" in astronomy, and spherical representations.

3.2.2.6. Graphical representations

Let I be a set with n elements, and D a dissimilarity on I. Let E be a metric space, and Δ a dissimilarity on E.

We shall say that the n points $M_1, M_2, \ldots, M_n$ are a representation of (I, D) into (E, Δ) if and only if
$$\forall (i,j) \in I^2, \ \Delta(M_i, M_j) = D_{ij}$$
The analogous definition also holds for similarities.

The representation depends on Δ, which has to be specified.

- The representation most commonly used is the euclidean representation: set $E = \mathbb{R}^n$ and Δ the distance associated to the inner product norm.
- If $E = \mathbb{R}^n$, another choice for Δ could be the L_1-distance.

Some dissimilarities will permit a euclidean representation, others a L_1-representation, and some will permit neither of them.

In fact the set of possible representations is wide. We will make a survey of some cases with the objective of pointing out the type of SF underlying each model.

Let $M_1, M_2, \ldots, M_n$ be n points lying on a Euclidean sphere (representation often used for categorical variables). In that case E is the sphere (with radius one) of $\mathbb{R}^n$. There are numerous eligible choices for the dissimilarity Δ, because the relative position of any two points can be described in many ways. We could use for instance the Euclidean distance (length of the chord M_iM_j) or the geodetical distance (length of the shortest arc M_iM_j).

In fact any quantity tending to zero as M_i tends to M_j can be choosen as a dissimilarity between M_i and M_j. Any quantity increasing as M_i tends to M_j can be choosen as a similarity.

Phrased differently, the dissimilarity between two points M_i and M_j on the euclidean sphere can be expressed in a variety of mathematical notions, some of them we shall now review. Using θ, the central angle $\widehat{M_iM_j}$, we can define

$d_1 = 1 - \cos\theta$

$d_2 = \theta$

$d_3 = $ arc θ's sagitta

$d_4 = $ the length of the chord M_iM_j

$d_5 = \sin\theta$

$d_6 = \tan\frac{\theta}{2}$

$s_1 = \cos\theta$

$s_2 = \frac{\pi}{2} - \theta$

$s_3 = $ arc θ's apothem

Here d_1 to d_6 may be viewed as measures of dissimilarities while s_1 to s_3 can be considered as measures of similarities. Furthermore, there is a linear link between s_1 and d_1, s_2 and d_2, s_3 and d_3, as between s_1^2 and d_5^2. Between s_1 and d_4, there is a quadratic link, an homographic one between s_1^2 and d_6^2.

It is worth remarking that, though it is of common use, with categorical variables, given a similarity S and a dissimilarity D, to produce a Euclidean representation on a sphere, this is absolutely not justified. On the one hand, it has been proved that most indices do not permit a Euclidean representation (Fichet, Le Calvé 1984). On the other hand, this representation makes use of s_1 and d_4, linked by a quadratic function, though a linear function links the given indices D and S.

It would be much more appropriate to use the couples (s_1, d_1), or (s_2, d_2), or else (s_3, d_3), more especially as by using them, most of the indices for categorical data can be exactly represented (Beninel, 1987).

3.3. The $W^M(D^p)$ forms

3.3.1. Definitions and properties

We will now concern ourselves with a very important family of SF, which we will call W-forms. To define a W-form, we first have to choose a point M of I that will act as origin for the form. Then a W-form will be a $n \times n$-matrix whose terms are linear combinations of D_{Mi}, D_{Mj}, D_{ij} hence the name "W-form".

Definition 3.3.1. Let M be a point belonging to I and D a dissimilarity on I. We call W-form of D evaluated at point M, denoted $W^M(D)$, the $n \times n$-matrix whose elements are
$$\forall (i,j) \in I^2, \ W^M(D)_{ij} = \frac{1}{2}\left(D_{Mi} + D_{Mj} - D_{ij}\right).$$

For p belonging to $\mathbb{R}_+$ we will also consider the following form, called "W-form of D^p evaluated at point M"
$$\forall (i,j) \in I^2, \ W^M(D^p)_{ij} = \frac{1}{2}\left(D_{Mi}^p + D_{Mj}^p - D_{ij}^p\right). \qquad \square$$

It is a known property that if D is a dissimilarity, then, for any positive p, D^p is also a dissimilarity.

Some remarks concerning the W-forms can be of interest.

- The multiplicative coefficient -1 for D_{ij}^p is not sufficient in order for $W^M(D^p)$ to be a decreasing function in D, because of the positive terms D_{Mi}^p and D_{Mj}^p. Thus Theorem 3.2.1 cannot be applied to W^p-forms.

- It is easy to show that usually $W^M(D^p)$ is not a similarity, because of the condition $W_{ii} \geq W_{ij}$ failing to be true (this condition is equivalent to $D_{Mi}^p \geq D_{Mj}^p - D_{ij}^p$).

- On the other hand, whenever D^p is a distance, then, for any M, $W^M(D^p)$ is an index of similarity.

- However, when D_{Mi} is a constant with respect to i, $W^M(D^p)$, as a function of D, is decreasing. It happens, for instance, for many indices defined on "presence-absence" variables, where there exists a point O (the null variable) such that $D_{Oi} = 1$ and $0 \leq D_{ij} \leq 1$. Then, for any p the $W^O(D^p)$ are indices of similarity. In that peculiar but important case, $W^O(D^p)$ can be rewrited as

$$\forall (i,j) \in I^2, \; W^O(D^p)_{ij} = 1 - \frac{1}{2} D^p_{ij}.$$

This property is no longer true when we consider the $W^M(D^p)$-form at any point M different from O: the $W^M(D^p)$-forms generally are not similarities.

From the definition of $W^M(D^p)$ we derive

$$\forall (i,j) \in I^2, \; W^M(D^p)_{ii} = D^p_{Mi}$$
$$D^p_{ij} = W^M(D^p)_{ii} + W^M(D^p)_{jj} - 2W^M(D^p)_{ij}$$

Finally, M and N being two points in I, we can explicit the relation between the W^M-form and the W^N-form:

$$\forall (i,j) \in I^2, \; W^M(D^p)_{ij} = W^N(D^p)_{ij} + X_i + X_j$$
$$\text{with} \qquad X_i = \frac{1}{2}\left(D^p_{Mi} - D^p_{Ni}\right)$$

- A similarity S and a dissimilarity D are said to be "W-associated" if and only if

$$\forall (i,j) \in I^2, \; D_{ij} = S_{ii} + S_{jj} - 2\,S_{ij}$$

This is not a one-to-one relation: a dissimilarity D can be W-associated to several similarities. Let S and T be two such similarities, W-associated to D. Then S and T are related by:

$$\forall (i,j) \in I^2, \; S_{ij} = T_{ij} + Z_i + Z_j$$
$$\text{with} \qquad Z_i = \frac{1}{2}(S_{ii} - T_{ii})$$

Finally we shall note that $W^M(D)$ is a $n \times n$-matrix with a null-line and a null-column (because $\forall i \in I$, $W^M(D^p)_{Mi} = 0$). We will frequently use the restriction of $W^M(D)$, i.e. the $(n-1) \times (n-1)$-matrix obtained by cutting both null-line and null-column.

Both following cases $p = 1$ and $p = 2$ are of great importance:
- $p = 1$
$$\forall (i,j) \in I^2,\ W^M(D)_{ij} = \frac{1}{2}\left(D_{Mi} + D_{Mj} - D_{ij}\right)$$

$W^M(D_{ij})$ is a kind of measures how the three points M, i, and j deviate from the straight line.

- $p = 2$
$$\forall (i,j) \in I^2,\ W^M(D^2)_{ij} = \frac{1}{2}\left(D^2_{Mi} + D^2_{Mj} - D^2_{ij}\right)$$

$W^M(D^2_{ij})$ derives from the Cosine Law: in a triangle ABC, we have
$$BC^2 = AC^2 + AB^2 - 2\,AB\,AC\,\cos A$$
which leads to $\overrightarrow{AB} \cdot \overrightarrow{AC} = \frac{1}{2}(AB^2 + AC^2 - BC^2)$ and $W^M(D^2)_{ij}$ can be viewed, if D is Euclidean, as the inner product $\overrightarrow{Mi} \cdot \overrightarrow{Mj}$.

3.3.2. The $W^M(D^2)$ form

We investigate now the properties of the $W^M(D^2)$-forms.

The following theorem is generally known as Frechet's theorem (1935). With some alterations in the expressing, it can also be fastened to Gauss, Minkowsky, or Schoënberg.

Theorem 3.3.1. *A dissimilarity matrix D can be considered as a distance matrix between n points of a Euclidean space if and only if there exists a point M such that the matrix $W^M(D^2)$ is non-negative definite (NND).*

The dimension of the representative space is equal to the rank of the matrix. If there exists an M such that $W^M(D^2)$ is NND, then $W^M(D^2)$ is NND for any M.

Proof. Consider the spectral form of $W^M(D^2)$:
$$\forall (i,j) \in I^2,\ W^M(D^2)_{ij} = \sum_{k=1}^{r} \lambda_k\, X_i^k\, X_j^k.$$

Then
$$W^M(D^2)_{ii} = \sum_{k=1}^{r} \lambda_k \left(X_i^k\right)^2$$

and, as
$$(D^2)_{ij} = W^M(D^2)_{ii} + W^M(D^2)_{jj} - 2\,W^M(D^2)_{ij},$$

it follows that

$$\forall (i,j) \in I^2, \ (D^2)_{ij} = \sum_{k=1}^{r} \lambda_k \left(X_i^k\right)^2 + \sum_{k=1}^{r} \lambda_k \left(X_j^k\right)^2 - 2 \sum_{k=1}^{r} \lambda_k X_i^k X_j^k$$
$$= \sum_{k=1}^{r} \lambda_k \left(X_i^k - X_j^k\right)^2$$

and this is the square of a Euclidean distance if and only if $\lambda_k \geq 0$.

Conversely, assume there exists a representation such that

$$\forall (i,j) \in I^2, \ (D^2)_{ij} = \sum_{k=1}^{r} \mu_k \left(Y_i^k - Y_j^k\right)^2 \quad \text{with} \quad \mu_k \geq 0.$$

Then, as

$$(D^2)_{Oi} = \sum_{k=1}^{r} \mu_k \left(Y_i^k\right)^2,$$

it follows that

$$W^O(D^2)_{ij} = \frac{1}{2} \sum_{k=1}^{r} \mu_k \left[\left(Y_i^k\right)^2 + \left(Y_{ij}^k\right)^2 - \left(Y_i^k - Y_j^k\right)^2 \right]$$
$$= \sum_{k=1}^{r} \mu_k Y_i^k Y_j^k$$

and since the μ_k are positive, this is a NND matrix.

If there exists a Euclidean representation, for any M the matrix of the inner products $\overrightarrow{Mi} \cdot \overrightarrow{Mj}$ is NND, and according to the Cosine Law this matrix is none other than $W^M(D^2)$. ∎

Torgerson form

We can consider the value of W in any point, provided we know its relative distances to all others. To overrule the arbitrary choice of point M, Torgerson (1958) suggested taking the value at the average point, i.e. the gravity centre G. We then need to compute the values of D_{Gi}^2.

Using Koenig's theorem, and defining

$$\forall i \in I, \ D_{i\bullet}^2 = \frac{1}{n} \sum_j D_{ij}^2 \quad \text{and} \quad D_{\bullet\bullet}^2 = \frac{1}{n} \sum_j D_{i\bullet}^2$$

we easily get

$$D_{Gi}^2 = D_{i\bullet}^2 - \frac{1}{2} D_{\bullet\bullet}^2$$

so that
$$\forall (i,j) \in I^2, \; W^G(D^2)_{ij} = \frac{1}{2}\left(D^2_{i\bullet} + D^2_{j\bullet} - D^2_{ii} - D^2_{\bullet\bullet}\right).$$

Now, in a Euclidean space, the centre of gravity belongs to that same space, so that $W^G(D^2)$ is NND if and only if $W^M(D^2)$ is NND, and we can apply the previous theorem to $W^G(D^2)$.

But we could prefer other choices. For example, instead of G, we could choose the average $W^{\bullet}(D^2)$ of the $W^M(D^2)$. That would mean defining a point H such that

$$\forall i \in I, D^2_{Hi} = D{i\bullet}^2 \quad \text{and} \quad W^{\bullet}(D^2) = W^H(D^2).$$

There is no difficulty in proving that W^M, W^G, and W^H are simultaneously NND.

If D is not Euclidean, it is worth remarking that in that case G would be the very wrong choice:

On one hand, the positivity of D^2_{Gi} is no longer secured. Since D^2_{Gi} is defined as

$$D^2_{Gi} = D^2_{i\bullet} - \frac{1}{2}D^2_{\bullet\bullet}.$$

It is evident that this quantity can be negative.

But the main point leading to the rejection of G in that case is that the distances were calculated by means of Koenig's theorem ... which stands only if D is a Euclidean distance, and thus $W^G(D^2)$ has no meaning!

These remarks do not apply to $W^M(D^2)$ and $W^{\bullet}(D^2)$.

If D is Euclidean, the main interest of G comes from the fact that when G is the origin, the factorial plane corresponds to the maximum of the inertia. When the origin is an arbitrary M, the factorial plane corresponds to the maximum of the moment of order 2 about M.

When considering subsets of I in the analysis, the W^G-form is not easy to use: adding one point to the data leads to an $(n+1) \times (n+1)$-matrix, whose elements all have to be recalculated, since the mean point G has changed.

On the other hand, the new matrix W^M is obtained by adding one row and one column to the former W^M: the W^M-form on a subset of I is a submatrix of the W^M-form on I.

This last property does not hold for the Torgerson-form, which is thus unfitted for mathematical induction on n.

3.3.3. Transformations of D

What can be done when D is not a Euclidean distance or when it is not even a distance? A first approach could consist in looking for approximative representations: this is the field of *multidimensional scaling* (in its broadest sense). We will select another way of approach, consisting in transforming D so that we get the wanted property: this is the field of changes of metrics.

Among the possible transformations, our choice will be in favour of those which least modify the informations upon D. Thus we will select monotonous functions because they are order-preserving.

Both following methods of transformations are often used:

- the additive constant: by adding a positive constant to every term of the matrix D (or to D^2) with the exception of the diagonal we obtain a distance or a euclidean distance.
- the D^α functions (generally with $0 \leq \alpha \leq 1$).

The case of the additive constant has already been largely considered. We shall only note that the constant often happens to be very large with respect to the values of d_{ij}, so that the distortion is important.

We will now consider the power functions D^α, and first define how to choose a in order for D^α to be a distance. Then we shall consider getting a euclidean distance.

Theorem 3.3.2.

a) The set of all α's such that D^α is a distance is a closed set.

b) If D is a semi-definite dissimilarity, let us put

$$k = \sup_{i,j} d_{ij} \qquad q = \inf_{i,j}\{d_{ij} : d_{ij} \neq 0\} \qquad \alpha = \frac{\ln 2}{\ln k - \ln q}$$

then α belongs to the set referred to in a)

c) If D is not semi-definite, D^α is a distance if and only if $\alpha = 0$.

Proof.

a) We need only prove that if D is a distance, D^α is a distance, for $\alpha \leq 1$. The triangle inequality holding for D also holds for D^α due to the inequality

$$\forall (a, b) \in \mathbb{R}_+, \forall \alpha \in [0, 1], \ a^\alpha + b^\alpha \geq (a + b)^\alpha.$$

b) This property follows from the fact that the triangle inequality holds for any dissimilarity Δ such that

$$\forall (i, j) \in I^2, \ i \neq j, \qquad \frac{1}{2} \leq \delta_{ij} \leq 1.$$

Let us put $k = \sup_{i,j} d_{ij}$. Let Δ be defined by $\delta_{ij} = \frac{1}{k} d_{ij}$. Then

$$\forall (i, j) \in I^2, \ 0 \leq \delta_{ij} \leq 1.$$

- If D is definite, then
$$\inf_{i,j} d_{ij} > 0.$$
Let us put
$$\alpha = \frac{\ln 2}{\ln(\sup d_{ij}) - \ln(\inf d_{ij})}$$
hence
$$\left(\frac{1}{k}\inf_{i,j} d_{ij}\right)^\alpha = \frac{1}{2}.$$

So Δ^α, with values between $\frac{1}{2}$ and 1, is a distance, and the same is true for D^α.

This value of α is not the supremum of all α's. The supremum is given by the formula
$$\alpha' = \sup_\alpha \{\inf_{i,j,k}\left(d_{ij}^\alpha + d_{jk}^\alpha - d_{ik}^\alpha\right) \geq 0\}.$$

- If D is semi-definite, let us define $I^0 = \{i : \exists j \text{ such that } d_{ij} = 0\}$ and $I^+ = I - I^0$. Then D restricted to I^+ is definite, and the precedent theorem holds. Furthermore, on I^0 D_{ij} is constant (and so is D_{ij}^α, and the proof of b) is achieved.

c) If D is not semi-proper, there exists i, j, k with $d_{ij} = 0$ and $d_{ik} \neq d_{jk}$. The triangle inequality applied to D^α holds only if $d_{ik}^\alpha = d_{jk}^\alpha$ and thus $\alpha = 0$. ∎

D^α and the Euclidean distances

It will be equivalent to prove that D is a Euclidean distance or that $W^M((f(D)^2)$ is NND. To that purpose the following property can be of great use:

Theorem 3.3.3. (generalized Schur's lemma)

Assume f to be a real function, such that

(C1) *The expansion of f into a serie about $t = 0$ exists, with radius of convergence r,*
$$f(t) = \sum_k c_k t^k, \quad \text{with } c_k \geq 0, \text{ continuous in } t \text{ at the point } t = 1.$$

Assume $A = [a_{ij}]$ to be a real matrix such that A is symmetric, NND, and $a_{ij} \leq r$. Let us define $B = [b_{ij}] = [f(a_{ij})]$ and $A^{*n} = [a_{ij}^n]$. Then

1) *the matrix B is NND,*
2) *the matrix B is not positive definite (PD) iff there exists an X such that,*
$$\forall n \in \mathbb{N}, \, X^t A^{*n} X = 0 \quad \text{(condition \textbf{C2})}.$$

Proof. (Joly, Le Calvé (1986)) Since condition (C1) refers to f and condition (C2) to A, it follows that

- for any f and any g both satisfying (C1)
$$\text{"}f(A) \text{ is PD"} \iff \text{"}g(A) \text{ is PD"},$$
- If $\lim_{n \to +\infty} A^{*n} = I$, then (C2) holds for A.
- If S is a similarity matrix with positive terms, then (C2) holds for S. ∎

Corollary 3.3.1. Assume S to be a NND similarity matrix. Then the matrices defined by their general term as follows are NND:

$$\frac{1}{1-s_{ij}} \qquad 1-\sqrt{1-s_{ij}} \qquad \forall \alpha \in [0,1],\ 1-\left(1-s_{ij}^{\alpha}\right)$$

$$\forall \alpha \geq 1,\ \frac{1}{1-s_{ij}^{\alpha}} \qquad e^{s_{ij}} \qquad \ln\frac{1+s_{ij}}{1-s_{ij}} \qquad \text{Arcsin } s_{ij}$$

For practical applications, it will be useful to remember that if D is a Euclidean distance, then Arccos D and D^{α} for $0 \leq a \leq 1$ are Euclidean too. This last characteristic of D^{α} implies that the set of α numbers such that D^{α} is Euclidean is a non empty set, since D^0 is Euclidean, and is a closed interval. Hence there exists a power of D that is Euclidean. Looking for the supremum of all these α's ($\alpha \leq 1$) seems an interesting method, competing with the additive constant technique. We don't know the infimum of this supremum, at least not for any general n.

Definition 3.3.2. A dissimilarity is said to be *"quasi-hypermetric"* if its square root is a Euclidean distance.

Corollary 3.3.2. *The following dissimilarities are quasi-hypermetric with full rank: Jaccard, Sokal-Sneath-Anderberg, Czenakowski-Dice, Rogers-Tanimoto, Russel-Rao, Ochiaï.*

The proof, and the definitions of these indices (for categorical "presence-absence" variables), can be found in Fichet, Le Calvé (1984) or in Gower, Legendre (1986). (See also Appendix).

3.4. The $W^M(D)$ form

3.4.1. Geometrical interpretations and properties

In Section 3.3, the definition of the matrix $W^M(D)$ was introduced by the relation

$$\forall (i,j) \in I^2,\ i \neq j,\ W^M(D)_{ij} = \frac{1}{2}\left(D_{Mi} + D_{Mj} - D_{ij}\right).$$

Let us define the *metric segment* $[AB]$ as

$$[AB]_{\text{met}} = \{M : D_{AM} + D_{MB} = D_{AB}\}.$$

In an affine space, we define a *vector segment* by

$$[AB]_{\text{vec}} = \{M : \overrightarrow{AM} = k\overrightarrow{MB},\ 0 \leq k \leq 1\}.$$

In a Euclidean space, both definitions are equivalent. But in a normed space, they differ if the norm on the vector space is not the one inducing the metric. For example, if, on $\mathbb{R}^2$, we define the L^1-norm, the metric segment $[AB]$ consists of all points within the rectangle with vertices A and B and whose sides are parallel to the axes.

If M belongs to the metric segment $[ij]_{\text{met}}$, $W^M(D)_{ij}$ is null and, in the general case, $W^M(D)_{ij}$ can thus measure how M deviates from the metric segment $[ij]$.

For another interpretation, let us suppose that D is a distance on a space E and consider the set A, the intersection of the metric segments $[Mi]$ and $[Mj]$:

$$A = [Mi] \cap [Mj]$$

The set A is non-empty since M belongs to A. Then,

$$\forall N \in A, \quad W^M(D)_{ij} = D_{MN} + W^N(D)_{ij}.$$

It follows that

$$\forall N \in A, \quad D_{MN} \leq W^M(D)_{ij}$$

Then $W^M(D)_{ij}$ can be considered as the length of the greatest metric segment included in both $[Mi]$ and $[Mj]$, and thus defines a kind of "metric similarity" between these two metric segments. Though $W^M(D)_{ij}$ is not an inner product, it sometimes plays a very analogous part.

Lastly, assume $X, Y, \ldots$ to be categorical variables. They can be viewed as characteristic functions (indicators) of some sets $X, Y, \ldots$. If we consider the Hamming metric $D_{X,Y} = |X \Delta Y|$, then

$$W^\emptyset(D)_{X,Y} = |X \cap Y|.$$

Theorem 3.4.1 lists some properties of $W^M(D)$.

Theorem 3.4.1. *Let D be a dissimilarity. Then*

- D is a distance iff
$$\forall M \in I, \forall (i,j) \in I^2, W^M(D)_{ij} \geq 0$$

- if D is a chain, then for every triple i, j, k one and only one of the three quantities $W^i(D)_{jk}$, $W^j(D)_{ik}$, $W^k(D)_{ij}$ is null.

- D is a Hamming metric if and only if $\forall M \in I$, $W^M(D)_{ij}$ is an integer, and if $W^\emptyset(D)_{ij} = |X_i \cap X_j|$

- D is quasi-hypermetric if and only if $W^M(D)$ is NND

- If D is an ultrametric, then $W^i_{jk} = W^j_{jk} = \frac{1}{2} D_{ik}$

- If D is an additive tree metric, then for any arbitrary triple (i, j, k), there exists an M such that $W^M_{ij} = W^M_{jk} = W^M_{ik} = 0$.

3.4.2. About metric projection

We defined earlier the notion of "metric segment". In the same way we can define the metric projection of a point onto a subset.

Definition 3.4.1. Given a subset A, we shall call *metric projection of i onto A* a point i^* such that

$$D_{ii^*} = \inf_{k,\ell \in A} W^i(D)_{k\ell} = \inf_{k,\ell \in A} \{\frac{1}{2}(D_{ik} + D_{i\ell} - D_{k\ell})\} \qquad \square$$

If there exists a point j such that $D_{ij} = D_{ii^*}$, j can be considered as i^*.

If there exists such a point j, and if D is a proper dissimilarity, then j is unique.

If there exists no j such that $D_{ij} = D_{ii^*}$, by adding to I the point i^* we define I^* by $I^* = I \cup \{i^*\}$, and by lining D with one column and one row, we get D^*.

The metric projection of i onto the total set I is called *"the foot of i"*.

On I^* (I completed with all the feet), we define δ_{ij} *"distance"* between the feet, such that

$$\forall (i,j) \in I^2,\ D_{ij} = D_{ii^*} + \delta_{ij} + D_{jj^*}$$

Then we can establish the following theorem.

Theorem 3.4.2. *With the above notations,*

- δ_{ij} *is a distance.*
- *the binary relation defined on I^2 by: "$iRj \Leftrightarrow \delta_{ij} = 0$" is an equivalence relation.*
- $(iRj) \iff (\forall k \in I, \forall x \in I,\ W^k_{ij} = W^k_{ix} = W^k_{jx})$.

We shall remark that δ is a distance, even when D fails to be one. Furthermore, the equivalence classes are all subsets on which $W^M(D)$ has a constant value (whatever D may be).

It should be noted that D is a star distance if and only if all points have the same foot. On an additive tree, the foot of a point is the node under which the point hangs. We deduce that D is *"representable"* by an additive tree if and only if, for any M, there exists a partition such that

$$\text{for any } i \text{ and } j \text{ belonging to distinct classes } W^M(D)_{ij} = D_{MM^*}.$$

(Le Calvé (1988)).

3.4.3. $W^M(D)$ and "$\mathcal{M}^1$-type" distance

Definition 3.4.2. A dissimilarity D is said to be an "$\mathcal{M}^1$-type" distance[†] iff the following equality holds:
$$D_{ij} = \sum_{k=1}^{r} |X_i^k - X_j^k|. \qquad \square$$

Such a distance is called "*city-block*" distance. It plays an important part in data analysis. Since rotating the axes causes the distances to change (rotations are not isometry), the axes have an intrinsic importance, which can be of interest in many problems. Furthermore the "$\mathcal{M}^1$-type" distance is the widest class of distances allowing easy to read representations, and permitting thus the best approximation (see Critchley, this volume). The absolute value makes the calculations difficult; it explains why there are so few available results. In particular, we know of no result similar to:

"D is Euclidean if and only if $W^M(D^2)$ is NND"

We will now establish such a result.

Definition 3.4.3. A symmetric $n \times n$-matrix M is said to be *realisable* if and only if it can be written
$$\forall (i,j) \in I^2, M_{ij} = \sum_k a_k X_i^k X_j^k \quad \text{with} \quad a_k \geq 0, \quad \text{and} \quad X_i^k \in \{0,1\}. \qquad \square$$

It may be noticed that this definition is very similar to the definition of a NND matrix. It was first used by Kelly (1972), with a simplificated form, owing to the fact that he was working with integers, and needed thus only to consider $a_k = 1$.

Theorem 3.4.3. *The dissimilarity D is of "$\mathcal{M}^1$-type" if and only if there exists an M such that $W^M(D)$ is realisable. If there exists an M such that $W^M(D)$ is realisable, then it is realisable for any M.*

Proof. Let us assume that $W^M(D)$ is realisable:
$$W^M(D)_{ij} = \sum_k a_k X_i^k X_j^k.$$

From the identity
$$D_{ij} = W^M(D)_{ii} + W^M(D)_{jj} - 2W^M(D)_{ij},$$
we deduce, as in the Euclidean case,
$$D_{ij} = \sum_k a_k \left[\{X_i^k\}^2 + \{X_j^k\}^2 - 2 X_i^k X_j^k \right]$$
$$= \sum_k a_k \left(X_i^k - X_j^k \right)^2$$
$$= \sum_k a_k |X_i^k - X_j^k|$$

[†] We use the $\mathcal{M}^1$ notation, for Minkowski spaces, and L_1 for normed spaces.

Conversely, if
$$D_{ij} = \sum_k a_k |X_i^k - X_j^k| = \sum_k a_k \left(X_i^k - X_j^k\right)^2,$$
it follows from the definition of $W^M(D)_{ij}$ that
$$W^M(D)_{ij} = \sum_k a_k X_i^k X_j^k.$$

∎

The following theorem strengthens the parallelism between NND matrices and realisable matrices.

Theorem 3.4.4. *Let f be a real function, whose expansion into a serie has positive coefficients and a convergence radius r, continuous at the point r. Then, if $A = [a_{ij}]$ is realisable, so is $B = [f(a_{ij})]$.*

Proofs of the Theorem 3.4.4 and of the below Corollary 3.4.1 can be found in Joly, Le Calvé (1992). The demonstration is analogous to that of Theorem 3.3.3.

Corollary 3.4.1. *The following indices of dissimilarity, defined for categorical variables, are "city-block" distances: Jaccard, Sokal-Sneath-Anderberg, Czenakowski-Dice, Rogers-Tanimoto, Russel-Rao, Ochiaï.*

This result completes the result of Corollary 3.3.2 and is a very strong incitation to use $\mathcal{M}^1$-type representation for categorical variables.

Appendix: Some indices of dissimilarity for categorical variables

Let I be a set of n individuals and J a set of p attributes. The $n \times p$-matrix X is a zero-one matrix defined by

$$X_{ik} = \begin{cases} 1 & \text{if the individual } i \text{ possesses the attribute } k, \\ 0 & \text{if the individual } i \text{ does not posses the attribute } k. \end{cases}$$

We define

n_{ij} to be the number of attributes common to i and j

$$n_{ij} = \sum_{k=1}^{p} X_{ik} X_{jk}$$

$n_{\bar{i}\bar{j}}$ to be the number of attributes missing both for i and j

$$n_{\bar{i}\bar{j}} = \sum_{k=1}^{p} (1 - X_{ik})(1 - X_{jk})$$

q_{ij} to be the number of disagreements between i and j

$$q_{ij} = \sum_{k=1}^{p} |X_{ik} - X_{jk}|$$

n_i to be the number of attributes the individual i possesses

$$n_i = \sum_{k=1}^{p} X_{ik}$$

We now list some of the most frequently used indices of similarity defined on categorical "presence-absence" variables.

III. SIMILARITY FUNCTIONS

RAO $\qquad S_1(i,j) = \dfrac{n_{ij}}{n}$

KULCYNSKI $\qquad S_2(i,j) = \dfrac{n_{ij}}{q_{ij}}$

JACCARD $\qquad S_3(i,j) = \dfrac{n_{ij}}{n_{ij} + q_{ij}}$

CZEKANOWSKI - DICE $\quad S_4(i,j) = \dfrac{n_{ij}}{n_{ij} + \frac{1}{2} q_{ij}}$

ANDERBERG $\qquad S_5(i,j) = \dfrac{n_{ij}}{n_{ij} + 2\, q_{ij}}$

ROGERS - TANIMOTO $\quad S_6(i,j) = (n_{ij} + n_{\bar{i}\bar{j}})(n_{ij} + 2\, q_{ij} + n_{\bar{i}\bar{j}})$

SOKAL - SNEATH $\qquad S_7(i,j) = (n_{ij} + n_{\bar{i}\bar{j}})(n_{ij} + \frac{1}{2}\, q_{ij} + n_{\bar{i}\bar{j}})$

Simple Matching $\qquad S_8(i,j) = \dfrac{n_{ij} + n_{\bar{i}\bar{j}}}{n}$

HAMMAN $\qquad S_9(i,j) = n_{ij} - q_{ij} + n_{\bar{i}\bar{j}}$

KULCYNSKI $\qquad S_{10}(i,j) = \dfrac{1}{2}\, \dfrac{n_{ij}}{n_i + n_{ij}\, n_j}$

ANDERBERG $\qquad S_{11}(i,j) = \dfrac{1}{4} \left(\dfrac{n_{ij}}{n_i} + \dfrac{n_{ij}}{n_j} + \dfrac{n_{\bar{i}\bar{j}}}{n_{\bar{i}}} + \dfrac{n_{\bar{i}\bar{j}}}{n_{\bar{j}}} \right)$

OCHIAI $\qquad S_{12}(i,j) = \dfrac{n_{ij}}{\sqrt{n_i\, n_j}}$

OCHIAI $\qquad S_{13}(i,j) = \dfrac{n_{ij}}{\sqrt{n_i\, n_j}}\, \dfrac{n_{\bar{i}\bar{j}}}{\sqrt{n_{\bar{i}}\, n_{\bar{j}}}}$

YULE $\qquad S_{14}(i,j) = \dfrac{n_{ij}\, n_{\bar{i}\bar{j}} - (n_i - n_{ij})(n_j - n_{ij})}{n_{ij}\, n_{\bar{i}\bar{j}} + (n_i - n_{ij})(n_j - n_{ij})}$

References

Al Ayoubi, B. (1991), Analyse des données de type M^1, Thèse, Université de Haute Bretagne, Rennes, France.

Beninel, F. (1987), Problèmes de représentations sphériques des tableaux de dissimilarité, Thèse de 3ème cycle, Université de Haute Bretagne, Rennes, France.

Fichet, B., Le Calvé, G. (1984), Structure géométrique des principaux indices de dissimilarité sur signes de présence-absence, *Statist. Anal. Données*, 9 (3), pp. 11–44.

Fréchet, M. (1935), Sur la définition axiomatique d'une classe d'espaces vectoriels distanciés applicables vectoriellement sur l'espace de Hilbert, *Ann. of Math.*, 36, pp. 705–718.

Gower, J.C., Legendre, P. (1986), Metric and Euclidean properties of dissimilarity coefficients, *J. Classification*, 3, pp. 5–48. [3].

Joly, S., Le Calvé, G. (1986), Metric and Euclidean properties of dissimilarity coefficients, *Statist. Anal. Données*, 11, pp. 30–50.

Joly, S., Le Calvé, G. (1992), Realisable 0-1 matrices and city-block distance, Rapport de recherche 92-1, Laboratoire Analyse des Données, Université de Haute Bretagne, Rennes, France.

Kelly, J.B. (1968), Products of zero-one matrices, *Canad. J. Math.*, 20, pp. 298–329.

Kelly, J.B. (1972), Hypermetric spaces and metric transforms, In: Shisha, O., ed, *Inequalities III*, Academic Press, New York, pp. 149–158.

Le Calvé, G. (1987), L_1-embeddings of a data structure (I, D), In: Dodge, Y., ed., *Statistical Data Analysis based on the L_1-norm and Related Methods*, North-Holland, Amsterdam, pp. 195–202.

Le Calvé, G. (1988), Similarities functions, In: Edwards, D., Raum, N.E., eds., *Proceedings of COMPSTAT 88*, Physica Verlag, Heidelberg, pp. 341–347.

Schoenberg, I.J. (1937), On certain metric spaces arising from Euclidean spaces by a change of metric and their imbedding in Hilbert space, *Ann. of Math.*, 38, pp. 787–793.

Schoenberg, I.J. (1938), Metric spaces and positive definite functions, *Trans. Amer. Math. Soc.*, 44, pp. 522–536.

Schur, J. (1911), Bemerkungen zur Theorie der beschrankter Bilinearformen mit unendlich vielen Veränderlichen, *J. Reine Angew. Math.*, 140, pp. 1–28.

Torgerson, W.S. (1958), *Theory and methods of scaling*. Wiley, New York.

Chapter 4.
An order-theoretic unification and generalisation of certain fundamental bijections in mathematical classification. I*

Frank Critchley[†]

Bernard Van Cutsem[††]

4.1. Introduction and overview

The primary objective of this and the following chapter is to show that one simple results unifies and generalises a number of familiar bijections in the mathematical classification literature. These include the classical bijections concerning ultrametrics on a finite set, obtained in Benzécri (1965), Hartigan (1967), Jardine, Jardine and Sibson (1967) and Johnson (1967), as well as their several later extensions reported in Jardine and Sibson (1971), Janowitz (1978), Barthélémy, Leclerc and Monjardet (1984) and the unpublished research report Critchley and Van Cutsem (1989).

The importance of such bijections cannot be overstated. They play a fundamental and indispensable role in the theory, methods and algorithms which underpin practical applications of mathematical classification.

The direct benefit of the present approach is that it brings a unity to the domain while at the same time providing a number of further generalisations. Generalisation is made in three senses. A real-valued dissimilarity on a finite set becomes a dissimilarity on an arbitrary set taking values in an arbitrary ordered set admitting a minimum. The symmetry requirement of dissimilarities is dropped and we work with the resulting predissimilarities. And bijections become (dual) order-isomorphisms, explained below. It also opens up a wide range of methodological advances, at least some of which we hope to explore in later work. These include the ability to treat multi-attribute dissimilarities, the provision of a mathematical framework in which to study the asymptotics of hierarchical cluster analysis, the natural accommodation of asymmetry, and the facility to extend to rectangular and multi-way data.

---

* *In* Van Cutsem, B. (Ed.), (1994) *Classification and Dissimilarity Analysis*, Lecture Notes in Statistics, Springer-Verlag, New York.

[†] University of Birmingham, U.K.

[††] Université Joseph Fourier, Grenoble, France.

We make explicit two obvious and general disclaimers. The present two chapters do not contain all the possible special cases of the general approach presented here. Neither is this approach the only possible one to the subject.

At the same time, we signal the following. In a related future paper, we plan to develop the present general approach by taking advantage of naturally occurring additional mathematical structure. This will enable us to capitalise on the fact that previous bijections are here shown to be (dual) order-isomorphisms. It will, in particular, permit the study of pyramids (Diday, 1984) and pseudo-hierarchies (Fichet, 1984) which involve a total order on the set S of objects to be classified. A variety of other natural structures is also possible upon the set S, the set E below (typically E is $S \times S \equiv S^2$ in mathematical classification), or the set L of admissible dissimilarity values, while the power set $\mathcal{P}(E)$ is, of course, always a Boolean lattice. Any such particular additional structure can be incorporated with advantage into the very general framework to be developed here.

Equally it would be both feasible and informative to explore the links between the present approach and others. These include the approaches based on preordonnances (Lerman, 1968, 1970), hypergraphs (Batbedat, 1988, 1990, 1991) and element-to-set dissimilarities (Zaks and Muchnik, 1989). We restrict ourselves here to remarking that the first and last of these appear to be intrinsically less general than the present approach. Some restriction on L (such as L is totally ordered) is necessary for predissimilarities to define a preordonnance. That is, a *total* preorder. And the fixed point approach of Genkin and Muchnik (1991) seems to make essential use of the fact that S is finite. After receiving an earlier version of the present work, Batbedat wrote (1991) indicating how his earlier paper (1988) can be generalised to the level of the present paper.

The organisation of this chapter is as follows. No previous knowledge of ordered sets is assumed. Instead, some self-contained notes on this subject are given in Section 4.2. Section 4.3 collects together for reference the definitions of the principal types of (pre)dissimilarity with which we shall be concerned and establishes some basic notation. Section 4.4 gives an overview of the seven existing bijections that will be unified and generalised. The single unifying and generalising result is given in Section 4.5. Section 4.6 to 4.9 then develop the necessary theory to show in Section 4.10 that this result does indeed embrace four of the bijections reviewed in Section 4.4. The other three bijections are dealt with in the following chapter.

In both this chapter and the following one, proofs are given in Appendices, while those parts of the chapters than can be omitted with little loss at a first reading are marked with an asterisk.

This chapter is based on Critchley and Van Cutsem (1992).

4.2. A few notes on ordered sets

4.2.1. Introduction

Dissimilarities, as usually defined, take values in the set $\mathbb{R}_+ \equiv [0, \infty)$ of real numbers greater than or equal to zero. Now this set has many properties: order-theoretic, arithmetic, topological, ... and so on. Janowitz (1978) observed that often nothing is lost (the same results remain true) and, moreover, insight is gained (into exactly which properties of $\mathbb{R}_+$ are being used) if we replace $\mathbb{R}_+$ by a more general set L. As in most of Janowitz' work, it will suffice for our present purposes to focus exclusively on certain of the order-theoretic properties of $\mathbb{R}_+$ and to reflect these in L. We view this as not only giving theoretical advantages, but also practical ones in terms of wider applicability as we indicate below.

First, we need some ordered sets terminology for which, with only a few exceptions, we follow Blyth and Janowitz (1972). Let L be a set and let R be a binary relation on L. Of the properties that R may enjoy, those most commonly encountered are as follows: R is said to be

(a) *reflexive* if $\forall \ell \in L$, $\ell R \ell$.

(b) *transitive* if $(\ell_1 R \ell_2$ and $\ell_2 R \ell_3) \Rightarrow (\ell_1 R \ell_3)$.

(c) *anti-symmetric* if $(\ell_1 R \ell_2$ and $\ell_2 R \ell_1) \Rightarrow \ell_1 = \ell_2$.

(d) *symmetric* if $\ell_1 R \ell_2 \Rightarrow \ell_2 R \ell_1$.

A relation R which satisfies (a), (b) and (d) is, of course, an *equivalence relation* on L. We shall be principally concerned here with relations R which satisfy (a), (b) and (c). Such a relation is called an *order relation* on L, or simply an *ordering* on L, and will usually be denoted by $\leq$. Note that some authors call this a partial order (cf. total order below). We also define $\ell_1 < \ell_2$ to mean $\ell_1 \leq \ell_2$ but $\ell_1 \neq \ell_2$.

By an *ordered set*, we mean a set L together with an ordering on it. So that self-dissimilarities can be sensibly defined, an ordered set must satisfy the following condition:

LMIN. L has a minimum element, denoted 0. That is, $0 \leq \ell$ for all $\ell \in L$.

Such an element is clearly unique, by antisymmetry.

It is of paramount importance to note that $\ell_1 \not\leq \ell_2$ and $\ell_2 < \ell_1$ are not logically equivalent in an ordered set. However, they are in an ordered set L which obeys the following condition:

LTO. L is *totally ordered*. That is, every pair of elements of L are *comparable*. In other words, $\forall (\ell_1, \ell_2) \in L^2$ either $\ell_1 \leq \ell_2$ or $\ell_2 \leq \ell_1$.

If neither $\ell_1 \leq \ell_2$ nor $\ell_2 \leq \ell_1$ holds, we say ℓ_1 and ℓ_2 are *incomparable* and write $\ell_1 \| \ell_2$.

As one example, which illustrates the practical advantages of moving from $\mathbb{R}_+$ to a more general ordered set L, consider the following situation. It is desired to simultaneously analyse the dissimilarity between objects in terms of several attributes. For example, between electoral candidates in terms of their foreign, economic and social policies. Then, if $\delta_i \in \mathbb{R}_+$ is their dissimilarity for the ith attribute for $i = 1, \ldots, p$ say, their multi-attribute dissimilarity is $\ell = (\delta_1, \ldots, \delta_p) \in \mathbb{R}_+^p$. Here we equip $\mathbb{R}_+^p$ with the order:

$$\ell \leq \ell' \Leftrightarrow \forall i = 1, \ldots, p : \delta_i \leq \delta_i'$$

where, on the right hand side, we use the standard ordering on the real numbers. Note that $\mathbb{R}_+^p$ ($p > 1$) obeys LMIN but not LTO. Thus multi-attribute dissimilarities can be incomparable. This and all other joint (or cross-attribute) information is, of course, ignored if p separate standard analyses, one for each attribute, are performed using $\mathbb{R}_+$. In contrast, such multi-attribute dissimilarities are easily analysed within the framework we describe below.

4.2.2. Duality and order-isomorphisms

Let R be a binary relation on a set L and let R^* denote its converse. That is, R^* is defined by $\ell_1 R^* \ell_2$ if and only if $\ell_2 R \ell_1$. It is readily seen that R^* is an ordering if (and only if) R is. We denote the converse of $\leq$ by $\geq$ and of $<$ by $>$. By the *dual* of an ordered set we mean the same set equipped with the converse order. When necessary, we use L^* to denote the dual of the ordered set L. This simple duality concept runs throughout ordered set theory giving, essentially at once, new dual definitions and results from old primal ones.

If P and Q are ordered sets, a mapping $\alpha : P \to Q$ is called *isotone* (resp. *antitone*) if $p_1 \leq p_2 \Rightarrow \alpha(p_1) \leq \alpha(p_2)$ (resp. $\alpha(p_1) \geq \alpha(p_2)$). Now the inverse of an isotone bijection need not be isotone. For a counter example, see Blyth and Janowitz (1972, p.6). We shall say that the ordered set P and Q are *order-isomorphic* if and only if there is an isotone bijection $\alpha : P \to Q$ such that α^{-1} is isotone. We say that $\alpha : P \to Q$ and $\alpha^{-1} : Q \to P$ establish this order-isomorphism. Thus, P and Q are order-isomorphic if and only if there is a mapping $\alpha : P \to Q$ which is onto and satisfies $p_1 \leq p_2 \Leftrightarrow \alpha(p_1) \leq \alpha(p_2)$.

Similarly, we say that P and Q are *dually order-isomorphic* if and only if there is an antitone bijection $\alpha : P \to Q$ whose inverse is also antitone. Clearly, P and Q are dually order-isomorphic $\Leftrightarrow P$ and Q^* are order-isomorphic.

(Dual) order-isomorphisms play a key role in this paper. They involve bijections which, moreover, preserve ordered set structure.

4.2.3. Semi-lattices and lattices*

Let L be any ordered set. A *filter* of L is any nonempty subset $\tilde{L}$ with the property that
$$\left(\ell \in \tilde{L} \text{ and } \tilde{\ell} \geq \ell\right) \Rightarrow \tilde{\ell} \in \tilde{L},$$
and a *principal filter* of L is any filter of the form
$$[\ell) \equiv \{\tilde{\ell} \in L : \tilde{\ell} \geq \ell\}.$$
We say that ℓ generates the principal filter $[\ell)$. Dually, an *ideal* of L is a filter of L^*, and a *principal ideal* of L is a principal filter of L^*. We denote by $(\ell]$ the principal ideal $\{\tilde{\ell} \in L : \tilde{\ell} \leq \ell\}$ generated by ℓ. A filter or ideal of L is called proper if it is a proper subset of L.

We say that an ordered set L is a *join semi-lattice* if it satisfies the following condition.

JSL. The set intersection of any two principal filters of L is a principal filter of L.

In this case we define the *join* of ℓ_1 and ℓ_2, denoted $\ell_1 \vee \ell_2$, by $[\ell_1) \cap [\ell_2) = [\ell_1 \vee \ell_2)$.

Dually, an ordered set is a *meet semi-lattice* if L^* is a join semi-lattice, in which case the *meet* of ℓ_1 and ℓ_2, denoted $\ell_1 \wedge \ell_2$, is defined by $(\ell_1] \cap (\ell_2] = (\ell_1 \wedge \ell_2]$.

Let $\tilde{L}$ be a non-empty subset of an ordered set L. Then $\ell \in L$ is called an *upper bound* of $\tilde{L}$ in L if $\forall \tilde{\ell} \in \tilde{L}$, $\tilde{\ell} \leq \ell$ and the (necessarily unique) *least upper bound* (or *supremum*) of $\tilde{L}$ in L if it is an upper bound of $\tilde{L}$ and such that, for every upper bound $\bar{\ell}$ of $\tilde{L}$ in L, $\ell \leq \bar{\ell}$. When it exists, this supremum is denoted by $\vee \tilde{L}$. It is easy to see that an ordered set L is a join semi-lattice if and only if every two element subset of L admits a supremum, in which case $\vee\{\ell_1, \ell_2\} = \ell_1 \vee \ell_2$. Thus, by induction, in a join semi-lattice every finite subset has a supremum and, moreover, $\vee\{\ell_1, \ell_2, \ldots, \ell_n\} = \ell_1 \vee \ell_2 \vee \ldots \vee \ell_n$. There are, of course, the dual notions of a *lower bound* of $\tilde{L}$ and its *greatest lower bound*, denoted $\wedge \tilde{L}$.

An ordered set L which forms, with respect to its ordering, both a join semi-lattice and a meet semi-lattice is called a *lattice*. For example, every totally ordered set, such as $\mathbb{R}_+$, is a lattice in which the join and meet of two elements are their maximum and minimum respectively. In particular, we have here that $\{\ell_1 \vee \ell_2, \ell_1 \wedge \ell_2\} = \{\ell_1, \ell_2\}$. More generally, $\mathbb{R}_+^p$ defined above is a lattice in which
$$\ell_1 \vee \ell_2 = (\max(\delta_{11}, \delta_{21}), \ldots, \max(\delta_{1p}, \delta_{2p}))$$
$$\text{and } \ell_1 \wedge \ell_2 = (\min(\delta_{11}, \delta_{21}), \ldots, \min(\delta_{1p}, \delta_{2p})).$$
Note that here $\ell_1 \vee \ell_2$ and $\ell_1 \wedge \ell_2$ are in general different from both ℓ_1 and ℓ_2.

A join semi-lattice is called *join-complete* if *every* non-empty subset admits a least upper bound. Clearly, if L is join-complete, then L contains a (necessarily unique) maximum element. The notion of meet-completeness is defined in a similar way. A lattice which is both join- and meet-complete is simply called *complete*. We say that a lattice is *bounded* if it has both a maximum and a minimum element. In particular, every complete lattice is bounded. Clearly, every finite lattice is complete.

4.2.4. Residual and residuated mappings*

Let P and Q be ordered sets. A mapping $\alpha : P \to Q$ is called *residual* if its preimage

$$\alpha^{\leftarrow}([q)) \equiv \{p \in P : \alpha(p) \in [q)\}$$

of every principal filter $[q)$ of Q is a principal filter of P, while a mapping $\alpha^{+} : Q \to P$ is called *residuated* if its preimage of every principal ideal of P is a principal ideal of Q.

Residuated and residual mappings enjoy many important properties: see especially the book by Blyth and Janowitz (1972) on Residuation Theory. In our present context, it turns out that residual maps play the more important role. We therefore denote by $\mathrm{Res}(P,Q)$ the set of all residual maps from P to Q and by $\mathrm{Res}^{+}(Q,P)$ the set of all residuated maps from Q to P.

Residual and residuated maps occur in natural pairings: see, for example, Section 2 of chapter 1 of Blyth and Janowitz (1972). Let α be a mapping from an ordered set P to an ordered set Q. Then it can be shown that the following two propositions are equivalent:

(1) α is residual.

(2) (a) α is isotone

and

(b) $\exists$ a (necessarily unique) isotone map $\alpha^{+} : Q \to P$ such that

$$\alpha^{+} \circ \alpha \leq id_P \text{ and } \alpha \circ \alpha^{+} \geq id_Q$$

where id_P denotes the identity map on P and mappings are ordered pointwise.

This unique α^{+} is residuated. Moreover, an isotone map $\alpha : P \to Q$ is residual if and only if $\forall q \in Q$, $\{p \in P : \alpha(p) \geq q\}$ is non-empty and admits a minimum element. Indeed, the unique α^{+} of (2) above is defined by

$$\forall q \in Q, \alpha^{+}(q) = \min\{p \in P : \alpha(p) \geq q\}.$$

Moreover, the mapping $\rho : \alpha \to \alpha^{+}$ establishes a natural dual order-isomorphism between $\mathrm{Res}(P,Q)$ and $\mathrm{Res}^{+}(Q,P)$. If $\alpha^{+} \in \mathrm{Res}(Q,P)$, then its associated residual map $\alpha = \rho^{-1}(\alpha^{+})$ is (well-) defined by

$$\forall p \in P, \alpha(p) = \max\{q \in Q : \alpha^{+}(q) \leq p\}.$$

Note also that by characterisation (2) above:

$$\alpha(p) \geq q \Leftrightarrow p \geq \alpha^{+}(q).$$

4.3. Predissimilarities

In this section we collect together for reference the definitions of the principal types of dissimilarity with which we shall be concerned. In fact, it costs nothing and will prove advantageous later, to be rather more general than that in three ways. We also establish some basic notation.

From now on S will denote an arbitrary nonempty set and L an arbitrary nonempty ordered set. Whenever, as here, we work directly with dissimilarities, we necessarily also suppose LMIN. In the usual approach, a dissimilarity is defined on a finite set taking values, as we noted above, in $\mathbb{R}_+$. We define more generally a dissimilarity on S taking values in L. Thus an L-dissimilarity on S, or simply a dissimilarity, is here defined to be a symmetric function $f : S \times S \to L$ which vanishes on the diagonal (that is, $\forall a \in S, f(a,a) = 0$). The set of all such is denoted $\mathcal{F}_\Delta$. Note that we write $f(a,b)$ in place of the more correct but cumbersome $f((a,b))$.

Secondly, it will cost nothing, and have the benefits of greater generality and mathematical ease, to drop the symmetry assumption. We therefore define an L-predissimilarity on S, or simply a predissimilarity, to be any function $f : S \times S \to L$ which vanishes on the diagonal. The set of all such is denoted $\mathcal{F}_0$.

Finally, although essential to the definition of predissimilarities, it turns out to be very advantageous to work at the higher level of generality in which the vanishing on the diagonal condition is dropped, knowing that it can always be added back in as a special case whenever required. (The same is true, of course, of the symmetry condition.) Thus, we let $\mathcal{F}$ denote the set of all functions $f : S \times S \to L$.

We summarise these definitions, and make a number of others, in the first two columns of Table 4.1 to which frequent reference will be made. Note that "even" and "definite" defined there become "semi-propre" and "propre" in the French literature. The final column of this table should be ignored at present. We remark that the definition of an ultrametric given there is rather general, when we recall that L is any ordered set obeying LMIN. When L is also totally ordered, so that $\max(\ell_1, \ell_2)$ is defined, we see that this reduces to the usual definition of an ultrametric in the mathematical classification literature, and that a definite ultrametric is then an ultrametric in the usual mathematical sense.

For later use, $B_f(a, \ell)$, or simply $B(a, \ell)$, denotes, for any predissimilarity $f \in \mathcal{F}_0$, and for any $a \in S$, $\ell \in L$, the closed ball centre a, radius ℓ determined by f. That is $\{b \in S : f(a,b) \le \ell\}$. Note that $a \in B(a, \ell)$ by definition. We also note here that, throughout this chapter and the following one, weak and strict set inclusion are consistently denoted by $\subseteq$ and $\subset$ respectively. Also, for any set E, $\mathcal{P}(E)$ denotes the power set of E. That is, the set of all subsets of E. These are ordered by inclusion. That is, $D \le D'$ means $D \subseteq D'$. When $E = S \times S$, we can and do identify $\mathcal{P}(E)$ with the set of all binary relations on S.

Name of $f \in \widetilde{\mathcal{F}}$	Definition of $\widetilde{\mathcal{F}} \subseteq \mathcal{F}$	Definition of $\widetilde{\mathcal{M}} \subseteq \mathcal{M}$
(a) Predissimilarity	$\mathcal{F}_o = \{f \in \mathcal{F} : \forall a \in S, f(a,a) = 0\}$	$\mathcal{M}_r = \{m \in \mathcal{M} : m \text{ is reflexive}\}^{(1)}$
(b) Symmetric	$\mathcal{F}_s = \{f \in \mathcal{F} : f(a,b) = f(b,a)\}$	$\mathcal{M}_s = \{m \in \mathcal{M} : m \text{ is symmetric}\}$
(c) Transitive	$\mathcal{F}_t = \{f \in \mathcal{F} : (f(a,b) \leq \ell \text{ and } f(b,c) \leq \ell) \Rightarrow f(a,c) \leq \ell\}$	$\mathcal{M}_t = \{m \in \mathcal{M} : m \text{ is transitive}\}$
(d) Dissimilarity	$\mathcal{F}_\Delta = \mathcal{F}_o \cap \mathcal{F}_s$	$\mathcal{M}_{rs} = \mathcal{M}_r \cap \mathcal{M}_s$
(e) Preultrametric	$\mathcal{F}_{ot} = \mathcal{F}_o \cap \mathcal{F}_t$	$\mathcal{M}_{rt} = \mathcal{M}_r \cap \mathcal{M}_t$
(f) Ultrametric	$\mathcal{F}_u = \mathcal{F}_o \cap \mathcal{F}_s \cap \mathcal{F}_t$	$\mathcal{M}_\sim = \mathcal{M}_r \cap \mathcal{M}_s \cap \mathcal{M}_t$
(g) Definite predissimilarity	$\mathcal{F}'_o = \{f \in \mathcal{F}_o : f(a,b) = 0 \Rightarrow a = b\}$	$\mathcal{M}'_r = \{m \in \mathcal{M}_r : m(0) \text{ is the identity equivalence relation}\}$
(h) Definite dissimilarity	$\mathcal{F}'_\Delta = \mathcal{F}_\Delta \cap \mathcal{F}'_o$	$\mathcal{M}'_{rs} = \mathcal{M}_{rs} \cap \mathcal{M}'_r$
(i) Definite ultrametric	$\mathcal{F}'_u = \mathcal{F}_u \cap \mathcal{F}'_o$	$\mathcal{M}'_\sim = \mathcal{M}_\sim \cap \mathcal{M}'_r$
(j) Right-even predissimilarity	$\mathcal{F}_{RE} = \{f \in \mathcal{F}_o : f(a,a') = 0 \Rightarrow a \overset{Rf}{\sim} a'\}^{(2)}$	$\mathcal{M}_{RE} = \{m \in \mathcal{M}_r : m(0) \text{ is } \overset{Rf}{\sim}\}$
(k) Left-even predissimilarity	$\mathcal{F}_{LE} = \{f \in \mathcal{F}_o : f(b',b) = 0 \Rightarrow b \overset{Lf}{\sim} b'\}^{(2)}$	$\mathcal{M}_{LE} = \{m \in \mathcal{M}_r : m(0) \text{ is } \overset{Lf}{\sim}\}$
(l) Even predissimilarity	$\mathcal{F}_E = \mathcal{F}_{LE} \cap \mathcal{F}_{RE}$	$\mathcal{M}_E = \mathcal{M}_{LE} \cap \mathcal{M}_{RE}$

Table 4.1

The subsets of $\mathcal{F}$ of principal interest and the corresponding subsets of $\mathcal{M}$

In the above table, $\mathcal{F}$ is the set of all maps $f : S \times S \to L$ and $\mathcal{M}$ is the set of all level maps $m : L \to \mathcal{P}(S \times S)$.

Notes on Table 4.1:

(1) We identify $\mathcal{P}(S \times S)$ with the set of all binary relations on S. A map $m : L \to \mathcal{P}(S \times S)$ is called reflexive (respectively, symmetric, transitive, or an equivalence) if $m(\ell)$ has this property for each $\ell \in L$.

(2) The equivalence relations $\overset{Rf}{\sim}$ and $\overset{Lf}{\sim}$ on S, called right- and left-indistinguishability by f, are defined by

$$a \overset{Rf}{\sim} a' \text{ if } \forall b \in S, f(a,b) = f(a',b)$$

and

$$b \overset{Lf}{\sim} b' \text{ if } \forall a \in S, f(a,b) = f(a,b').$$

4.4. Bijections

4.4.0. Overview

Table 4.2 gives an overview of the seven bijections reported in the literature that will be unified and generalised. The rows and column divisions of the table correspond to increasing generality of S and L respectively, (note that LMIN is necessarily assumed here). The new terms introduced in this table, such as "indexed hierarchy", are defined in Sections 4.4.1 to 4.4.7 below, where the numbered bijections are stated formally without proof. Where relevant, variants of these bijections or developments are also given. Along the way, we pick up three clues that there is a single structure underlying these apparently rather disparate results.

4.4.1. Indexed hierarchies (S finite, $L = \mathbb{R}_+$)*

When S is finite, an *indexed hierarchy (on S)* (Benzécri, 1965; see also Benzécri et al. 1973) is a pair $(\mathcal{H}, g)$, where $\mathcal{H}$ is a collection of subsets of S and g is a function from $\mathcal{H}$ to $\mathbb{R}_+$ which satisfy:

H0. The empty set $\emptyset \notin \mathcal{H}$.

H1. $\forall (H, H') \in \mathcal{H}^2$, $H \cap H' \in \{\emptyset, H, H'\}$.

H2. $S \in \mathcal{H}$.

H3. $\forall a \in S$, $\{a\} \in \mathcal{H}$.

I1. $H \subset H' \Rightarrow g(H) < g(H')$.

I2. $g(H) = 0 \Leftrightarrow H = \{a\}$ for some $a \in S$.

For each $f \in \mathcal{F}'_u$, we define $u_1(f)$ to be $(\mathcal{H}_f, g_f)$ where

$$\mathcal{H}_f = \{B_f(a, \ell) : a \in S, \ell \in \mathbb{R}_+\} \text{ and}$$

$$\forall H \in \mathcal{H}_f, g_f(H) = \max\{f(a, b) : a \in H, b \in H\}.$$

For each indexed hierarchy $(\mathcal{H}, g)$, we define $v_1(\mathcal{H}, g) \in \mathcal{F}_0$ by

$$\forall (a, b) \in S^2, (v_1(\mathcal{H}, g))(a, b) = \min\{g(H) : a \in H, b \in H, H \in \mathcal{H}\}.$$

Theorem 4.4.1. *Let S be finite and let $L = \mathbb{R}_+$. Then u_1 and its inverse v_1 establish a bijection between $\mathcal{F}'_u$ and the set of all indexed hierarchies on S.*

With an eye to later generalisations, we make two remarks.

	$L = \mathbb{R}_+$	L obeys JSL	L arbitrary
S finite	1. {definite ultrametrics} $\leftrightarrow$ {indexed hierarchies} 2. {ultrametrics} $\leftrightarrow$ {dendrograms} 3. {dissimilarities} $\leftrightarrow$ {numerically stratified clusterings}	7. {dissimilarities} $\leftrightarrow$ {residual maps $m : L \to \Sigma(S)$}	
S arbitrary	4. {ultrametrics} $\leftrightarrow$ {indexed regular generalised hierarchies} 5. {ultrametrics} $\leftrightarrow$ {generalised dendrograms} 6. {predissimilarities} $\leftrightarrow$ {"prefilters"}		SEE NOTE (3)

Table 4.2

The seven bijections to be unified and generalised

In the above table, the symbol $A \leftrightarrow B$ denotes a particular bijection between the sets A and B reported in the literature and reviewed here in Secion 4.4.

Notes on Table 4.2 :

(1) Bijections 4, 5 and 6 generalise bijections 1, 2 and 3 respectively.

(2) Bijection 7 generalises bijection 3.

(3) All seven bijections are generalised, in this chapter or in the following one, to the case where both S and L are arbitrary.

Remark 4.4.1. An alternative, equivalent way to define g_f is as follows:

$$\forall H \in \mathcal{H}_f, g_f(H) = \min\{\ell : H = B_f(a, \ell) \text{ for some } a \in S\}.$$

To see this, let $H = B(a_0, \ell_0)$ and suppose that $f(a, b)$ attains its maximum over $a \in H$ and $b \in H$ when $a = a^*$ and $b = b^*$. Note that, by transitivity of f, $f(a^*, b^*) \leq \ell_0$. Then clearly, $B(a_0, \ell_0) = B(a_0, f(a^*, b^*))$. Since f is ultrametric, $a^* \in B(a, \ell) \Rightarrow B(a, \ell) = B(a^*, \ell)$, as shown by Benzécri (and proved in greater generality in the following chapter). The proof is completed by noting that:

$$H = B(a, \ell) \text{ for some } a \in S \Rightarrow H = B(a^*, \ell), \text{ by the above}$$
$$\Rightarrow \ell \geq f(a^*, b^*), \text{ as } b^* \in H. \qquad \square$$

Remark 4.4.2. It is immediate from H1 and I1 that an alternative, equivalent way to define $v_1(\mathcal{H}, g)$ is as follows:

$$\forall (a, b) \in S^2, (v_1(\mathcal{H}, g))(a, b) = g(H_{ab})$$

where $H_{ab} = \cap \{H \in \mathcal{H} : a \in H, b \in H\}$. $\qquad \square$

4.4.2. Dendrograms (S finite, $L = \mathbb{R}_+$)*

In 1967 essentially the following bijection was published in separate papers by Hartigan, by Jardine, Jardine and Sibson, and by Johnson. Let $Eq(S)$ denote the set of all equivalence relations on S. Then, with S finite, a *dendrogram (on S)* is a function $m : \mathbb{R}_+ \to Eq(S)$ satisfying:

D1. m is isotone. That is, $\ell_1 \leq \ell_2 \Rightarrow m(\ell_1) \subseteq m(\ell_2)$.

D2. $\exists \bar{\ell} \in \mathbb{R}_+$ such that $m(\bar{\ell}) = S \times S$.

D3. $\forall \ell \in \mathbb{R}_+, \exists \varepsilon > 0$ such that $m(\ell) = m(\ell + \varepsilon)$.

For each $f \in \mathcal{F}_u$, define $u_2(f) : \mathbb{R}_+ \to \mathcal{P}(S \times S)$ by

$$\forall \ell \in \mathbb{R}_+, (u_2(f))(\ell) = \{(a, b) \in S^2 : f(a, b) \leq \ell\}.$$

For each dendrogram m on S, define $v_2(m) \in \mathcal{F}_0$ by

$$\forall (a, b) \in S^2, (v_2(m))(a, b) = \inf\{\ell \in \mathbb{R}_+ : (a, b) \in m(\ell)\}.$$

Because of D3, "inf" can be replaced by "min" in the definition of v_2.

Theorem 4.4.2. *Let S be finite and let $L = \mathbb{R}_+$. Then u_2 and its inverse v_2 establish a bijection between $\mathcal{F}_u$ and the set of all dendrograms on S.*

4.4.3. Numerically stratified clusterings (S finite, $L = \mathbb{R}_+$)*

Let $\Sigma(S)$ denote the set of all reflexive, symmetric relations on S. With S finite, a *numerically stratified clustering (NSC) (on S)* is a function $m : \mathbb{R}_+ \to \Sigma(S)$ satisfying conditions called NSC1, NSC2 and NSC3 (Jardine and Sibson, 1971) which are formally identical to D1, D2 and D3 respectively. The only difference lies in the range of m. Moreover, we define u_3 and v_3 in precisely the same way as u_2 and v_2 respectively with the appropriate change of domains (u_3 is defined on $\mathcal{F}_\Delta$ not $\mathcal{F}_u$ while v_3 is defined on the NSC's not on the dendrograms). These formal similarities are the first clue that there is a single general structure in play here.

Theorem 4.4.3. *Let S be finite and let $L = \mathbb{R}_+$. Then u_3 and its inverse v_3 establish a bijection between $\mathcal{F}_\Delta$ and the set of all numerically stratified clusterings on S.*

4.4.4. Indexed regular generalised hierarchies (S arbitrary, $L = \mathbb{R}_+$)*

Let $\Pi(S)$ denote the set of all partitions of S. Critchley and Van Cutsem (1989) defined an *indexed regular generalised hierarchy (IRGH) (on S)* to be a pair $(\mathcal{H}, g)$, where $\mathcal{H}$ is a collection of subsets of S and g is a function from $\mathcal{H}$ to $\mathbb{R}_+$ satisfying:

H0. $\emptyset \notin \mathcal{H}$.

H1. $\forall (H, H') \in \mathcal{H}^2$, $H \cap H' \in \{\emptyset, H, H'\}$.

GH2. $\forall a \in S, S = \cup \{H \in \mathcal{H} : a \in H\}$.

GH3. $\exists$ a terminal partition of S in $\mathcal{H}$. That is, $\exists \Pi \in \Pi(S)$ such that $\forall A \in \Pi$,

   (i) $A \in \mathcal{H}$

   and (ii) $(H \subseteq A$ and $H \in \mathcal{H}) \Rightarrow H = A$.

Note that, given H1, Π is unique.

I1. $H \subset H' \Rightarrow g(H) < g(H')$.

GI2. $g(H) = 0 \Leftrightarrow H$ belongs to the terminal partition of S in $\mathcal{H}$.

And the regularity conditions that $\forall (a, b) \in S^2$ and $\forall \ell \in \mathbb{R}_+$:

RH. $H_{ab} \in \mathcal{H}$ where $H_{ab} = \cap \{H \in \mathcal{H} : a \in H, b \in H\}$.

RH1. $H_a(\ell) \in \mathcal{H}$ where $H_a(\ell) = \{b \in S : g(H_{ab}) \leq \ell\}$.

RH2. $g(H_a(\ell)) \leq \ell$.

An IRGH is called *definite* if the terminal partition is the set of all singletons of S. That is, if H3 holds.

For each $f \in \mathcal{F}_u$, we define $u_4(f)$ to be $(\mathcal{H}_f, g_f)$ where

$$\mathcal{H}_f = \{B_f(a, \ell) : a \in S, \ell \in \mathbb{R}_+\}$$

IV. UNIFICATION AND GENERALISATION OF FUNDAMENTAL BIJECTIONS 99

and
$$\forall H \in \mathcal{H}_f, g_f(H) = \min\{\ell : H = B(a, \ell) \text{ for some } a \in S\}.$$

(Critchley and Van Cutsem show that the relevant minimum in the definition of $g_f(H)$ does indeed exist.) For each IRGH $(\mathcal{H}, g)$ we define $v_4(\mathcal{H}, g) \in \mathcal{F}_0$ by

$$\forall (a, b) \in S^2, (v_4(\mathcal{H}, g))(a, b) = g(H_{ab}).$$

Now, if S is finite, $(\mathcal{H}, g)$ is an indexed hierarchy if and only if it is a definite IRGH, (Critchley and Van Cutsem, 1989). Recalling Remark 4.4.1 and Remark 4.4.2, we see that u_4 and v_4 extend u_1 and v_1 respectively. Thus the following result does indeed generalise Benzécri's bijection (Theorem 4.4.1).

Theorem 4.4.4. Let $L = \mathbb{R}_+$. Then u_4 and its inverse v_4 establish a bijection between $\mathcal{F}_u$ and the set of all indexed regular generalised hierarchies on S. Their relevant restrictions are again mutually inverse and establish a bijection between $\mathcal{F}'_u$ and the set of all definite indexed regular generalised hierarchies on S.

4.4.5. Generalised dendrograms (S arbitrary, $L = \mathbb{R}_+$)*

Following Critchley and Van Cutsem (1989), a *generalised dendrogram (on S)* is a function $m : \mathbb{R}_+ \to Eq(S)$ satisfying:

GD1. $\forall \ell \in \mathbb{R}_+, m(\ell) = \cap\{m(\ell + \varepsilon) : \varepsilon > 0\}$.

GD2. $\cup\{m(\ell) : \ell \in \mathbb{R}_+\} = S \times S$.

A generalised dendrogram satisfying

DD. $m(0) = \{(a, a) : a \in S\}$

is said to be *definite*.

Clearly, every dendrogram on S is a generalised dendrogram on S. (For D2 $\Rightarrow$ GD2, while (D1 and D3) $\Rightarrow$ GD1.) Equally, it is clear that the converse is true *if S is a finite set*. The mappings u_5 and v_5 are formally identical to u_2 and v_2 respectively, with the appropriate change of domain of v_5. Again, "inf" can be replaced by "min" in the definition of v_5, here because of GD1. The following result does indeed therefore generalise Theorem 4.4.2.

However, it is not hard to see that, in general, a generalised dendrogram on S need not be a dendrogram on S. Whereas GD1 $\Rightarrow$ D1 always, neither D2 nor D3 is implied by both GD1 and GD2. For example, the function $c : \mathbb{R}_+ \to Eq(S)$ defined by

$$\forall \ell \in \mathbb{R}_+, c(\ell) = \{(a, a) : a \in \mathbb{R}\} \cup \{(a, b) \in \mathbb{R}^2 : \max(a, b) \le \ell\}$$

is a generalised dendrogram on $\mathbb{R}$ but is not a dendrogram on $\mathbb{R}$. Both conditions D2 and D3 fail. The same example with $\mathbb{R}$ replaced by the natural numbers provides a counterexample even for S countably infinite. Although D3 now holds, D2 still fails.

We now state the generalisation of Theorem 4.4.2.

Theorem 4.4.5. Let $L = \mathbb{R}_+$. Then u_5 and its inverse v_5 establish a bijection between $\mathcal{F}_u$ and the set of all generalised dendrograms on S. Their relevant restrictions are again mutually inverse and establish a bijection between $\mathcal{F}'_u$ and the set of all definite generalised dendrograms on S.

4.4.6. Prefilters (S arbitrary, $L = \mathbb{R}_+$)*

Critchley and Van Cutsem (1989) introduced the idea of a "prefilter" (on S). In the light of our subsequent order-theoretic generalisations, this is perhaps a rather unfortunate choice of name. However we keep it for historical consistency. No confusion should arise thereby. A "prefilter" on (S) is a function $\phi : S \times \mathbb{R}_+ \to \mathcal{P}(S)$ such that $\forall a \in S$ and $\forall \ell \in \mathbb{R}_+$:

PF0. $a \in \phi(a, 0)$.

PF1. $\phi(a, \ell) = \cap \{\phi(a, \ell + \varepsilon) : \varepsilon > 0\}$.

PF2. $\cup \{\phi(a, \ell) : \ell \in \mathbb{R}_+\} = S$.

A prefilter ϕ on S is called:

Symmetric if $b \in \phi(a, r)$ implies $a \in \phi(b, r)$.

Definite if $\forall a \in S, \phi(a, 0) = \{a\}$.

Right Even if $a' \in \phi(a, 0)$ implies $\forall \ell \in \mathbb{R}_+, \phi(a, \ell) = \phi(a', \ell)$.

To every prefilter ϕ we associate its transpose prefilter ϕ^T defined by $b \in \phi^T(a, \ell) \Leftrightarrow a \in \phi(b, \ell)$. A prefilter ϕ is called *left even* if ϕ^T is right even, and *even* if it is both left and right even.

For any $f \in \mathcal{F}_0$, let $u_6(f) : S \times \mathbb{R}_+ \to \mathcal{P}(S)$ be defined by:

$$\forall a \in S, \forall \ell \in \mathbb{R}_+ : (u_6(f))(a, \ell) = B_f(a, \ell) \equiv \{b \in S : f(a, b) \leq \ell\}$$

and for any prefilter ϕ, let $v_6(\phi) \in \mathcal{F}_0$ be defined by:

$$\forall (a, b) \in S^2, (v_6(\phi))(a, b) = \inf\{\ell \in \mathbb{R}_+ : b \in \phi(a, \ell)\}.$$

Here "inf" can be replaced by "min" because of PF1.

At least in the symmetric case, we may expect a close relation between prefilters and generalised dendrograms, by analogy with that between numerically stratified clusterings and dendrograms noted above. This is indeed the case. Axiom PF0 is simply the analogue of the "vanishing on the diagonal" property which characterises predissimilarities within the set of all functions from $S \times S$ to L. Recalling that in the generalised dendrogram m corresponding to an ultrametric f:

$$m(\ell) = \{(a, b) \in S^2 : f(a, b) \leq \ell\},$$

we see that axioms PF1 and PF2 are, in an obvious sense, the sectionings by each $a \in S$ of GD1 and GD2 respectively; and that u_6 and v_6 are defined by corresponding sectionings of u_2 and v_2. Thus the following result generalises Theorem 4.4.3 and strengthens the first clue that there is a single underlying result in play here.

The second clue, also illustrated in the following theorem, is the ability to obtain multiple new bijections simply by appropriate restrictions of a basic one.

Theorem 4.4.6. *Let $L = \mathbb{R}_+$. Then u_6 and its inverse v_6 establish a bijection between $\mathcal{F}_0$ and the set of all prefilters on S. Moreover, their relevant restrictions are again mutually inverse and establish a bijection between the symmetric (resp. definite or right even or left even or even) members of $\mathcal{F}_0$ and the symmetric (resp. definite or right even or left even or even) prefilters on S.*

4.4.7. Residual maps (S finite, L obeys LMIN and JSL)*

Janowitz (1978), recalled in Barthélemy, Leclerc and Monjardet (1984), obtained a different generalisation of Theorem 4.4.3 to that presented in the previous section. He kept S finite but greatly relaxed the conditions on L which here becomes any ordered set obeying LMIN as it must (see Section 4.2.1) and JSL (see Section 4.2.3).

Replacing $\mathcal{F}_u$ by $\mathcal{F}_\Delta$ and "dendrogram" by "residual map from L to $\Sigma(S)$" (see Section 4.2.4), we define u_7 and v_7 in a formally identical way to u_2 and v_2. Note that here "inf" can be replaced by "min" in the definition of v_7 because the preimage by m of the principal filter $[\{(a,b)\}]$ of $\Sigma(S)$ is a principal filter of L.

Recall that $\mathrm{Res}(L, \Sigma(S))$ denotes the set of all residual maps $m : L \to \Sigma(S)$ (see Section 4.2.4). We order $\mathcal{F}_\Delta$ and $\mathrm{Res}(L, \Sigma(S))$ pointwise. That is,

$$f \leq f' \text{ if } \forall (a,b) \in S^2, f(a,b) \leq f'(a,b), \text{ while } m \leq m' \text{ if } \forall \ell \in L, m(\ell) \subseteq m'(\ell).$$

Theorem 4.4.7. *Let S be finite and suppose L satisfies LMIN and JSL.*

(a) *u_7 and its inverse v_7 establish a dual order-isomorphism between $\mathcal{F}_\Delta$ and $\mathrm{Res}\,(L, \Sigma(S))$.*

(b) *For each $m \in \mathrm{Res}(L, \Sigma(S))$, the associated residuated map $m^+ : \Sigma(S) \to L$ is given by*

$$m^+(\emptyset) = 0 \text{ and } \forall \emptyset \neq D \subseteq S \times S, \, m^+(D) = \max\{f(a,b) : (a,b) \in D\}.$$

(c) *In particular, defining $f^+ \in \mathcal{F}_\Delta$ by $f^+(a,b) = m^+(\{(a,b)\})$, we have $f^+ = f$.*

Remark 4.4.3. We note that Janowitz (1978, lemma 4.1) began by showing that, when S is finite and $L = \mathbb{R}_+$, m is residual if and only if m is a numerically stratified clustering. □

The fact that we can throw away so much of the structure of $\mathbb{R}_+$ and still obtain a bijection of the relevant kind is the third and final clue that there is a single underlying result. And the fact that we get not only a bijection but actually a dual order-isomorphism confirms Janowitz' claim that much is to be gained from taking an order-theoretic approach.

However, it turns out that residual maps are not the key to unlocking the single general result. Rather we need the concept of a level map, introduced in the following section.

4.5. The unifying and generalising result

We are now ready to state the single result which, it will be shown, unifies and generalises the seven bijections reviewed in Section 4.4.

This result concerns the functions $f : E \to L$ where E is *any* nonempty set and, as declared above, L is *any* nonempty ordered set. In applications we shall often take E to be $S \times S$, where S is the arbitrary nonempty set declared above. However, we gain both insight and generality by working with a completely general set E.

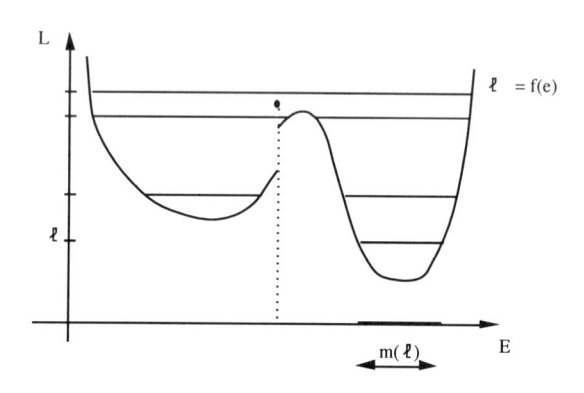

Figure 4.1

An example of the basic $f \leftrightarrow m$ bijection

This key result is at once both simple and general. Its essence is conveyed by Figure 4.1. There, E and L are taken to be subsets of $\mathbb{R}$ in order to have a picture but this is not essential to the argument. The Figure shows the graph $\{(e, f(e)) : e \in E\}$ of a function $f : E \to L$. The epigraph of f is the set $\{(e, \ell) : e \in E, \ell \in L, \ell \geq f(e)\}$. Given f, we can define a naturally associated map $m : L \to \mathcal{P}(E)$ which to each $\ell \in L$ associates the (possibly empty) corresponding *level set* $m(\ell) \equiv \{e \in E : f(e) \leq \ell\}$. Importantly, given a map $m : L \to \mathcal{P}(E)$ defined in this way, we can recover the original function $f : E \to L$ from it. This can be seen intuitively as follows. In Figure 4.1, a number of these level sets $m(\ell)$ have been translated parallel to the E axis to form one or more horizontal line segments in the epigraph of f at level ℓ. It is clear that by doing this for all the level sets $m(\ell)$ we fill out the whole of the epigraph of f. We then recover the graph of f (and hence f itself) as the lower boundary of this epigraph. Formally, we recover $f(e)$ as the least ℓ for which $e \in m(\ell)$. Thus we see, intuitively at least, that the function $f : E \to L$ and the so-called "level" map $m : L \to \mathcal{P}(E)$ are in a natural bijection.

We now formalise this in the general case and, further, introduce natural orders on the sets of functions f and m. Let $\mathcal{F}$ denote the set of all maps $f : E \to L$ and $\mathcal{N}$ the set of all maps $m : L \to \mathcal{P}(E)$. The function $u : f \to m_f$ associates to each $f \in \mathcal{F}$ the element $m_f \in \mathcal{N}$ defined by its level sets. That is:

$$\forall f \in \mathcal{F}, \forall \ell \in L : m_f(\ell) = \{e \in E : f(e) \leq \ell\}.$$

A map $m \in \mathcal{N}$ is called <u>level</u> if, in fact, $m \in \mathcal{M} \equiv u(\mathcal{F})$. Conversely, the function $v : m \to f$ associates to each level map $m \in \mathcal{M}$ <u>the</u> map $f_m \in \mathcal{F}$ such that $m = m_{f_m}$. This map f_m is indeed necessarily unique. For suppose that $f, f' \in \mathcal{F}$ satisfy $m_f = m_{f'}$ and let $e \in E$. As $\leq$ is reflexive, $e \in m_f(f(e))$. Hence, $e \in m_{f'}(f(e))$. That is, $f'(e) \leq f(e)$. Similarly, $f(e) \leq f'(e)$. It follows that $f = f'$, as $\leq$ is anti-symmetric and e was arbitrary.

The sets $\mathcal{F}$ and $\mathcal{N}$ are ordered pointwise. That is,

$$f \leq f' \Leftrightarrow \forall e \in E, f(e) \leq f'(e) \text{ and } m \leq m' \Leftrightarrow \forall \ell \in L, m(\ell) \leq m'(\ell),$$

where we recall that the members of $\mathcal{P}(E)$ are ordered by inclusion. Unless otherwise stated, every subset of an ordered set is supposed to be endowed with the order induced upon it.

Theorem 4.5.1. *The maps u and v establish a natural dual order-isomorphism between $\mathcal{F}$ and $\mathcal{M}$.*

The dual nature of this order-isomorphism is clear intuitively: the larger the function, the smaller its level sets.

Clearly, every restriction of u and v to a pair of corresponding subsets $\tilde{\mathcal{F}} \subseteq \mathcal{F}$ and $\tilde{\mathcal{M}} \subseteq \mathcal{M}$ (that is, $\tilde{\mathcal{M}} = u(\tilde{\mathcal{F}})$; equivalently $\tilde{\mathcal{F}} = v(\tilde{\mathcal{M}})$) induces a natural dual order-isomorphism between them.

The obvious question now is how to characterise when a map $m : L \to \mathcal{P}(E)$ is level. A first direct answer is given in the following theorem. For each $e \in E$ and for each $m \in \mathcal{N}$, we define $L_{e,m}$ to be $\{\ell \in L : e \in m(\ell)\}$. The following lemma is immediate.

Lemma 4.5.1.

(a) Let $m \in \mathcal{N}$. Then $\forall e \in E$, $L_{e,m} = m^{\leftarrow}([\{e\}))$.

(b) Let $m \in \mathcal{M}$. Then $\forall e \in E$, $L_{e,m} = [f_m(e))$.

Theorem 4.5.2. *Let $m \in \mathcal{N}$. Then the following three propositions are equivalent:*

(1) *m is level.*

(2) *(a) m is isotone and (b) $\forall e \in E$, $L_{e,m}$ has a minimum element.*

(3) *$\forall e \in E$, $L_{e,m}$ is a principal filter of L.*

Alternative answers to this characterisation question, which are of greater interest from our current perspective, are provided in the following sections. We first introduce some additional properties that the ordered set L may enjoy. In the same spirit as Janowitz (1978), we seek to isolate precisely which conditions on L, and possibly E, have been used implicitly in the existing mathematical classification literature where $L = \mathbb{R}_+$. We identify conditions under which the level maps can variously be characterised within $\mathcal{N}$ as the (to-be-defined) L-stratifications on E, the generalised L-stratifications on E

or the residual maps from L to $\mathcal{P}(E)$. In this last case, the links with the associated residuated maps are explored. These developments are then shown to generalise bijections 2, 3, 5 and 7 of Section 4.4. The other bijections (numbers 1, 4 and 6) are dealt with in the following chapter.

4.6. Further properties of an ordered set*

Table 4.3 lists together for reference a number of properties that an ordered set L map possess. The first three have been given in Section 4.2. The others will be used in the following sections. In connection with these latter, we note that $\forall \ell \in L,]\ell)$ denotes the (possibly empty) set $[\ell)\setminus\{\ell\}$, and that $]\ell) =]\ell')$ does not in general imply that $\ell = \ell'$.

LMIN. L has a minimum element, denoted 0.

LTO. L is totally ordered. That is, every pair of elements of L are comparable.

JSL. L is a join semi-lattice. That is, the set intersection of any two principal filters of L is a principal filter of L. Equivalently, every two element subset of L admits a supremum.

LUR. L is upper regular. That is,

$\forall \ell \in L,]\ell)$ nonempty $\Rightarrow \wedge]\ell)$ exists and is ℓ.

LWF. L is well-filtered. That is, every filter of L is either of the form $[\ell)$, or of the form $]\ell)$, for some $\ell \in L$.

LFILT. L is filter complete. That is, every proper filter of L admits an infimum.

Table 4.3

Some properties that ordered sets may enjoy

We collect various simple implications between these conditions on L into the following portemanteau proposition.

Proposition 4.6.1.

Let L be a nonempty ordered set. Then:

(a) LMIN $\Leftrightarrow L$ is a principal filter.

(b) LTO $\Rightarrow$ JSL.

(c) LWF $\Rightarrow$ LMIN.

(d) LFILT $\Rightarrow [\forall \ell \in L,]\ell)$ nonempty $\Rightarrow \wedge]\ell)$ exists$]$.

(e) (LFILT and LTO) $\Rightarrow$ [For each proper, non-principal, filter K of L, $K =]\ell_K)$ where $\ell_K = \wedge K$].

(f) (LFILT and LTO and LMIN) $\Rightarrow$ LWF.

(g) (LUR and LWF) $\Rightarrow$ LFILT.

We have the following useful result, which is now immediate from parts (c), (f) and (g) of the above.

Theorem 4.6.1. *Let L be a nonempty ordered set. Then:*

$$(\text{LTO and LUR}) \Rightarrow [\text{LWF} \Leftrightarrow (\text{LFILT and LMIN})].$$

Remark 4.6.1.

(a) Despite Proposition 4.6.1(d), LFILT does not imply LUR. Consider, for example, the set $L = \{0, 1\}$ with $0 < 1$. Then LFILT holds, but LUR fails since $\wedge]0) = 1$.

(b) The counterexamples presented in Remark 4.7.1 below show, inter alia, that the conditions LTO, LUR and LWF are logically independent. No two of them imply the third.

We are now ready to define the (generalised) stratifications and to explore when these are precisely the level maps.

4.7. Stratifications and generalised stratifications

A map $m \in \mathcal{N}$ is called an *L-stratification of E*, or simply a *stratification*, if it satisfies the following three conditions:

S1. m is isotone.

S2. $\exists \bar{\ell} \in L$ such that $m(\bar{\ell}) = E$.

S3. $\forall \ell \in L,]\ell)$ nonempty $\Rightarrow \exists \ell' \in]\ell)$ such that $m(\ell') = m(\ell)$.

The set of all stratifications is denoted $\mathcal{N}_S$. Note that these three conditions respectively parallel axioms D1 to D3 of a dendrogram, and also axioms NSC1 to NSC3 of a numerically stratified clustering.

Further $m \in \mathcal{N}$ is called an *generalised L-stratification of E*, or simply a *generalised stratification*, if it obeys the following two conditions:

GS1. m is upper semi-continuous. That is,

$$\forall \ell \in L,]\ell) \text{ nonempty} \Rightarrow m(\ell) = \cap \{m(\ell') : \ell' \in]\ell)\}.$$

GS2. m covers E. That is $\cup \{m(\ell) : \ell \in L\} = E$.

The set of all generalised stratifications is denoted $\mathcal{N}_G$. These two conditions parallel axioms GD1 and GD2 of a generalised dendrogram, and PF1 and PF2 of a "prefilter".

The following result is immediate from the definitions.

Proposition 4.7.1.

(a) *Every stratification is a generalised stratification. That is,* $\mathcal{N}_S \subseteq \mathcal{N}_G$.

(b) *Every upper semi-continuous map is isotone.*

(c) *Every level map* $m : L \to \mathcal{P}(E)$ *covers* E.

We can now identify sufficient conditions for $\mathcal{M} = \mathcal{N}_G$.

Theorem 4.7.1.

(a) LUR $\Rightarrow \mathcal{M} \subseteq \mathcal{N}_G$.

(b) LWF $\Rightarrow \mathcal{M} \supseteq \mathcal{N}_G$.

Thus,

(c) *(LUR and LWF)* $\Rightarrow \mathcal{M} = \mathcal{N}_G$.

It is natural to ask whether the sufficient conditions LUR and LWF are also necessary for the level maps to be precisely the generalised stratifications. A partial answer is provided in the following theorem. This theorem also begins our study of the case $\mathcal{M} = \mathcal{N}_S$. We need the following lemma.

Lemma 4.7.1.

(a) *Suppose LTO. Then, (m level* $\Rightarrow$ *m is upper continuous)* $\Rightarrow$ *LUR.*

(b) $\mathcal{M} \supseteq \mathcal{N}_S \Rightarrow$ *LMIN.*

(c) *Suppose LTO. Then,* $\mathcal{M} \supseteq \mathcal{N}_S \Rightarrow$ *LFILT.*

Together with Theorem 4.6.1 and Proposition 4.7.1(a), this lemma yields at once:

Theorem 4.7.2.

Suppose LTO. Then:

(a) $\mathcal{M} \subseteq \mathcal{N}_S \Rightarrow \mathcal{M} \subseteq \mathcal{N}_G \Rightarrow$ *LUR.*

(b) $\mathcal{M} \supseteq \mathcal{N}_G \Rightarrow \mathcal{M} \supseteq \mathcal{N}_S \Rightarrow$ *LFILT and LMIN.*

Thus:

(c) $\mathcal{M} = \mathcal{N}_G \Leftrightarrow$ *(LUR and LWF).*

(d) $\mathcal{M} = \mathcal{N}_S \Leftrightarrow \mathcal{M} = \mathcal{N}_S = \mathcal{N}_G \Leftrightarrow$ *(LUR and LWF and* $\mathcal{N}_G \subseteq \mathcal{N}_S$*).*

Partly with part (d) of this theorem in mind, we now explore when $\mathcal{N}_G \subseteq \mathcal{N}_S$ (and hence $\mathcal{N}_G = \mathcal{N}_S$) holds. By Proposition 4.7.1(b), every generalised stratification obeys S1. We also have the following two simple lemmas.

Lemma 4.7.2. *Suppose either (i) JSL and E is finite, or (ii) L is a complete lattice. Let $m \in \mathcal{N}$. Then: (m isotone and m covers E) $\Rightarrow$ m obeys S2.*

For each $\ell \in L$, $(\ell[$ denotes the (possibly empty) set $(\ell]\setminus\{\ell\}$. Let $m \in \mathcal{N}$. For each $\ell \in L$ for which $(\ell[$ is nonempty, we define $m(\ell-)$ to be $\cup\{m(\ell') : \ell' \in (\ell[\}$. Thus, whenever it is defined, $m(\ell-) \subseteq m(\ell)$ for an isotone map m. We say that $\ell \in L$ is a *jump point* of an isotone map $m \in \mathcal{N}$ if $(\ell[$ is nonempty and, in fact, $m(\ell-) \subset m(\ell)$. Recalling Theorem 4.5.2, we now have:

Lemma 4.7.3. *Suppose LTO. Let $m \in \mathcal{M}$. Then:*

(a) *ℓ is a jump point of m $\Leftrightarrow$ $(\ell[$ is nonempty and $\exists e \in E$ such that $\ell = f_m(e)$.*

(b) *$m(\ell) \subset m(\ell') \Leftrightarrow \exists$ a jump point $\tilde{\ell}$ of m with $\ell < \tilde{\ell} \le \ell'$.*

If L is totally ordered, the question of whether or not $m(\ell-)$ is defined is easily resolved. In this case, if LMIN, $m(\ell-)$ is defined for all ℓ other than the minimum; otherwise, it is defined for every $\ell \in L$.

In view of Lemma 4.7.3(a), we see that if LTO and if E is finite, then every level map has a finite number of jump points.

We are now ready to state the next theorem.

Theorem 4.7.3. *(LTO and LUR and LWF and E is finite) $\Rightarrow$ $\mathcal{M} = \mathcal{N}_S = \mathcal{N}_G$.*

For ease of reference, we draw together the main results of this section (from Theorem 4.7.1, 4.7.2 and 4.7.3) in the following single theorem.

Theorem 4.7.4.

(a) (LUR and LWF) $\Rightarrow$ $\mathcal{M} = \mathcal{N}_G$.

(b) LTO $\Rightarrow$ $[\mathcal{M} = \mathcal{N}_G \Leftrightarrow$ (LUR and LWF)].

(c) LTO $\Rightarrow$ $[\mathcal{M} = \mathcal{N}_S \Leftrightarrow \mathcal{M} = \mathcal{N}_S = \mathcal{N}_G \Leftrightarrow$ (LUR and LWF and $\mathcal{N}_G = \mathcal{N}_S$)].

(d) (LTO and LUR and LWF and E is finite) $\Rightarrow$ $[\mathcal{M} = \mathcal{N}_S = \mathcal{N}_G]$.

Remark 4.7.1.

(a) When L is totally ordered, parts (b) and (c) of Theorem 4.7.4 give necessary and sufficient conditions for the level maps to be precisely the generalised stratifications, or the stratifications, respectively.

(b) In the general case, part (a) of Theorem 4.7.4 gives two conditions which are together sufficient for $\mathcal{M} = \mathcal{N}_G$. Counterexamples 1 and 2 below show that neither condition on its own is sufficient for the conclusion.

(c) Part (d) of Theorem 4.7.4 gives four conditions which are together sufficient for $\mathcal{M} = \mathcal{N}_S = \mathcal{N}_G$. This conclusion is false if any one of these conditions is dropped, as the following four counterexamples show. These examples also demonstrate that the four conditions here are logically independent.

No.	LUR	LWF	LTO	E is finite	Conclusion
1	×	✓	✓	✓	$\mathcal{M} \not\subseteq \mathcal{N}_G$
2	✓	×	✓	✓	$\mathcal{M} \not\supseteq \mathcal{N}_S$
3	✓	✓	×	✓	$\mathcal{M} \not\subseteq \mathcal{N}_S$
4	✓	✓	✓	×	$\mathcal{M} \not\subseteq \mathcal{N}_S$

Table 4.4

The structure of the four counterexamples

Note: ✓ and × denote the truth and falsity of a condition respectively.

(d) We introduce four counterexamples. Which of the four conditions of part (d) of Theorem 4.7.4 they satisfy, and the relevant conclusion to be drawn from each example, is summarised in Table 4.4. In these examples when L is a subset of $\mathbb{R}$ it receives the usual order.

EXAMPLE 1. – $L = \{0\} \cup [1, \infty)$. $E = \{e_1, e_2\}$. $f(e_1) = 0$. $f(e_2) = 1$. m_f does not satisfy condition GS1 when $\ell = 0$.

EXAMPLE 2. – $L = (0, 1]$. $E = \{e_1, e_2\}$. If $\ell = 1, m(\ell) = E$. Else, $m(\ell) = \{e_1\}$. Then, $m \in \mathcal{N}_S$ but, since $L_{e_1,m}$ has no minimum, $m \notin \mathcal{M}$.

EXAMPLE 3. – $L = \{0, \ell_1, \ell_2\}$ where $0 < \ell_1$, $0 < \ell_2$ and $\ell_1 \| \ell_2$. $E = \{e_1, e_2\}$. $f(e_1) = \ell_1$. $f(e_2) = \ell_2$. m_f does not satisfy S2 or S3.

EXAMPLE 4. – $L = [0, \infty)$. $E = L$. $f =$ the identity function. m_f does not satisfy S2 or S3. For a countably infinite counterexample, E can be taken to be the integers or the rationals. In this case S2 (but not S3) fails.

4.8. Residual maps

We explore here the relations between level and residual maps, being particularly interested in cases where the subset $\mathcal{M}$ of all level maps in $\mathcal{N}$ coincides with the subset $\mathcal{M}_R$ of all residual maps in $\mathcal{N}$.

Let $m \in \mathcal{N}$. Then we recall (Section 4.2.4) that m is residual means that m is an isotone map with the property that the inverse image of *every* principal filter of $\mathcal{P}(E)$ is a principal filter of L. In contrast, by Lemma 4.5.1(a) and Theorem 4.5.2, m is level means that m is an isotone map which enjoys this property for those particular principal filters of $\mathcal{P}(E)$ generated by the singletons. Thus we have shown:

Proposition 4.8.1.

$\mathcal{M}_R \subseteq \mathcal{M}$.

Next we note that residual maps need not exist!

IV. UNIFICATION AND GENERALISATION OF FUNDAMENTAL BIJECTIONS

Proposition 4.8.2. $\mathcal{M}_R$ *is nonempty* $\Leftrightarrow$ LMIN.

Thus we see that a level map need not be residual. When LMIN fails, it suffices to take *any* level map $m : L \to \mathcal{P}(E)$. When LMIN holds, we can use a counterexample due to Leclerc (1984, p.11). Let $L = \{0, \ell_1, \ell_2, \ell_3, \ell_4, 1\}$ have the Hasse diagram shown in Figure 4.2. In any such diagram, $\tilde{\ell} \leq \ell$ is represented by joining the point representing $\tilde{\ell}$ to that representing ℓ by an increasing line segment. And we agree not to include any superfluous line segments that arise through the transitivity of $\leq$. Thus, for example, in this particular case $\ell_1 \leq \ell$ for $\ell = \ell_3, \ell_4$ or 1 but a line segment is not drawn in for the last of these as $\ell_1 \leq \ell_3$ and $\ell_3 \leq 1$ are already indicated. Further, let $E = \{a, b, c, d\}$ and define $f : E \to L$ via $f(a) = \ell_1$, $f(b) = \ell_2$, $f(c) = \ell_3$ and $f(d) = \ell_4$. Then m_f is level, by definition, but not residual since, for example:

$$m_f^{\leftarrow}([\{a, b\})) = \{\ell_3, \ell_4, 1\}$$

is not a principal filter of L. To see this, the set $m(\ell)$ is indicated by the side of each ℓ in Figure 4.2.

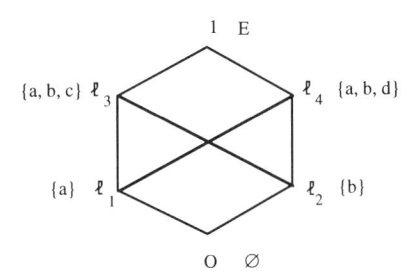

Figure 4.2

Leclerc's counterexample

When every level map is residual, that is when $\mathcal{M}_R = \mathcal{M}$, we say that E and L are *adapted*. This situation can be characterised as follows. A map $m \in \mathcal{M}$ is said to be *join complete* (with respect to E and L) if:

MJC. $\forall \emptyset \neq D \subseteq E, \vee\{f_m(e) : e \in D\}$ exists,

in which case it is denoted by $\ell_m(D)$.

Theorem 4.8.1.

(a) *If LMIN, then a level map is residual if and only if it is join complete.*

(b) *Thus* $\mathcal{M} = \mathcal{M}_R$, *that is* E *and* L *are adapted, if and only if*

       (i) LMIN

and   (ii) $\forall m \in \mathcal{M}$, m *is join complete.*

(c) *In particular, $\mathcal{M} = \mathcal{M}_R$ if*

either (1) *E is finite and L is a join semilattice with a minimum element.*
or (2) *L is a complete lattice.*

With reference to Leclerc's counterexample, we see that E is finite and LMIN holds there but that JSL fails since, for example, $[\ell_1) \cap [\ell_2) = \{\ell_3, \ell_4, 1\}$ is not a principal filter of L (as remarked above).

We note also that condition (2) in the above theorem is not as strong as it looks at first sight. In fact, it is equivalent to requiring that L be a join complete join semilattice with a minimum. See, for example, Blyth and Janowitz (1972, Theorem 4.2, p.29).

4.9. On the associated residuated maps

Given a residual map $m : L \to \mathcal{P}(E)$ it is natural to enquire about its associated residuated map $m^+ : \mathcal{P}(E) \to L$. Given a residuated map $m^+ : \mathcal{P}(E) \to L$, there is a naturally induced map $f^+ \in \mathcal{F}$ defined by restricting it to the singletons. That is, $f^+(e) \equiv m^+(\{e\})$. It is then natural to ask how, for any residual map $m \in \mathcal{M}_R$, this function f^+ obtained from its associated residuated map relates to the function f_m which it defines via $f_m = v(m)$.

These questions are answered in the following proposition. We denote by $\mathcal{M}_R^+$ the set of all residuated maps $m^+ : \mathcal{P}(E) \to L$ endowed with the pointwise order:

$$m_1^+ \leq m_2^+ \iff \forall D \subseteq E, m_1^+(D) \leq m_2^+(D).$$

Recall (Section 4.2.4) that there is a canonical dual order-isomorphism $\rho : m \to m^+$ between $\mathcal{M}_R$ and $\mathcal{M}_R^+$. We have the following result:

Proposition 4.9.1.

(a) *$\mathcal{M}_R^+$ is nonempty $\Leftrightarrow$ LMIN.*

(b) *Suppose LMIN. Then the canonical dual order-isomorphism $\rho : \mathcal{M}_R \to \mathcal{M}_R^+$ is given by:*

(i) $\rho : m \to m^+$ where

$$\forall D \subseteq E, m^+(D) \equiv \min\{\ell : D \subseteq m(\ell)\} = \begin{cases} 0 & \text{if } D = \emptyset \\ \ell_m(D) & \text{if } D \neq \emptyset \end{cases}$$

where $\ell_m(D) = \vee\{f_m(e) : e \in D\}$, as defined after condition MJC above.

And:

(ii) $\rho^{-1} : m^+ \to m$ where

$$\forall \ell \in L, m(\ell) \equiv \max\{D \in \mathcal{P}(E) : m^+(D) \leq \ell\} = \cup\{D : m^+(D) \leq \ell\}.$$

(c) *Suppose LMIN. Then for every residual map $m \in \mathcal{M}_R$, the function f^+ obtained by restricting its associated residuated map to the singletons coincides with the function f_m it defines via $f_m = v(m)$.*

Whenever L has a minimum, we define the maps $\varepsilon : v(\mathcal{M}_R) \to \mathcal{M}_R^+$ and $\eta : \mathcal{M}_R^+ \to \mathcal{F}$ by $\varepsilon = \rho o(u \mid_{v(\mathcal{M}_R)})$ and $\eta : m^+ \to f^+$. We find the following.

Theorem 4.9.1. *Suppose* LMIN. *Then* $\eta = \varepsilon^{-1}$. *In particular, we have:*

(a) $\eta(\mathcal{M}_R^+) = v(\mathcal{M}_R)$. *We denote this set by* $\mathcal{F}_R$.

And:

(b) *The following diagram commutes:*

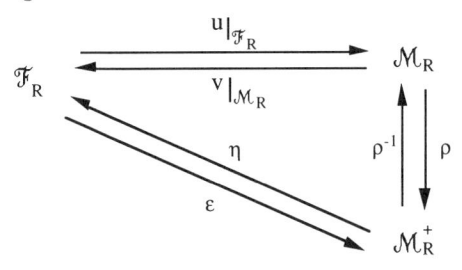

$$\begin{array}{ccc}
\mathcal{F}_R & \xrightarrow{u|_{\mathcal{F}_R}} & \mathcal{M}_R \\
 & \xleftarrow{v|_{\mathcal{M}_R}} & \\
 & \searrow^{\eta} & \updownarrow \rho^{-1} \,\, \rho \\
 & \searrow_{\varepsilon} & \\
 & & \mathcal{M}_R^+
\end{array}$$

Moreover, we have:

(c) *The mappings shown in the above diagram establish natural dual order-isomorphisms between* $\mathcal{F}_R$ *and* $\mathcal{M}_R$, *and between* $\mathcal{M}_R$ *and* $\mathcal{M}_R^+$, *and, hence, a natural order-isomorphism between* $\mathcal{F}_R$ *and* $\mathcal{M}_R^+$.

4.10. Some applications to mathematical classification

The general interest of the central result Theorem 4.5.1 lies in the fact that it shows that the ordered set $\mathcal{M}$ can be substituted for the ordered set $\mathcal{F}$. Any particular problem can be studied either in $\mathcal{F}$ or in $\mathcal{M}$ as may be most advantageous. In this section, we show how the problem of characterising certain classes of functions in $\mathcal{F}$ which appear naturally in mathematical classification can be readily resolved using their corresponding level maps. And how four of the seven bijections reported in Section 4.4 can be unified and generalised. Some other applications are noted in remarks following Corollary 4.10.1. The other three bijections and some further applications are discussed in the following chapter, also with reference to Theorem 4.5.1.

In order to discuss predissimilarities, we suppose throughout this section that LMIN holds and that $E = S \times S$ for some necessarily nonempty set S. We note in passing that we could take $E = S^k$ for some integer $k \geq 2$. This would permit us to introduce dissimilarity measures on subsets of k elements of S. The case $k = 3$ has been considered by Joly and Le Calvé (1992). Although we do not pursue this here, it is one of the benefits of the present rather general approach that such possibilities are so easily accommodated within a single framework. See also the following chapter for other possibilities.

We refer back now to Table 4.1 which, in the current special case LMIN and $E = S \times S$, identifies twelve subsets $\widetilde{\mathcal{F}}$ of $\mathcal{F}$. Identifying $\mathcal{P}(E)$ here, as we may, with the set of all binary relations on S, a map $m \in \mathcal{N}$ is called reflexive (respectively symmetric, transitive, or an equivalence) if $m(\ell)$ has this property for each $\ell \in L$. Using this terminology, twelve subsets $\widetilde{\mathcal{M}}$ of $\mathcal{M}$ were defined in the third column of Table 4.1, it being implicit that each of these was to be associated with the subset $\widetilde{\mathcal{F}}$ appearing in the same row of the table. Indeed, when we observe that in each row $\widetilde{\mathcal{M}} = u(\widetilde{\mathcal{F}})$, the following result follows at once from Theorem 4.5.1:

Corollary 4.10.1. *Suppose* LMIN *and* $E = S \times S$. *Then the relevant restrictions of the maps u and v establish a natural dual order-isomorphism between $\widetilde{\mathcal{F}}$ and $\widetilde{\mathcal{M}}$ for each of the twelve rows (a) to (l) of Table 4.1.*

This result is general in as much as S and L are arbitrary, the sole restriction LMIN being necessary for predissimilarities to be defined. Moreover, it can be extended or used in several ways. One can assert that *any* collection of predissimilarities is dually order-isomorphic with its image under the map u. And that the intersection or union of any collection of subsets of $\mathcal{F}$ is dually order-isomorphic to the same intersection or union of their images under u. Since L is so general and, in particular, not necessarily totally ordered, we have a framework in which we can handle multi-attribute dissimilarities. Since L is not necessarily finite, this framework can also handle asymptotics, for example of hierarchical cluster analysis methods. Since the elements of $\mathcal{F}_0$ are not necessarily dissimilarities, we can also analyse problems of asymmetry. Finally, as we develop a little in the following chapter, since E is so general we can handle rectangular and multi-way data.

We hope to explore at least some of these methodological innovations in future papers. For now, we finish this paper by identifying bijections 2, 3, 5 and 7 of Section 4.4 as special cases of Corollary 4.10.1. First of all we remark that the nonnegative reals $\mathbb{R}_+$ endowed with the usual order enjoy all the properties of ordered sets listed in Table 4.3 above.

BIJECTIONS 2 and 3. (S finite, $L = \mathbb{R}_+$).

Using part (d) of Theorem 4.7.4, we see that parts (f) and (d) of Corollary 4.10.1 imply respectively the bijections reported in Theorems 4.4.2 and 4.4.3.

BIJECTION 5. (S arbitrary, $L = \mathbb{R}_+$).

Using part (a) of Theorem 4.7.4, we see that parts (f) and (i) of Corollary 4.10.1 imply the bijections reported in Theorem 4.4.5.

BIJECTION 7. (S finite, L obeys LMIN and JSL).

Using part (c)(1) of Theorem 4.8.1, we see that part (d) of Corollary 4.10.1 is precisely the dual order-isomorphism reported in part (a) of Theorem 4.4.7. Moreover, parts (b) and (c) of Theorem 4.4.7 are special cases of the corresponding parts of Proposition 4.9.1 respectively. Finally, we note that Janowitz' result reported in Remark 4.4.3 is implicit in Theorems 4.7.4 (d) and 4.8.1 (c)(1).

Bijections 1, 4 and 6 are dealt with in the next chapter.

Acknowledgements

It is a pleasure to gratefully acknowledge the assistance of B. Leclerc and B. Monjardet in opening our eyes to the advantages of the order-theoretic approach. And also the support provided by:

(a) a grant from the Royal Society under the European Science Exchange Programme (FC)

(b) two periods as Professeur Associé à l'Université Joseph Fourier (FC)

(c) a grant from the Warwick Research and Innovations Fund (FC)

(d) a sabbatical semester granted by l'Université Joseph Fourier (BVC)

and (e) travel grants from the British Council (both authors).

Appendix A: Proofs

We give here the proofs of the results stated in Section 4.5 and following. Where necessary, we state as Facts well-known results about ordered sets, along with an appropriate reference where a proof may be found.

Proof of Theorem 4.5.1

By definition of $\mathcal{M}$, u is onto. By the proof in the paragraph preceding Theorem 4.5.1, u is $1-1$. Further, by definition of v, $f = v(m) \Leftrightarrow m = u(f)$. Thus, $v = u^{-1}$. It only remains to note that u (resp. v) is antitone as $\leq$ is transitive (resp. reflexive). ∎

Proof of Lemma 4.5.1

(a) $\ell \in L_{e,m} \Leftrightarrow \{e\} \subseteq m(\ell)$.

(b) $\ell \in L_{e,m} \Leftrightarrow f_m(e) \leq \ell$. ∎

Fact A1

Let P and Q be ordered sets and let $f : P \to Q$ be any mapping. Then the following are equivalent:

(1) *f is isotone.*

(2) *the preimage of every principal ideal of Q is either empty or is an ideal of P.*

(3) *the preimage of every principal filter of Q is either empty or is a filter of P.*

Proof

See Blyth and Janowitz (1972, Theorem 2.1, p.5). ∎

Fact A2

A filter is a principal filter if and only if it has a minimum element.

Proof

Obvious. ∎

Proof of Theorem 4.5.2

We show that $(1) \Rightarrow (2) \Rightarrow (3) \Rightarrow (1)$.

$(1) \Rightarrow (2)$... As m is level, $\exists f_m \in \mathcal{F}$ such that $\forall \ell \in L, m(\ell) = \{e \in E : f_m(e) \leq \ell\}$. Thus, as $\leq$ is transitive, m is isotone. Property (2)(b) is immediate from Lemma 4.5.1 (b).

$(2) \Rightarrow (3)$. Recalling Lemma 4.5.1 (a) and property 2(a), Fact A1 shows that $L_{e,m}$ is either empty or is a filter. Recalling 2(b) and Fact A2, $L_{e,m}$ is therefore a principal filter.

IV. UNIFICATION AND GENERALISATION OF FUNDAMENTAL BIJECTIONS   115

(3) ⇒ (1). $\forall e \in E$, we can write $L_{e,m} = [\ell_{e,m})$. Then, defining $f \in \mathcal{F}$ by $f(e) = \ell_{e,m}$, we find $m = m_f$. ∎

Proof of Proposition 4.6.1

Note first that: L nonempty $\Leftrightarrow L$ is a filter of L.

(a) Immediate from the above note and Fact A2.

(b) Immediate.

(c) By the above note, L is a filter of L. Clearly $L =]\ell)$ is impossible. Hence $L = [\ell)$ with $\ell = 0$.

(d) Clearly, $]\ell)$ is nonempty $\Leftrightarrow]\ell)$ is a filter, while $]\ell) \subset L$ always.

(e) Clearly, $k \geq \ell_K$ for each $k \in K$. But $k = \ell_K$ is impossible. Hence $K \subseteq]\ell_K)$. Conversely,

$$\ell \in]\ell_K) \;\Rightarrow\; \ell \text{ is not a lower bound for } K, \text{ by definition of } \ell_K.$$
$$\Rightarrow\; \exists k \in K \text{ such that } \ell > k, \quad \text{by LTO.}$$
$$\Rightarrow\; \ell \in K, \quad \text{as } K \text{ is a filter.}$$

(f) Let K be a filter of L. Now $L = [0)$, by LMIN. Suppose then that K is proper. If K is a principal filter, $K = [\min K)$. If not, $K =]\ell_K)$ by (e).

(g) Let K be a proper, non-principal, filter of L. By LWF, $K =]\ell)$ for some $\ell \in L$. By LUR, this ℓ is $\wedge K$. □

Proof of Theorem 4.7.1

(a) Let $m \in \mathcal{M}$. By Proposition 4.7.1 (c), it suffices to show GS1. Let $\ell \in L$ with $]\ell)$ nonempty. Then, $\forall e \in E$:

$$e \in \cap\{m(\ell') : \ell' > \ell\} \Leftrightarrow \ell \geq f_m(e),$$

the forwards implication holding by LUR.

(b) Let $m \in \mathcal{N}_G$. Then m is isotone by Proposition 4.7.1 (b) and hence, by Theorem 4.5.2, it suffices to show that each $L_{e,m}$ has a minimum element. Let $e \in E$. Using Lemma 4.5.1 (a), the fact that m is isotone and Fact A1, $L_{e,m}$ is either empty or is a filter of L. The first of these is precluded by GS2. Using LWF, $L_{e,m}$ is therefore either of the form $[\ell)$ or a nonempty set of the form $]\ell)$. But the second of these is impossible by GS1. ∎

Proof of Lemma 4.7.1

(a) For each $\ell_0 \in L$, define the constant function $f^{\ell_0} \in \mathcal{F}$ by $f^{\ell_0}(e) = \ell_0$ for all $e \in E$. Writing $m^{\ell_0} = u(f^{\ell_0})$, we have that:

$$\forall \ell \in L, m^{\ell_0}(\ell) = \begin{cases} E & \text{if } \ell \geq \ell_0 \\ \emptyset & \text{else (that is, by LTO, if } \ell < \ell_0). \end{cases}$$

Suppose $\exists (\ell, \ell_0) \in L^2$ with $\ell < \ell_0$. Then $m^{\ell_0}(\ell) = \emptyset$ implies $\exists \ell' \in]\ell)$ with $\ell' < \ell_0$, as m^{ℓ_0} is upper semi-continuous by hypothesis. Clearly, ℓ is a lower bound for $]\ell)$. Let k be any lower bound of $]\ell)$. It suffices to show that $k \leq \ell$. Suppose the contrary. That is, by LTO, suppose $l < k$. Then, arguing as above, $\exists \ell' \in]\ell)$ with $\ell' < k$ which is impossible.

(b) Define $m \in \mathcal{N}$ by $\forall \ell \in L, m(\ell) = E$ so that $L_{e,m} = L$ for each $e \in E$. Then m is a stratification and hence, by hypothesis, level. Thus, by Theorem 4.5.2, $L_{e,m} \equiv L$ has a minimum.

(c) Let K be a proper, non-principal filter of L and note that, as L is totally ordered, $\ell \notin K \Leftrightarrow \forall k \in K, \ell < k$. Suppose, if possible, that $\wedge K$ does not exist. Then, define $m \in \mathcal{N}$ by

$$\forall \ell \in L, m(\ell) = \left\{ \begin{array}{ll} E, & \text{if } \ell \in K \\ \emptyset, & \text{else} \end{array} \right\} \text{ so that } \forall e \in E, L_{e,m} = K.$$

Clearly, m satisfies conditions S1 and S2 of a stratification. Consider now S3. Let $\ell \in L$ with $]\ell)$ nonempty. If $\ell \in K$, then $m(\ell') = m(\ell)$ $\forall \ell' > \ell$, as K is a filter. If $\ell \notin K$, then ℓ is a lower bound of K. As K has no greatest lower bound, $\exists \ell' \in L$ such that $\ell < \ell'$ and, moreover, $\ell' \notin K$ so that $m(\ell') = m(\ell)$. Thus m is a stratification and hence level. Thus, by Theorem 4.5.2, K has a minimum, yielding the desired contradiction. ∎

Proof of Lemma 4.7.2

By GS2, $\forall e \in E, \exists \ell_e \in L$ such that $e \in m(\ell_e)$. But $\vee \{\ell_e : e \in E\}$ exists by hypothesis. Call it $\bar{\ell}$. Then, as m is isotone, $m(\bar{\ell}) = E$. ∎

Proof of Lemma 4.7.3

(a) Let $\ell \in L$ with $(\ell[$ nonempty. Then:

$$m(\ell-) \subset m(\ell) \Leftrightarrow \exists e \in E \text{ such that } \forall \ell' < \ell, \ell' < f_m(e) \leq \ell, \text{ using LTO}$$
$$\Leftrightarrow \exists e \in E \text{ with } f_m(e) = \ell.$$

(b) Similarly, $m(\ell) \subset m(\ell') \Leftrightarrow \exists e \in E$ such that $\ell < f_m(e) \leq \ell'$. The result now follows, using (a). ∎

Proof of Theorem 4.7.3

In view of parts (c) and (d) of Theorem 4.7.2, $\mathcal{M} = \mathcal{N}_G$ holds by hypothesis and it suffices to prove $\mathcal{N}_G \subseteq \mathcal{N}_S$. Let $m \in \mathcal{N}_G$. Then m satisfies S1, by Proposition 4.7.1 (b), and hence S2, by Proposition 4.6.1 (b) and Lemma 4.7.2. Consider now S3. Suppose then that there exists $\ell \in L$ with $]\ell)$ nonempty. We note that m is level and, by Lemma 4.7.3 (a), has a finite number of jump points. Thus J is finite, where

$$J = \{\ell' : \ell' \in]\ell) \text{ and } \ell' \text{ is a jump point of } m\}.$$

IV. UNIFICATION AND GENERALISATION OF FUNDAMENTAL BIJECTIONS   117

If J is empty, $m(\ell') = m(\ell)$ for all $\ell' > \ell$ using Lemma 4.7.3 (b). Otherwise, $\underline{\ell} \equiv \min J$ exists. Suppose, if possible:

$$\exists \ell' \in L \text{ with } \ell < \ell' < \underline{\ell}. \tag{$*$}$$

Then:
$$\begin{aligned} m(\ell) &= m(\underline{\ell}), \text{ using GS1, LTO and } (*). \\ &\supset m(\underline{\ell}-), \text{ as } \underline{\ell} \text{ is a jump point of } m. \\ &= m(\ell), \text{ using GS1, LTO and } (*) \text{ again.} \end{aligned}$$

This contradiction shows that $\nexists \ell' \in L$ with $\ell < \ell' < \underline{\ell}$. Now, by definition of $\underline{\ell}$, $\tilde{\ell}$ is not a jump point of m for all $\ell < \tilde{\ell} \leq \ell'$. Thus, using Lemma 4.7.3 (b) again, $m(\ell') = m(\ell)$ for all $\ell < \ell' < \underline{\ell}$. ∎

Proof of Proposition 4.8.2

Suppose $\exists m \in \mathcal{M}_R$. Note that $[\emptyset) = \mathcal{P}(E)$ so that $m^{\leftarrow}([\emptyset)) = L$. Thus L is a principal filter. Equivalently, by Fact A2, LMIN. Conversely, if LMIN, then the map defined by $\forall \ell \in L, m(\ell) = E$ is clearly residual. ∎

Proof of Theorem 4.8.1

(a) Let $m \in \mathcal{M}$. Recalling the proof of Proposition 4.8.2,

m is residual $\Leftrightarrow$ LMIN and $\forall \emptyset \neq D \subseteq E, m^{\leftarrow}([D))$ is a principal filter of L.

It suffices now to note that $m^{\leftarrow}([D)) = \cap \{L_{e,m} : e \in D\}$ where $L_{e,m} = [f_m(e))$.

(b) Immediate, from Proposition 4.8.1 and part (a).

(c) Immediate, from part (b). ∎

Proof of Proposition 4.9.1

(a) Immediate from the fact that ρ is a bijection and Proposition 4.8.2.

(b) (i) Immediate from the definition of ρ and of $\ell_m(D)$.

   (ii) This follows from the definition of ρ^{-1} and from recalling the general result (Section 4.2.4) that $m^+(D) \leq \ell \Leftrightarrow D \leq m(\ell)$.

(c) Immediate from b(i). ∎

Proof of Theorem 4.9.1

The fact that $\eta = \varepsilon^{-1}$ follows at once from $f^+ = f_m$ (Proposition 4.9.1) and establishes (a) and (b). The two dual order-isomorphisms in part (c) follow from Theorem 4.5.1 and the general theory of ordered sets (Section 4.2.4) respectively. Their composition is therefore an order-isomorphism. ∎

References

Barthélémy, J.P., Leclerc, B., Monjardet, B. (1984), Ensembles ordonnés et taxonomie mathématique, *Ann. Discrete Mathematics*, 23, pp. 523–548.

Batbedat, A. (1988), Les isomorphismes HTS et HTE (après la bijection de Benzécri/ Johnson), *Metron*, 46, pp. 47–59.

Batbedat, A. (1990), *Les approches pyramidales dans la classification arborée*, Masson, Paris.

Batbedat, A. (1991), Ensembles ordonnés, dissimilarités, hypergraphes (finis on infinis), Personal communication.

Benzécri, J.P. (1965), Problèmes et méthodes de la taxinomie, Rapport de recherche de l'Université de Rennes I, France.

Benzécri, J.P., et al. (1973), *L'analyse des données. I. La taxinomie*, Dunod, Paris.

Blyth, T.S., Janowitz, M.F. (1972), *Residuation theory*, Pergamon Press, Oxford.

Critchley, F., Van Cutsem, B. (1989), Predissimilarities, prefilters and ultrametrics on an arbitrary set. Rapport de Recherche commun de l'Université de Warwick, U.K. et du Laboratoire TIM3-IMAG, Grenoble, France.

Critchley, F., Van Cutsem, B. (1992), An order-theoretic unification and generalisation of certain fundamental bijections in mathematical classification – I, Joint Research Report, University of Warwick, U.K., and Université Joseph Fourier, Grenoble, France.

Diday, E. (1984), Une représentation visuelle des classes empiétantes: les pyramides, Research Report n° 291, INRIA, Rocquencourt, France.

Fichet, B. (1984), Sur une extension de la notion de hiérarchie et son équivalence avec certaines matrices de Robinson, Journées de Statistique, Montpellier, France.

Hartigan, J.A. (1967), Representations of similarity matrices by trees, *J. Amer. Statist. Assoc.*, 62, pp. 1140–1158.

Janowitz, M.F. (1978), An order theoretic model for cluster analysis, *SIAM J. Appl. Math.*, 34, pp. 55–72.

Jardine, C.J., Jardine, N., Sibson, R. (1967), The structure and construction of taxonomic hierarchies, *Math. Biosci.*, 1, 171–179.

Jardine, N., Sibson, R. (1971), *Mathematical Taxonomy*, Wiley, London.

Johnson, S.C. (1967), Hierarchical clustering schemes, *Psychometrika*, 32, pp. 241–254.

Joly, S., Le Calvé, G. (1992), 3-way distances, *J. Classification*, (to appear).

Lerman, I.C. (1968), Analyse du problème de la recherche d'une hiérarchie de classifications, Report 22, Maison des Sciences de l'Homme, Paris, France.

Lerman, I.C. (1970), *Les bases de la classification automatique*, Gauthier-Villars, Paris.

Zaks, Y.M., Muchnik, I.B. (1989), Monotone systems for incomplete classification of a finite set of objects, *Automat. Remote Control*, 4, pp. 155–164.

Chapter 5.
An order-theoretic unification and generalisation of certain fundamental bijections in mathematical classification. II*

Frank Critchley[†]

Bernard Van Cutsem[††]

5.1. Introduction and overview

In chapter 4 of this book, the simple but general Theorem 4.5.1 was established and shown to have a variety of important applications in mathematical classification. In particular, it was seen to unify and generalise four bijections reported in the literature concerning dendrograms and generalised dendrograms, numerically stratified clusterings and residual maps.

The present chapter continues this development. In particular, an extended version of a particular case of Theorem 4.5.1 is given and is seen to unify and generalise three further bijections involving "prefilters", indexed hierarchies and indexed regular generalised hierarchies. Other applications of this extension, for example to rectangular and multiway data, are briefly noted as is a certain natural duality which occurs there. The application to "prefilters" yields three characterisations of ultrametrics taking values in an arbitrary nonempty ordered set, and a further three characterisations if that set is totally ordered. Finally, following Jardine and Sibson (1971), it is easy to establish conditions under which a subset of predissimilarities admits a unique subdominant member for any given initial predissimilarity. In particular, it is shown that the ultrametrics have this property. This greatly generalises the familiar fundamental result underlying single linkage hierarchical cluster analysis. That case deals with dissimilarities which are symmetric by definition, take values in the real numbers and are defined on a finite nonempty set. In the present case, we work with predissimilarities which need not be symmetric, which take values in an arbitrary nonempty ordered set (thereby accommodating multiattribute dissimilarities) and which are defined on any nonempty set (thereby providing a natural framework in which to study asymptotic behaviour).

The organisation of this chapter is as follows. Section 5.2 recalls the key result (Theorem 4.5.1) of the previous chapter before extending the relevant special case of it

---

* *In* Van Cutsem, B. (Ed.), (1994) *Classification and Dissimilarity Analysis*, Lecture Notes in Statistics, Springer-Verlag, New York.

[†] University of Birmingham, U.K.

[††] Université Joseph Fourier, Grenoble, France.

appropriately to our present purposes. Three other aspects of this extension are noted in Section 5.3. Section 5.4 generalises bijection 6 on "prefilters", using the concept of a reflexive, level foliation. Section 5.5 gives characterisations of ultrametrics in terms of this concept. Section 5.6 introduces Benzécri structures and shows that they generalise both Benzécri's indexed hierarchies and the indexed regular generalised hierarchies of Critchley and Van Cutsem. These structures are then used in Section 5.7 to obtain the desired generalisations of the final two bijections (numbers 1 and 4). The final Section 5.8 deals with subdominants.

Wherever we judge that it will not unduly hinder the reader, we avoid duplication by referring the reader to chapter 4 for definitions, results, etc. In particular, formal statements of the bijections referred to above, and to the literature where they may be found, appear in that chapter. We also draw attention to the twelve pairs of subsets defined in Table 4.1, to the layout and numbering of the seven bijections indicated in Table 4.2, and to the list of properties that an ordered set may enjoy given in Table 4.3. Thus, for example, we will use $\mathcal{F}_0$ for the set of all predissimilarities on an arbitrary nonempty set S and taking values in an arbitrary nonempty ordered set L, and will write LMIN to denote the condition that L has a minimum. We recall also that the adjectives "even" and "definite" applied to certain types of predissimilarity correspond to "semi-propre" and "propre" in French.

As in chapter 4, proofs are given in an Appendix and parts of the chapter that can be omitted with little loss on a first reading are indicated by an asterisk.

This chapter is based on Critchley and Van Cutsem (1992).

5.2. The case $E = A \times B$ of Theorem 4.5.1

We first recall the key result, Theorem 4.5.1 of chapter 4, before extending it in the particular case $E = A \times B$.

Let E be any nonempty set and let L be any nonempty ordered set. Let $\mathcal{F}$ denote the set of all maps $f : E \to L$, and $\mathcal{N}$ the set of all maps $m : L \to \mathcal{P}(E)$. The function $u : f \to m_f$ associates to each $f \in \mathcal{F}$ the element $m_f \in \mathcal{N}$ defined by its level sets. That is,
$$\forall f \in \mathcal{F}, \ \forall \ell \in L, \ m_f(\ell) = \{e \in E : f(e) \leq \ell\}.$$
A map $m \in \mathcal{N}$ is called level if, in fact, $m \in \mathcal{M} \equiv u(\mathcal{F})$. Conversely the function $v : m \to f_m$ associates to each level map $m \in \mathcal{M}$ the necessarily unique map $f_m \in \mathcal{F}$ which generates it via $m = u(f)$. The set $\mathcal{F}$ and $\mathcal{M}$ are ordered pointwise. That is:
$$f \leq f' \Leftrightarrow \forall e \in E, f(e) \leq f'(e) \quad \text{while} \quad m \leq m' \Leftrightarrow \forall \ell \in L, m(\ell) \subseteq m'(\ell)$$
where we recall that the members of the power set $\mathcal{P}(E)$ are ordered by inclusion. Unless otherwise stated, every subset of an ordered set is taken to be endowed with the order induced upon it. The following simple but general result is now essentially immediate.

Theorem 4.5.1. *The maps u and v establish a natural dual order-isomorphism between $\mathcal{F}$ and $\mathcal{M}$.*

Clearly, every restriction of u and v to a pair of corresponding subsets $\widetilde{\mathcal{F}} \subseteq \mathcal{F}$ and $\widetilde{\mathcal{M}} \subseteq \mathcal{M}$ (that is, $\widetilde{\mathcal{M}} = u(\widetilde{\mathcal{F}})$; equivalently $\widetilde{\mathcal{F}} = v(\widetilde{\mathcal{M}})$) induces a natural dual order-isomorphism between them. Thus, for example, if $E = S \times S$ and LMIN, the set $\mathcal{F}_0 \equiv \{f \in \mathcal{F} : \forall a \in S, f(a,a) = 0\}$ of all L-predissimilarities on S is naturally dually order-isomorphic to the set $\mathcal{M}_r$ of all level maps that are reflexive. Whenever $E = S \times S$, we identify $\mathcal{P}(E)$ with the binary relations on S in the obvious way and say that a map $m \in \mathcal{N}$ is reflexive (respectively symmetric, transitive, or an equivalence) if $m(\ell)$ has this property for each $\ell \in L$.

In the previous chapter, bijections 2, 3, 5 and 7 were unified and generalised by Theorem 4.5.1 by identifying conditions under which the set $\mathcal{M}$ of all level maps in $\mathcal{N}$ coincided precisely with the L-stratifications of E, the generalised L-stratifications of E, or the residual maps $m : L \to \mathcal{P}(E)$. In this chapter, we deal with bijections 1, 4 and 6 by a different route. Throughout this section we suppose that $E = A \times B$ for some necessarily nonempty sets A and B, with general elements denoted a and b respectively. The essence of the approach here is to produce a *section* of $f \in \mathcal{F}$ or of $m \in \mathcal{M}$ for each $a \in A$.

Let $\mathcal{F}_A$ denote the set of all families of functions from B to L indexed by A. We define the map $\sigma_A : \mathcal{F} \to \mathcal{F}_A$ by $\sigma_A(f) = \{f_a\}_{a \in A}$ where $f_a(b) \equiv f(a,b)$. We call the map $f_a : B \to L$ the a-section of f. Clearly, σ_A is a bijection which induces natural order on $\mathcal{F}_A$ by

$$\{f_a\}_{a \in A} \leq \{f'_a\}_{a \in A} \Leftrightarrow f \leq f'.$$

Similarly, let $\mathcal{M}_A$ denote the set of all families of level maps from L to $\mathcal{P}(B)$ indexed by A. For each $m \in \mathcal{M}$, we define its a-section to be the map $m_a : L \to \mathcal{P}(B)$ given by

$$\forall \ell \in L, m_a(\ell) = \{b \in B : (a,b) \in m(\ell)\}.$$

As $L_{b,m_a} = L_{(a,b),m}$, each m_a is level by Theorem 4.5.2. The map $\varsigma_A : \mathcal{M} \to \mathcal{M}_A$ defined by $\varsigma_A(m) = \{m_a\}_{a \in A}$ is clearly a bijection which induces a natural order on $\mathcal{M}_A$ by

$$\{m_a\}_{a \in A} \leq \{m'_a\}_{a \in A} \Leftrightarrow m \leq m'.$$

Relatedly, a function $\phi : A \times L \to \mathcal{P}(B)$ is called a foliation if each $\phi(a, \cdot)$ covers B, (that is if $\forall a \in A, \cup \{\phi(a,\ell) : \ell \in L\} = B$), and a level foliation if, in fact, each $\phi(a, \cdot) : L \to \mathcal{P}(B)$ is level. The set of all these latter is denoted Φ_A. Clearly, the sets Φ_A and $\mathcal{M}_A$ can be identified via $\phi(a,\ell) = m_a(\ell)$ and this we do.

Finally, the maps $u_A : \mathcal{F}_A \to \mathcal{M}_A$ and $v_A : \mathcal{M}_A \to \mathcal{F}_A$ are defined by $u_A : \{f_a\}_{a \in A} \to \{\tilde{u}_B(f_a)\}_{a \in A}$ and $v_A : \{m_a\}_{a \in A} \to \{\tilde{v}_B(m_a)\}_{a \in A}$ where $\tilde{u}_B$ and $\tilde{v}_B$ are the maps u and v of Theorem 4.5.1 in the special case where $E = B$. With these definitions, the following simple but general result is an immediate corollary of Theorem 4.5.1.

Corollary 5.2.1. *Suppose $E = A \times B$. Then the following diagram commutes:*

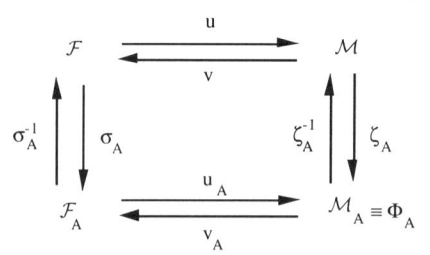

$$\begin{array}{ccc}
\mathcal{F} & \xrightleftharpoons[v]{u} & \mathcal{M} \\
\sigma_A^{-1} \Big\uparrow \Big\downarrow \sigma_A & & \zeta_A^{-1} \Big\uparrow \Big\downarrow \zeta_A \\
\mathcal{F}_A & \xrightleftharpoons[v_A]{u_A} & \mathcal{M}_A \equiv \Phi_A
\end{array}$$

Moreover, (compositions of) the mappings shown establish natural:

(a) *order-isomorphisms between $\mathcal{F}$ and $\mathcal{F}_A$, and between $\mathcal{M}$ and $\mathcal{M}_A$.*

(b) *dual order-isomorphisms between the other four pairs of sets: $\mathcal{F}$ and $\mathcal{M}$, $\mathcal{F}$ and $\mathcal{M}_A$, $\mathcal{F}_A$ and $\mathcal{M}$, $\mathcal{F}_A$ and $\mathcal{M}_A$.*

This simple result is rather general, and can be extended or used in several ways, in the same senses as noted for Corollary 4.10.1. Here we content ourselves with simply noting that, of course, the relevant restrictions of the maps shown establish natural (dual) order-isomorphisms between all pairs of all corresponding quartets of subsets $(\widetilde{\mathcal{F}}, \widetilde{\mathcal{M}}, \sigma_A(\widetilde{\mathcal{F}}), \zeta_A(\widetilde{\mathcal{M}}))$ where $\widetilde{\mathcal{M}} = u(\widetilde{\mathcal{F}})$; equivalently, $\widetilde{\mathcal{F}} = v(\widetilde{\mathcal{M}})$.

The aspect of this result which allows us to unify and generalise the bijections based on "prefilters" and on (generalisations of) indexed hierarchies, is that it brings the set Φ_A naturally into play. In particular, any subset of $\mathcal{F}$ is dually order-isomorphic to the corresponding subset of Φ_A. Before pursuing this, we pause to note several other aspects.

5.3. Other aspects of the case $E = A \times B$ *

Although we do not develop them in the present paper, we note here three other aspects of the case $E = A \times B$.

5.3.1. Duality

There is a natural duality here in which the roles of A and B are reversed throughout. Thus, for example, each $f : A \times B \to L$ can be associated with its dual $f^0 : B \times A \to L$ defined by $f^0(b, a) = f(a, b)$. Clearly, $f^{00} = f$. This gives dual definitions and results from primal ones, in the usual way. Thus, for example, we have at once the dual result, Corollary $5.2.1^0$ say, that the following diagram commutes and the mappings shown

V. UNIFICATION AND GENERALISATION OF FUNDAMENTAL BIJECTIONS   125

establish natural (dual) order-isomorphisms between all pairs of the sets $\mathcal{F}$, $\mathcal{M}$, $\mathcal{F}_B$ and $\mathcal{M}_B$:

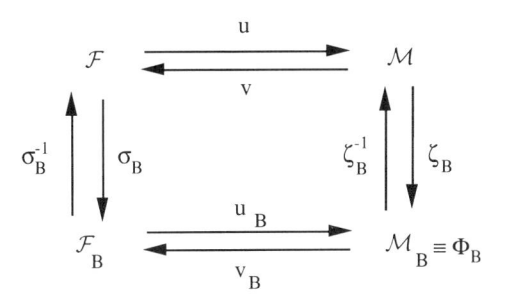

$$
\begin{array}{ccc}
\mathcal{F} & \xrightleftharpoons[v]{u} & \mathcal{M} \\
\sigma_B^{-1} \updownarrow \sigma_B & & \zeta_B^{-1} \updownarrow \zeta_B \\
\mathcal{F}_B & \xrightleftharpoons[v_B]{u_B} & \mathcal{M}_B \equiv \Phi_B
\end{array}
$$

Here, the definitions of terms with a subscript B are the duals of those with a subscript A given above. Thus, for example, $\mathcal{F}_B$ is the set of all families of functions from A to L indexed by B and σ_B maps $f \in \mathcal{F}$ to $\{f_b\}_{b \in B}$ where $f_b(a) = f(a,b)$. We call f_b the b-section of f. Similarly $m_b : L \to \mathcal{P}(A)$ defined by

$$\forall \ell \in L,\ m_b(\ell) = \{a \in A : (a,b) \in m(\ell)\}$$

is called the b-section of m, and ζ_B maps $m \in \mathcal{M}$ to $\{m_b\}_{b \in B}$.

This duality also underlies the notions of left- and right-evenness of a predissimilarity. Let $f \in \mathcal{F}$. Then we say that a and a' are right-indistinguishable by f if $\forall b \in B, f(a,b) = f(a',b)$. Dually, we say b and b' are left-indistinguishable by f if $\forall a \in A, f(a,b) = f(a,b')$. Clearly, these are equivalence relations on A and B and we write them as $a \overset{Rf}{\sim} a'$ and $b \overset{Lf}{\sim} b'$ respectively. Suppose for the rest of Section 5.3.1 that $A = B = S$, that LMIN holds, and that f is a predissimilarity. Then, as defined in Table 4.1, $f \in \mathcal{F}_0$ is called

(i) right-even if $f(a,a') = 0 \Rightarrow a \overset{Rf}{\sim} a'$

and

(i)⁰ left-even if $f(b',b) = 0 \Rightarrow b \overset{Lf}{\sim} b'$.

Let $f \in \mathcal{F}_0$ and let $m = u(f)$. We see from Corollary 5.2.1 that propositions (i) to (iv) below are equivalent. From Corollary 5.2.1⁰, so are their duals (i)⁰ to (iv)⁰:

(i) f is right even.   (i)⁰ f is left even.
(ii) $m(0)$ is $\overset{Rf}{\sim}$.   (ii)⁰ $m(0)$ is $\overset{Lf}{\sim}$.
(iii) $f_a(a') = 0 \Rightarrow f_a = f_{a'}$.   (iii)⁰ $f_b(b') = 0 \Rightarrow f_b = f_{b'}$.
(iv) $a' \in m_a(0) \Rightarrow m_a = m_{a'}$.   (iv)⁰ $b' \in m_b(0) \Rightarrow m_b = m_{b'}$.

5.3.2. Multiway data

We consider here the practical scope of the framework presented in Section 5.2. Clearly this framework can handle two mode data in which, for example, $f(a,b)$ measure the preference of assessor a from a panel of assessors A for candidate b from some pool B of candidates. However it is clearly not limited to this particular sort of rectangular data array. Rather, any such two way table of data can be accommodated simply by regarding it as the image of $A \times B$ under f. That is, as $\{f(a,b) : a \in A, b \in B\}$. Thus a labels rows, b labels columns, f_a carries the information in row a, f_b that in column b, etc. In particular, the duality of Section 5.3.1, becomes the familiar one between rows and columns.

Moreover, the framework can be extended in the obvious way to cover higher-way data. Thus, for example, three way data associated with a function $f : A_1 \times A_2 \times A_3 \to L$ can be accommodated by taking $A = A_1$, $B = A_2 \times A_3$ and then $A = A_2$, $B = A_3$. Of course, duality is lost in the higher-way case. This has important consequences. For example, the decomposition of the particular f given above is not unique. We could instead take $A = A_1 \times A_2$, $B = A_3$ and then $A = A_1$, $B = A_2$. For a survey of multiway data from the exploratory data analysis point of view, see Coppi and Bolasco (1989).

5.3.3. Residual maps

The links of Corollary 5.2.1 with residual maps are explored in the following proposition. Recall that E and L are said to be adapted when every level map $m : L \to \mathcal{P}(E)$ is residual. In this case, the level maps and the residual maps coincide.

Proposition 5.3.1. *Suppose $E = A \times B$ and LMIN. Then:*

(a) *$m \in \mathcal{M}$ is residual $\Rightarrow \forall a \in A, m_a : L \to \mathcal{P}(B)$ is residual.*

(b) *$A \times B$ and L are adapted $\Rightarrow B$ and L are adapted.*

Of course, there is a dual result.

The counterexample below shows that the converses of both parts of this proposition are false in general. This underlines the fact, noted in Section 4.4.7, that the fundamental notion here is of level rather than residual maps.

Counterexample

Let L be the interval $[0, 1)$ of the reals endowed with the usual order. Let $E = A \times B$ where A is the interval $(0,1)$ and $B = \{1\}$. Then, as L is a join semi-lattice with a minimum and B is finite, B and L are adapted by Theorem 4.8.1(c). For each $a \in A$, define $m_a : L \to \mathcal{P}(B)$ by:

$$\forall \ell \in L, m_a(\ell) = \begin{cases} \emptyset & \text{if } \ell < 1 - a. \\ B & \text{if } \ell \geq 1 - a. \end{cases}$$

Then $L_{m_a}(1) = [1-a, 1)$ so that each m_a is level, by Theorem 4.5.2, and hence residual, as B and L are adapted. Let $m \equiv \zeta_A^{-1}(\{m_a\}_{a \in A})$ be the level map from L to $\mathcal{P}(A \times B)$ generated by the $\{m_a\}_{a \in A}$. Then

$$\{f_m(a,b) : (a,b) \in E\} \equiv \{(1-a) : 0 < a < 1\}$$

does not have a least upper bound in L. Thus, by Theorem 4.8.1(a), m is not residual and, in particular, $A \times B$ and L are not adapted. □

5.4. Prefilters

We return now to our principal agenda of identifying existing bijections within the single general framework of Corollary 5.2.1. From now on we assume that $A = B = S$ and that LMIN holds, since these assumptions are made in those parts of mathematical classification that we shall be considering.

A level foliation $\phi : S \times L \to \mathcal{P}(S)$ is called reflexive (respectively symmetric, transitive, or an equivalence) if the corresponding map $m = \zeta_S^{-1}(\phi)$ has this property. Here and henceforth we take ζ_S, Φ_S, etc. to be ζ_A, Φ_A, etc. as in Corollary 5.2.1, rather than ζ_B, Φ_B, etc. from the dual result.

Taking $L = \mathbb{R}_+$, we see from part (a) of Theorem 4.7.4 that a "prefilter" as defined in Section 4.4.6 is precisely a reflexive level foliation. The following result does indeed therefore generalise Theorem 4.4.6. It is immediate from Corollary 4.10.1 and Corollary 5.2.1.

Corollary 5.4.1. *Suppose $E = S \times S$ and LMIN. Then,*

(a) *the predissimilarities and the reflexive level foliations are naturally dually order-isomorphic via $f \leftrightarrow \phi = (\zeta_S \circ u)f$.*

(b) *this induces, by restriction, natural dual order-isomorphisms between each subset $\widetilde{\mathcal{F}}$ of $\mathcal{F}_0$ given in Table 4.1 and the corresponding reflexive level foliations in $(\zeta_S \circ u)(\widetilde{\mathcal{F}})$.*

We now use this result to characterise the ultrametrics in a variety of ways.

5.5. Ultrametrics and reflexive level foliations

5.5.1. The main result

Recall (Section 4.3) that a dissimilarity is a symmetric predissimilarity, that an ultrametric is a transitive dissimilarity, and that this rather general definition of an ultrametric reduces to the usual one when L is totally ordered. We have the following characterisations of ultrametrics in terms of their corresponding reflexive level foliations.

Theorem 5.5.1. *Suppose LMIN and $E = S \times S$. Let f be an L-predissimilarity on S and let $\phi = (\zeta_S \circ u)f$ be the corresponding reflexive level foliation.*

(a) *The following four propositions are equivalent:*

(1) *f is an ultrametric on S.*

(2) *$\{b, c\} \subseteq \phi(a, \ell) \Rightarrow c \in \phi(b, \ell)$.*

(3) *for each $\ell \in L$, $a \stackrel{\ell}{\sim} b$ if $b \in \phi(a, \ell)$ defines an equivalence relation on S.*

(4) *$b \in \phi(a, \ell) \Rightarrow \phi(a, \ell) = \phi(b, \ell)$.*

(b) *Suppose now that LTO holds. We define the following hierarchical condition on Range (ϕ):*

(HIER). $\forall H \in \text{Range}(\phi), \forall H' \in \text{Range}(\phi), H \cap H' \in \{\emptyset, H, H'\}$.

Then the following seven propositions are equivalent: (1) to (4) above and:

(5) (i) HIER *and* (ii) $\phi(a, \ell) \subseteq \phi(a', \ell') \Rightarrow \phi(a', \ell') = \phi(a, \ell')$.

(6) (i) HIER *and* (ii) $\phi(a, \ell) \subset \phi(a', \ell') \Rightarrow \ell < \ell'$.

(7) (i) HIER *and* (ii) $b \in \phi(a, \ell) \Leftrightarrow a \in \phi(b, \ell)$.

This theorem summarises a variety of useful characterisations of ultrametricity. In particular, it both illumines and simplifies the formulation of two decompositions of an ultrametric obtained when S is finite and $L = \mathbb{R}_+$. These are the decomposition based on a maximal simplex of minimal diameter introduced in Van Cutsem (1983a) and the ziggurat decomposition proposed in Critchley (1983).

In Section 5.5.2, we offer a number of remarks on Theorem 5.5.1. In Section 5.5.3, we derive variants of Theorem 5.5.1 under stronger initial assumptions upon f.

5.5.2. Remarks on Theorem 5.5.1*

We make the following remarks.

Remark 5.5.1. Perhaps the most important thing to note is that part (b) of Theorem 5.5.1 fails completely if L is not totally ordered, as the following example shows. Let $L = \{(i, j) : i, j \in \{0, 1\}\}$ where $0 < 1$ and we use the product order. Let $S = \{a, b, c\}$ and let $f : S^2 \to L$ be the ultrametric defined by $f(a, b) = (1, 0)$, $f(b, c) = (0, 1)$ and $f(a, c) = (1, 1)$. Let $H = \phi(b, (1, 0))$ and $H' = \phi(b, (0, 1))$. Then $H = \{a, b\}$ and $H' = \{b, c\}$ so that HIER fails. □

Remark 5.5.2. However, without assuming LTO, we still have $I \Rightarrow 5(\text{ii})$ and $7(\text{ii})$ where I is any of conditions 1 to 4. Inspection of the proof of Theorem 5.5.1 shows that $I \Rightarrow 5(\text{ii})$, while clearly $I \Rightarrow 4 \Rightarrow 7(\text{ii})$. □

Remark 5.5.3. Recalling that condition 7(ii) is equivalent to the symmetry of f (Corollary 5.4.1(b)), we see from Theorem 5.5.1 that, if L is totally ordered:

$$f \text{ symmetric} \Rightarrow (\text{HIER} \Leftrightarrow f \text{ transitive}).$$

This characterisation of HIER fails completely if the symmetry condition is dropped even if LTO is retained. In this case, HIER neither implies nor is implied by the transitivity of f. This is demonstrated by the following counter-examples in which $L = \mathbb{R}_+$ and $S = \{a, b, c\}$. Note that these counter-examples are smallest possible since both HIER and f transitive necessarily hold when $|S| = 1$ or 2. Consider the following predissimilarity matrices in which $f(a, b)$ occurs in row a and column b.

Example 1	a	b	c
a	0	1	2
b	1	0	1
c	1	1	0

Example 2	a	b	c
a	0	2	1
b	2	0	1
c	2	2	0

In Example 1, HIER holds but f is not transitive.

In Example 2, f is transitive but HIER fails. □

Remark 5.5.4. Provided condition LTO is retained, property 6(ii) is logically equivalent to the weaker version of property 5(ii) obtained by restricting attention to $\ell = \ell'$. That is:
$$\text{LTO} \Rightarrow (6(\text{ii}) \Leftrightarrow 5^-(\text{ii}))$$
where $5^-(\text{ii})$ is the condition: $\phi(a, \ell) \subseteq \phi(a', \ell) \Rightarrow \phi(a', \ell) = \phi(a, \ell)$.

In particular, LTO $\Rightarrow (5(\text{ii}) \Rightarrow 6(\text{ii}))$.

A proof is given in Appendix B. □

Remark 5.5.5. Suppose here that L is totally ordered. Then we see from Theorem 5.5.1 that

HIER $\Rightarrow$ The three conditions 5(ii), 6(ii) and 7(ii) are logically equivalent.

This is no longer the case if the condition HIER is dropped. Of course, we still have 5(ii) $\Rightarrow$ 6(ii) by the previous remark, but no other implication of the form $A(\text{ii}) \Rightarrow B(\text{ii})$ holds with A and B being distinct members of $\{5, 6, 7\}$. This is shown by the following counterexamples in which $L = \mathbb{R}_+$ and $S = \{a, b, c\}$. Note that these counterexamples are again smallest possible as it is easily shown that 5(ii), 6(ii) and 7(ii) are equivalent when $|S| = 1$ or 2. Consider the following predissimilarity matrices:

Example 1	a	b	c
a	0	1	2
b	2	0	0
c	2	0	0

Example 2	a	b	c
a	0	1	2
b	2	0	1
c	1	2	0

Example 3	a	b	c
a	0	3	1
b	3	0	1
c	1	1	0

In the following array, a tick ($\checkmark$) denotes that the Example in that row has the property in that column. A cross ($\times$) denotes that it does not.

	5(ii)	6(ii)	7(ii)
Example 1	$\checkmark$	$\checkmark$	$\times$
Example 2	$\times$	$\checkmark$	$\times$
Example 3	$\times$	$\times$	$\checkmark$

□

5.5.3. Variants of Theorem 5.5.1

The following strengthenings of Theorem 5.5.1 are possible under the stated additional assumptions on f.

Proposition 5.5.1. *Suppose* LMIN *and* $E = S \times S$. *Let* f *be an L-predissimilarity on* S *and let* $\phi = (\zeta_{SOU})f$ *be the corresponding reflexive level foliation. In the notation of Theorem 5.5.1, we have:*

(a) *If* f *is definite, then:* $1 \Leftrightarrow 2 \Leftrightarrow 3 \Leftrightarrow 4 \Leftrightarrow 5(ii)$.

  Thus, *if* LTO *also holds, we also have* $5(ii) \Leftrightarrow 6 \Leftrightarrow 7$.

(b) *If* f *is symmetric, then:*
$$\text{LTO} \Rightarrow \left(1 \Leftrightarrow 2 \Leftrightarrow 3 \Leftrightarrow 4 \Leftrightarrow \text{HIER} \Rightarrow 5(ii) \Rightarrow 6(ii)\right).$$

(c) *If* f *is definite and symmetric, then:*
$$\text{LTO} \Rightarrow \left(1 \Leftrightarrow 2 \Leftrightarrow 3 \Leftrightarrow 4 \Leftrightarrow \text{HIER} \Leftrightarrow 5(ii) \Rightarrow 6(ii)\right).$$

With reference to this result, neither of the one-way implications HIER $\Rightarrow$ 5(ii) and 5(ii) $\Rightarrow$ 6(ii) of part (b) can be reversed. Nor can the one-way implication 5(ii) $\Rightarrow$ 6(ii) of part (c) be reversed. This is demonstrated by the following examples in which $L = \mathbb{R}_+$ and $S = \{a, b, c, d\}$. These examples are again smallest possible, since it is easy to show that for $|S| \leq 3$, $5(i) \Leftarrow 5(ii) \Leftarrow 6(ii)$ for any symmetric f. Consider the following predissimilarity matrices.

Example 1	a	b	c	d
a	0	0	0	1
b	0	0	1	0
c	0	1	0	0
d	1	0	0	0

Example 2	a	b	c	d
a	0	1	2	3
b	1	0	3	2
c	2	3	0	1
d	3	2	1	0

In the first example, 5(ii) holds but 5(i) fails.

In the second example, where, we note, f is definite, 6(ii) holds but 5(ii) fails.

5.6. On generalisations of indexed hierarchies

5.6.1. Introduction

It is clear from part (c) of Proposition 5.5.1 that we are close to a major extension of Benzécri's bijection between definite ultrametrics and indexed hierarchies. This bijection, recalled as Theorem 4.4.1, deals with the case where S is finite and $L = \mathbb{R}_+$. Critchley and Van Cutsem's bijection between ultrametrics and indexed regular generalised hierarchies, recalled in Theorem 4.4.4, retains $L = \mathbb{R}_+$ but allows S to be arbitrary and does away with the definiteness requirement. In the following section we will go further and relax the conditions on L. However, it is already clear from Remark 5.5.1 that some condition on L, such as LTO, will be needed. In the present preliminary section, we will show that an additional condition, labelled LSU below, will also be required.

The organisation of this section is as follows. In Section 5.6.2, we introduce the concept of a Benzécri structure and develop some of its elementary properties. In Section 5.6.3, we establish that indexed hierarchies and indexed regular generalised hierarchies are special cases of Benzécri structures, coinciding with them under the appropriate conditions on S and L. In Section 5.6.4, we show the necessity of a condition labelled LSU, and establish a range of cases in which this condition holds.

5.6.2. Benzécri structures

We define a *Benzécri structure*, on an arbitrary nonempty set S and an arbitrary ordered set L admitting a minimum, to be a pair $(\mathcal{H}, g)$ where $\mathcal{H}$ is a collection of subsets of S and g is a function from $\mathcal{H}$ to L which satisfy the following conditions:

H1. $\forall (H, H') \in \mathcal{H}^2$, $H \cap H' \in \{\emptyset, H, H'\}$.

GH2. $\forall a \in S$, $S = \cup \{H \in \mathcal{H} : a \in H\}$.

I1. $\forall (H, H') \in \mathcal{H}^2$, $H \subset H' \Rightarrow g(H) < g(H')\}$.

R1. $\forall (a, b) \in S^2$, $H_{ab} \in \mathcal{H}$ where $H_{ab} \equiv \cap \{H \in \mathcal{H} : a \in H, b \in H\}$.

R2. $\forall a \in S$, $\forall \ell \in L$, $H_a(\ell) \in \mathcal{H}$ where $H_a(\ell) \equiv \{b \in S : g(H_{ab}) \leq \ell\}$.

R3. $\forall a \in S$, $\forall \ell \in L$, $g(H_a(\ell)) \leq \ell$.

R4. $\forall a \in S$, $g(H_{aa}) = 0$.

A Benzécri structure is called definite if $\forall a \in S$, $H_{aa} = \{a\}$.

This definition is rather similar to that of an indexed regular generalised hierarchy (IRGH for short), except that here the more general set L replaces $\mathbb{R}_+$. In particular, conditions H1, GH2 and I1 are common to both definitions, while R1 to R3 are simply a relabelling, for the sake of clarity, of conditions RH, RI1 and RI2 respectively of an IRGH. However, there are three important differences: axioms H0 and GH3 have been removed, while GI2 has been replaced by the regularity condition R4. This is for good reason. It will be shown that axioms H0 and GH3 of an indexed regular generalised hierarchy are *redundant* and that GI2 can be replaced by the *weaker* condition R4.

Recall also that the concept of an IRGH, defined by the following set of nine conditions:

$$\text{H0, H1, GH2, GH3, I1, GI2, RH, RI1 and RI2,}$$

generalises that of an indexed hierarchy (IH for short) which is defined by the six conditions:

$$\text{H0, H1, H2, H3, I1 and I2.}$$

We will also show that axiom H0 of an indexed hierarchy is redundant, that the sole role of H3 is to ensure definiteness and that I2 can be replaced by the weaker condition R4.

We establish first some elementary properties of Benzécri structures. Others are given in subsection 5.6.4 below. Let $\mathcal{H}$ be a collection of subsets of S. Then, with reference to part (d) of the following result, we recall that a terminal partition of S in $\mathcal{H}$ is a partition Π of S which satisfies:

$$\forall A \in \Pi, \text{ (i) } A \in \mathcal{H} \quad \text{and (ii)} \quad (H \in \mathcal{H} \text{ and } H \subseteq A) \Rightarrow H = A.$$

Moreover, when $\mathcal{H}$ satisfies H1, a terminal partition is unique if such exists.

Proposition 5.6.1. Let $(\mathcal{H}, g)$ be a Benzécri structure on S and L. Let $(a, b) \in S^2$, let $\ell \in L$ and let $H \in \mathcal{H}$. Then:

(a) $\{a\} \subseteq H_{aa} \subseteq H_{ab}$ and $\{a, b\} \subseteq H_{ab}$.

(b) $H_{ab} = H_{ba}$. In particular, $b \in H_a(\ell) \Leftrightarrow a \in H_b(\ell)$.

(c) $H_a(\ell) = \cup\{H_{ab} : b \in S, g(H_{ab}) \leq \ell\}$.

(d) $\{H_{aa} : a \in S\}$ is the terminal partition of S in $\mathcal{H}$. In particular, H0 holds.

(e) $H_{aa} \subseteq H_a(\ell)$. In particular, $a \in H_a(\ell)$.

(f) $H_a(0) = H_{aa}$ and $g(H_a(0)) = 0$.

(g) $g(H) = 0 \Leftrightarrow$ for some $a \in S$, $H = H_{aa}$.

We offer the following remarks on this result.

Remark 5.6.1. In any Benzécri structure:

(a) There is a certain symmetry, expressed in $H_{ab} = H_{ba}$.

(b) Conditions H0: "$\emptyset \notin \mathcal{H}$" and GH3: "$\exists$ a terminal partition of S in $\mathcal{H}$" of an IRGH are redundant, while GI2: "$g(H) = 0 \Leftrightarrow H$ belongs to the terminal partition of S in $\mathcal{H}$" is logically equivalent to the apparently weaker condition GI2$^-$: "$g(H) = 0 \Leftarrow H$ belongs to the terminal partition of S in $\mathcal{H}$" which, in turn, *is* condition R4. Similar remarks apply to IH conditions. These include the redundancy of H0, while I2 is equivalent to the apparently weaker condition I2$^-$: "$g(H) = 0 \Leftarrow H = \{a\}$ for some $a \in S$". The sole role of H3 is to ensure definiteness.

Note that an indexed hierarchy contains no regularity conditions. As we shall see, this is essentially because in an indexed hierarchy S is finite and therefore any subset $\mathcal{H}$ of $\mathcal{P}(S)$ consists of a finite number of elements. The following proposition rephrases condition H1 to bring out its essentially finite nature, in an obvious sense.

Proposition 5.6.2. *Let $\mathcal{H}$ be a collection of subsets of a nonempty set S. Then H1 is equivalent to the following condition:*

$\tilde{H}$1. *For each finite set $\{H_i : i = 1, \ldots, n\}$ of members of $\mathcal{H}$ whose intersection is nonempty, $\exists \underline{i}$ and $\overline{i}$ in $\{1, \ldots, n\}$ such that*

$$H_{\underline{i}} = \cap \{H_i : i = 1, \ldots, n\} \text{ and } H_{\overline{i}} = \cup \{H_i : i = 1, \ldots, n\}.$$

This result has important implications for Benzécri structures in the finite S case.

Proposition 5.6.3.

(a) *Let $\mathcal{H}$ be a collection of subsets of S and let g be a function from $\mathcal{H}$ to L. Suppose that conditions GH2 and H1 hold. Suppose also that S is finite. Then R1 holds. If I1 also holds, then so do R2 and R3.*

(b) *In particular, if S is finite, conditions R1, R2 and R3 of a Benzécri structure are redundant.*

We give now three examples which demonstrate the necessity of each of conditions R1 to R3 when S is not a finite set. In each case, we take $L = \mathbb{R}_+$. In Example 1, $(\mathcal{H}, g)$ does not obey R1 even though it satisfies H1, GH2 and I1. In fact, $(\mathcal{H}, g)$ satisfies the stronger conditions H0 to H3, I1 and I2 listed in the definition of an indexed hierarchy. In Example 2, $(\mathcal{H}, g)$ does not obey R2 even though it satisfies H0 to H3, I1, I2, R1 and R4. In Example 3, $(\mathcal{H}, g)$ does not obey R3 even though it satisfies H0 to H3, I1, I2, R1, R2 and R4. Noting that R4 doesn't make sense without R1, and that R3 doesn't make sense without R2, these examples therefore give a complete account of the logical independence of conditions R1 to R3.

Example 1

$S = \mathbb{R}$. $\mathcal{H} = \{\{a\} : a \in S\} \cup S \cup \{(-\ell, \ell) : \ell > 0\}$. $g : \mathcal{H} \to \mathbb{R}_+$ satisfies

$$g(\{a\}) = 0,\ g(S) = 1,\ \forall \ell > 0,\ g((-\ell, \ell)) = \ell/(\ell + 1).$$

Then $\mathcal{H}$ satisfies H0 to H3 and I1, I2 hold but, for any $\ell > 0$, $H_{-\ell\ell} \equiv [-\ell, \ell] \notin \mathcal{H}$ and so R1 fails. For a countably infinite example, replace $S = \mathbb{R}$ by $S = \mathbb{Q}$, the set of all rationals, and $(-\ell, \ell)$ by $(-\ell, \ell) \cap \mathbb{Q}$. Then for all positive $q \in \mathbb{Q}$, $H_{-qq} = [-q, q] \cap \mathbb{Q} \notin \mathcal{H}$.

□

Example 2

$S = \mathbb{R}$. $\mathcal{H} = \{\{a\} : a \in S\} \cup S \cup \{[-\ell, \ell] : \ell > 0\}$. $g : \mathcal{H} \to \mathbb{R}_+$ satisfies

$$g(\{a\}) = 0,\ g(S) = 1,\ \forall \ell > 0,\ g([-\ell, \ell]) = \ell/(\ell + 1).$$

Then H0 to H3, I1 and I2 clearly hold. Remarking that

$$\forall (a, b) \in S^2,\ H_{ab} = \begin{cases} \{a\} & \text{if } a = b \\ [-\ell_{ab}, \ell_{ab}] & \text{if } a \neq b \end{cases}$$

where $\ell_{ab} = \max\{|a|, |b|\}$, we see that R1 and R4 also hold. However, R2 fails since, for example,

$$\forall 0 < \varepsilon < 1,\ H_0(\varepsilon) = (-\ell_\varepsilon, \ell_\varepsilon) \notin \mathcal{H}$$

where $\ell_\varepsilon = \varepsilon/(1-\varepsilon)$.

For a countably infinite example replace $S = \mathbb{R}$ by $S = \mathbb{Q}$, each H in $\mathcal{H}$ by $H \cap \mathbb{Q}$, and so on, and then at the last step consider any rational ε in $(0,1)$. □

Example 3

$S = \mathbb{R}$. $\mathcal{H} = \{\{a\} : a \in S\} \cup S \cup \{[-n,n] : n \in \mathbb{N}\}$ where $\mathbb{N}$ is the set of strictly positive integers. $g : \mathcal{H} \to \mathbb{R}_+$ satisfies

$$g(\{a\}) = 0, \ g(S) = 1 + \varepsilon \text{ for some } \varepsilon > 0, \ g([-n,n]) = n/(n+1).$$

Then H0 to H3, I1 and I2 clearly hold, as do R1 and R4 when we remark that

$$\forall (a,b) \in S^2, \quad H_{ab} = \begin{cases} \{a\} & \text{if } a = b \\ [-n_{ab}, n_{ab}] & \text{if } a \neq b \end{cases}$$

where n_{ab} is the least n in $\mathbb{N}$ satisfying $n \geq \max\{|a|, |b|\}$. Moreover R2 also holds since, using Proposition 5.6.1(c), we have that:

$$\forall a \in S, \forall \ell \in \mathbb{R}_+, \ H_a(\ell) = \begin{cases} H_{aa} & \text{if } \ell = 0, \text{ or if } 0 < \ell < 1 \text{ and } n_{a0} > \ell/(1-\ell) \\ [-n_a^*, n_a^*] & \text{if } 0 < \ell < 1 \text{ and } n_{a0} \leq \ell/(1-\ell) \\ S & \text{if } \ell \geq 1 \end{cases}$$

when $n_a^* = \max\{n_{ab} : b \in S, n_{ab} \leq \ell/(1-\ell)\}$. However R3 fails since

$$\forall a \in S, \quad g(H_a(1)) = 1 + \varepsilon > 1.$$

Again, replacing $S = \mathbb{R}$ by $S = \mathbb{Q}$ and each H in $\mathcal{H}$ by $H \cap \mathbb{Q}$ in the obvious way yields a countably infinite counterexample. □

5.6.3. Special cases of Benzécri structures

We establish here that both indexed hierarchies and indexed regular generalised hierarchies are indeed special cases of Benzécri structures under the appropriate conditions. Thus the bijection to be established in Section 5.7 is indeed a generalisation of those of Benzécri (for IH's) and of Critchley and Van Cutsem (for IRGH's).

Proposition 5.6.4. *Let $\mathcal{H}$ be a collection of subsets of the nonempty set S and let g be a function from $\mathcal{H}$ to L. Then:*

(a) *If $L = \mathbb{R}_+$,*

  $(\mathcal{H}, g)$ is an indexed regular generalised hierarchy $\Leftrightarrow$ $(\mathcal{H}, g)$ is a Benzécri structure.

(b) *If $L = \mathbb{R}_+$ and if S is finite,*

  $(\mathcal{H}, g)$ is an indexed hierarchy $\Leftrightarrow$ $(\mathcal{H}, g)$ is a definite Benzécri structure.

5.6.4. The condition LSU

We begin by noting some further elementary properties of a Benzécri structure.

Proposition 5.6.5. *Let $(\mathcal{H}, g)$ be a Benzécri structure on S and L. Let $H \in \mathcal{H}$. Then:*

(a) (i) $H_{aa} \subseteq H \Leftrightarrow a \in H$ and (ii) $H_{ab} \subseteq H \Leftrightarrow a \in H$ and $g(H_{ab}) \leq g(H)$.

Moreover,

(b) *H can be written in any of the following forms:*
  (i) $H = \cup\{H_{aa} : H_{aa} \subseteq H\} = \cup\{H_{aa} : a \in H\}$.
  (ii) $H = \cup\{H_{ab} : H_{ab} \subseteq H\} = \cup\{H_{ab} : a \in H, g(H_{ab}) \leq g(H)\}$.
  (iii) $\forall a \in H, H = \{b \in S : H_{ab} \subseteq H\} = H_a(g(H))$.

And,

(c) $\{\tilde{\ell} \in L : H = H_{\tilde{a}}(\tilde{\ell}) \text{ for some } \tilde{a} \in S\}$ *admits a minimum, which is $g(H)$.*

In words, this result tells us that every H in a Benzécri structure $(\mathcal{H}, g)$ is the union of the members of the terminal partition which it contains (b(i)), is also the union of the sets H_{ab} which it contains (b(ii)) and, most importantly for our present purposes, can be written in the form $H_a(g(H))$ where a is any member of H. Together with condition R3, this then yields that the subset of L specified in (c) has a minimum which, moreover, is $g(H)$.

The fact that Benzécri structures, which subsume IH's and IRGH's, have this property means that the existence of a bijection of the desired form via $H_a(\ell) \leftrightarrow \phi(a, \ell)$ necessitates the following condition.

LSU. L and S are ultrametrically compatible. That is, LMIN and:

  $\forall$ ultrametrics $f \in \mathcal{F}_u, \forall a \in S, \forall \ell \in L$,

  $\{\tilde{\ell} \in L : \phi(a, \ell) = \phi(\tilde{a}, \tilde{\ell}) \text{ for some } \tilde{a} \in S\}$ has a minimum,

  where ϕ denotes $(\zeta_S ou)f$.

We establish now some general conditions under which LSU holds. To do this we define the following terms:

$\overline{\text{LSU}}$. L and S are ultrametrically boundedly join complete. That is, LMIN holds and

  $\forall f \in \mathcal{F}_u, \forall a \in S, \forall \ell \in L$:

  $\{f(a, b) : b \in S, f(a, b) \leq \ell\}$ has a least upper bound.

$\overline{\text{L}}$. L is boundedly join complete. That is, every nonempty subset of L that has an upper bound has a least upper bound.

We recall also the list of properties, given in Table 4.3, that L may enjoy; and Theorem 4.8.1 on conditions for E and L to be adapted. The following lemma indicates how the condition $\overline{\text{L}}$ relates to other properties of L. Part (a) is (essentially) classical.

Lemma 5.6.1. *Let L be a nonempty ordered set. Then:*

(a) *(LMIN and $\overline{L}$) $\Leftrightarrow$ L is a meet complete meet semi-lattice.*

(b) *LTO $\Rightarrow$ [(LMIN and $\overline{L}$) $\Leftrightarrow$ LWF].*

(c) *In particular, (LTO and LMIN) $\Rightarrow$ [$\overline{L}$ $\Leftrightarrow$ LWF].*

Also:

(d) *LFILT $\Rightarrow$ $\overline{L}$.*

(e) *The converse to (d) is false in general, but true if LTO.*

(f) *In particular, LTO $\Rightarrow$ [$\overline{L}$ $\Leftrightarrow$ LFILT].*

Remark 5.6.2. With reference to the relevant part of this lemma we note that $L = (0,1]$ shows:

(b) (LTO and $\overline{L}$) $\not\Rightarrow$ LWF. □

Proposition 5.6.6.

(a) *The following implications hold:*

(S finite, LMIN, JSL)

(L is a complete lattice)

(S^2 and L are adapted)

$\overline{\text{LSU}} \Rightarrow$ LSU.

(LUR, LWF) $\Rightarrow$ (LMIN, LFILT) $\Rightarrow$ (LMIN, $\overline{L}$)

(b) *If LTO holds,*

LWF $\Leftrightarrow$ (LMIN, LFILT) $\Leftrightarrow$ (LMIN, $\overline{L}$) $\Rightarrow$ $\overline{\text{LSU}}$ $\Leftrightarrow$ LSU.

We are now ready to extend Benzécri's bijection and that of Critchley and Van Cutsem.

5.7. Benzécri structures

Let $\mathcal{B}$ denote the set of all Benzécri structures on a nonempty set S and an ordered set L admitting a minimum. We introduce an order on $\mathcal{B}$ as follows:

$$(\mathcal{H}, g) \leq (\mathcal{H}', g') \Leftrightarrow \forall a \in S, \forall \ell \in L, H_a(\ell) \subseteq H'_a(\ell).$$

Let $\Phi_S^U \equiv (\zeta_S \text{ou})\mathcal{F}_u$ be the set of all reflexive level foliations that correspond to ultrametrics. Whenever LSU holds (see Proposition 5.6.6 of Section 5.6.4) we can and do define the map α on Φ_S^U by $\alpha(\phi) = (\mathcal{H}, g)$ where $\mathcal{H} = \text{Range}(\phi)$ and $g : \mathcal{H} \to L$ satisfies

$$\forall a \in S, \forall \ell \in L, g(\phi(a, \ell)) = \min\{\tilde{\ell} \in L : \phi(a, \ell) = \phi(\tilde{a}, \tilde{\ell}) \text{ for some } \tilde{a} \in S\}.$$

V. UNIFICATION AND GENERALISATION OF FUNDAMENTAL BIJECTIONS

The map β is defined on $\mathcal{B}$ by $\beta(\mathcal{H}, g) = \phi$ where $\phi: S \times L \to \mathcal{P}(S)$ is given by:

$$\forall a \in S, \forall \ell \in L, \phi(a, \ell) = H_a(\ell).$$

Recalling Proposition 5.6.4 of Section 5.6.3, the following result is indeed a generalisation of that of Benzécri for indexed hierarchies reported in Theorem 4.4.1, and of that of Critchley and Van Cutsem for indexed regular generalised hierarchies reported in Theorem 4.4.4.

Theorem 5.7.1. *Suppose LTO, LSU and $E = S \times S$. Then:*

(a) α *and* β *define a natural order-isomorphism between* Φ_S^U *and* $\mathcal{B}$.

And so:

(b) $\alpha \circ \zeta_s \circ u$ *and* $v \circ \zeta_s^{-1} \circ \beta$ *define a natural dual order-isomorphism between* $\mathcal{F}_u$ *and* $\mathcal{B}$.

Further:

(c) *The relevant restrictions of these maps define natural (dual) order-isomorphisms between the definite members of* $\mathcal{F}_u$, Φ_S^U *and* $\mathcal{B}$.

Remark 5.7.1. We offer the following remarks on Theorem 5.7.1:

(1) Condition LSU cannot be removed or relaxed as it is precisely what is required for α to be defined.

(2) The counterexample in Remark 5.5.1, we observe, satisfies LSU. Thus, even when LSU is retained, the theorem fails if LTO is removed.

(3) Let $\phi \in \Phi_S^U$ and let $(\mathcal{H}, g) = \alpha(\phi)$. Then it is shown in the proof of Theorem 5.7.1 that:

(i) $H_{ab} = \phi(a, f(a, b))$ and (ii) $g(H_{ab}) = f(a, b)$.

(4) Since $H_a(\ell) = \phi(a, \ell)$ it follows that, under the hypotheses of Theorem 5.7.1, the sets $\{H_a(\ell) : a \in S, \ell \in L\}$ enjoy all the properties listed in Theorem 5.5.1. In particular, in this case:

$\forall \ell \in L, \{H_a(\ell) : a \in S\}$ partition S.

(5) Part (a) of the theorem can of course be combined with other aspects of Corollary 5.2.1 to produce (dual) order-isomorphisms involving $\mathcal{B}$ other than that noted in part (b) above.

5.8. Subdominants

In this section on subdominants we follow the approach of Jardine and Sibson (1971, Section 8.3) which we find extends easily to the case where predissimilarities need not be symmetric, where the set S on which they are defined is not necessarily finite, and where the ordered set L admitting a minimum is not necessarily $\mathbb{R}_+$.

If such exists, a subdominant of a given predissimilarity $f_0 \in \mathcal{F}_0$ in a given subset $\widehat{\mathcal{F}}_0$ of $\mathcal{F}_0$ is a member g of $\widehat{\mathcal{F}}_0$ such that:

(i) $g \leq f_0$ and (ii) $\forall f \in \widehat{\mathcal{F}}_0, g \leq f \leq f_0 \Rightarrow g = f$.

Let $\emptyset \neq \widetilde{\mathcal{F}}_0 \subseteq \mathcal{F}_0$. Then $\widetilde{\mathcal{F}}_0$ is called bounded if:

$$\exists f_0 \in \mathcal{F}_0 \text{ such that } f \in \widetilde{\mathcal{F}}_0 \Rightarrow f \leq f_0.$$

Also, whenever the relevant suprema exist, we can and do define $\vee\widetilde{\mathcal{F}}_0$ to be the member of $\mathcal{F}_0$ given by:

$$\forall (a,b) \in S^2, \left(\vee\widetilde{\mathcal{F}}_0\right)(a,b) = \vee\{f(a,b) : f \in \widetilde{\mathcal{F}}_0\}.$$

Note that $\overline{\mathrm{L}}$ is a natural condition on L alone which is sufficient for $\vee\widetilde{\mathcal{F}}_0$ to be defined for every bounded subset $\widetilde{\mathcal{F}}_0$ of $\mathcal{F}_0$ and for every S. Proposition 5.6.6 identifies conditions under which $\overline{\mathrm{L}}$ holds. Finally, a subset $\widehat{\mathcal{F}}_0$ of $\mathcal{F}_0$ is called sup-closed if, for every bounded subset $\widetilde{\mathcal{F}}_0$ of $\widehat{\mathcal{F}}_0$, $\vee\widetilde{\mathcal{F}}_0$ is defined and belongs to $\widehat{\mathcal{F}}_0$. We have the following simple but important result.

Theorem 5.8.1. *Suppose* LMIN *and* $E = S \times S$. *Let* $\widehat{\mathcal{F}}_0$ *be a sup-closed subset of* $\mathcal{F}_0$. *Then:*

(a) $\forall f_0 \in \mathcal{F}_0, \hat{f}_0 \equiv \vee\{f \in \widehat{\mathcal{F}}_0 : f \leq f_0\}$ *is the unique subdominant of* f_0 *in* $\widehat{\mathcal{F}}_0$.

(b) $\hat{f}_0 = f_0 \Leftrightarrow f_0 \in \widehat{\mathcal{F}}_0$.

Thus to every sup-closed collection $\widehat{\mathcal{F}}_0$ of L-predissimilarities on S there is associated a well-defined subdominant method which maps the data f_0 to its unique subdominant $\hat{f}_0$. The most important example of this is given in the following result.

Proposition 5.8.1. *Suppose* LMIN *and* $E = S \times S$. *Suppose also* $\overline{\mathrm{L}}$. *Then the set* $\mathcal{F}_{ot}$ *of all ultrametric L-predissimilarities on S is sup-closed.*

We may therefore speak about "single-link" or "subdominant" hierarchical clustering for any nonempty set S and for any ordered set L that satisfies LMIN and $\overline{\mathrm{L}}$ (equivalently, (Lemma 5.6.1 (a)) is a meet complete meet semi-lattice). For examples of relevant data requiring such methods, see Section 4.2. We hope to develop these new methods in future work.

Finally we note that, assuming always LMIN and $\overline{\mathrm{L}}$, there are other sup-closed subsets of $\mathcal{F}_0$ of potential interest in defining subdominant methods. For example, generalising Jardine and Sibson (1971, §8.5 and §8.6), we define $\mathcal{F}_{B_k}$ to be the set of all $f \in \mathcal{F}_0$ such that $\forall \ell \in L, \forall R \subseteq S$ with $|R| = k$, $\forall (a,b) \in S^2$:

$$\left\{ \begin{array}{ll} \text{(i)} & \forall r \in R, f(a,r) \leq \ell \\ \text{(ii)} & \forall (r, \tilde{r}) \in R^2, f(r, \tilde{r}) \leq \ell \\ \text{(iii)} & \forall r \in R, f(r,b) \leq \ell \end{array} \right\} \Rightarrow f(a,b) \leq \ell$$

and $\mathcal{F}_{B_k^c}$ to be the set of all $f \in \mathcal{F}_0$ with the same property except that condition (ii) is removed. Then it is easy to show that both these sets are sup-closed and so are their intersections with the symmetric predissimilarities. Thus, both the fine and coarse k-clustering approach to overlapping hierarchical methods also generalise to any S and L. Equally, supposing that addition is defined on L (and is consistent with its order, that is: $\ell_1 \geq \tilde{\ell}_1$ and $\ell_2 \geq \tilde{\ell}_2$ implies $\ell_1 + \ell_2 \geq \tilde{\ell}_1 + \tilde{\ell}_2$) then the set

$$\mathcal{F}_M \equiv \{f \in \mathcal{F}_0 : \forall (a,b,c) \in S^3, f(a,b) + f(b,c) \geq f(a,c)\}$$

of all metric predissimilarities is defined and is easily seen to be sup-closed. In the case where S is finite and $L = \mathbb{R}_+$, Van Cutsem (1983b) discusses subdominants for several types of dissimilarities, together with algorithms to calculate them.

Acknowledgements

It is a pleasure to gratefully acknowledge the assistance of B. Leclerc and B. Monjardet in opening our eyes to the advantages of the order-theoretic approach and, in particular, for pointing out an improved version of Lemma 5.6.1. And also the support provided by:

(a) a grant from the Royal Society under the European Science Exchange Programme (FC)

(b) two periods as Professeur Associé à l'Université Joseph Fourier (FC)

(c) a grant from the Warwick Research and Innovations Fund (FC)

(d) a sabbatical semester granted by l'Université Joseph Fourier (BVC)

and (e) travel grants from the British Council (both authors).

APPENDIX B: Proofs

Proof of Proposition 5.3.1

(a) LMIN holds by hypothesis. Thus, by Theorem 4.8.1 (a),

$m \in \mathcal{M}$ is residual $\Leftrightarrow \forall \emptyset \neq D \subseteq E, \vee\{f_m(a,b) : (a,b) \in D\}$ exists.

Thus, recalling that $L_{b,m_a} = L_{(a,b),m}$ and taking $D = \{a\} \times \tilde{B}$, $m \in \mathcal{M}$ is residual $\Rightarrow \forall a \in A, \forall \emptyset \neq \tilde{B} \subseteq B, \vee\{f_{m_a}(b) : b \in \tilde{B}\}$ exists. This latter is equivalent to m_a being residual, using Theorem 4.8.1 (a) again.

(b) Let $\tilde{m} : L \to \mathcal{P}(B)$ be level. For all $a \in A$, put $m_a = \tilde{m}$ and let $m = \zeta_A^{-1}(\{m_a\}_{a \in A})$. Then $m : L \to \mathcal{P}(A \times B)$ is level by Corollary 5.2.1, and hence residual by hypothesis. Thus, by (a), $\tilde{m}$ is residual, as required. ∎

Proof of Theorem 5.5.1

(a) We prove $1 \Rightarrow 2 \Rightarrow 3 \Rightarrow 4 \Rightarrow 1$.

$\underline{1 \Rightarrow 2}$: $\{b,c\} \subseteq \phi(a,\ell) \Rightarrow f(b,a) \leq \ell$ and $f(a,c) \leq \ell$, using the symmetry of f.
$\Rightarrow f(b,c) \leq \ell$, as f is transitive.

$\underline{2 \Rightarrow 3}$: Let $\ell \in L$. The relation $\overset{\ell}{\sim}$ is clearly reflexive by definition. Suppose now $a \overset{\ell}{\sim} b$. Then $\{b,a\} \subseteq \phi(a,\ell)$ and hence, by 2, $b \overset{\ell}{\sim} a$. Finally, suppose $a \overset{\ell}{\sim} b$ and $b \overset{\ell}{\sim} c$. Then, as $\overset{\ell}{\sim}$ is symmetric, $\{a,c\} \subseteq \phi(b,\ell)$. Hence, by 2, $a \overset{\ell}{\sim} c$.

$\underline{3 \Rightarrow 4}$: It suffices to note that, by 3:

$$a \overset{\ell}{\sim} b \Rightarrow \{a \overset{\ell}{\sim} c \Leftrightarrow b \overset{\ell}{\sim} c\}.$$

$\underline{4 \Rightarrow 1}$: $b \in \phi(a, f(a,b))$, as $\leq$ is reflexive
$\Rightarrow \phi(a, f(a,b)) = \phi(b, f(a,b))$, by 4
$\Rightarrow f(b,a) \leq f(a,b)$, as $a \in \phi(a, f(a,b))$.

Similarly, $f(a,b) \leq f(b,a)$. Hence, f is symmetric, as $\leq$ is anti-symmetric. It remains to show that f is transitive. Suppose then that $f(a,b) \leq \ell$ and $f(b,c) \leq \ell$. Then $b \in \phi(a,\ell)$, by hypothesis.
$\Rightarrow \phi(a,\ell) = \phi(b,\ell)$, by 4.
$\Rightarrow f(a,c) \leq \ell$, as $c \in \phi(b,\ell)$ by hypothesis.

(b) Suppose now LTO. We prove that $4 \Rightarrow 5 \Rightarrow 6 \Rightarrow 7 \Rightarrow 1$.

$\underline{4 \Rightarrow 5}$: Suppose, if possible, HIER fails. Then $\exists H \equiv \phi(a,\ell)$ and $H' \equiv \phi(a',\ell')$ such that:

$$\exists b \in H \cap H' \text{ and } \exists c \in H \backslash H' \text{ and } \exists c' \in H' \backslash H.$$

V. UNIFICATION AND GENERALISATION OF FUNDAMENTAL BIJECTIONS   141

Using 4, $b \in H \cap H' \Rightarrow H = \phi(b, \ell)$ and $H' = \phi(b, \ell')$. Thus, $c \in H \backslash H' \Rightarrow \ell' < f(b, c) \le \ell$ and so $\ell' < \ell$. Similarly, $c' \in H' \backslash H \Rightarrow \ell < \ell'$. This contradiction establishes HIER. Suppose now $\phi(a, \ell) \subseteq \phi(a', \ell')$. Then:

$$a \in \phi(a, \ell), \text{ by definition.}$$
$$\Rightarrow a \in \phi(a', \ell'), \text{ by hypothesis.}$$
$$\Rightarrow \phi(a', \ell') = \phi(a, \ell'), \text{ by 4.}$$

$5 \Rightarrow 6$: Suppose $\phi(a, \ell) \subset \phi(a', \ell')$. Then $\phi(a, \ell) \subset \phi(a, \ell')$, using 5(ii). Hence $\ell < \ell'$, as L is totally ordered.

$6 \Rightarrow 7$: By Corollary 5.4.1(b), 7(ii) is equivalent to f being symmetric. Suppose, if possible, that f is not symmetric. Then $\exists (a, a') \in S^2$ such that $f(a, a') < f(a', a)$. Let $H = \phi(a, f(a, a'))$ and $H' = \phi(a', f(a, a'))$. Then $a' \in H \cap H'$ while $a \in H \backslash H'$. Thus, by HIER, $H' \subset H$ which contradicts 6(ii).

$7 \Rightarrow 1$: We prove (not 1) $\Rightarrow$ (not 7). The result is immediate if f is not symmetric. Suppose then that f is symmetric but that $\exists (a, b, c) \in S^3$ and $\ell \in L$ such that

$$f(a, b) \le \ell, f(b, c) \le \ell, \text{ but } f(a, c) > \ell.$$

Let $H = \phi(a, f(a, b))$ and $H' = \phi(c, f(b, c))$. Using the symmetry of f, $b \in H \cap H'$, $a \in H \backslash H'$ and $c \in H' \backslash H$. Thus HIER fails. ∎

Proof of Remark 5.5.4

Suppose first 6(ii). Suppose also that $\phi(a, \ell) \subseteq \phi(a', \ell)$. By 6(ii), strict inclusion is impossible and hence equality obtains. Conversely, suppose 5^-(ii). Suppose also that $\phi(a, \ell) \subset \phi(a', \ell')$. Then $\ell = \ell'$ is impossible by 5^-(ii), while $\ell > \ell'$ implies $\phi(a, \ell') \subseteq \phi(a, \ell) \subset \phi(a', \ell')$ which is again contrary to 5^-(ii). Hence $\ell < \ell'$ as L is totally ordered. ∎

Proof of Proposition 5.5.1

(a) Given Theorem 5.5.1 and Remark 5.5.2, it suffices to show 5(ii) $\Rightarrow$ 4. Let $b \in \phi(a, \ell)$. As f is definite, $\phi(b, 0) = \{b\}$. Thus $\phi(b, 0) \subseteq \phi(a, \ell)$, and so $\phi(a, \ell) = \phi(b, \ell)$ by 5(ii).

(b) Recalling that 7(ii) is equivalent to the symmetry of f, this is an immediate corollary of Theorem 5.5.1 and Remark 5.5.4.

(c) Immediate from (a) and (b). ∎

Proof of Proposition 5.6.1

(a) and (b) are immediate.

(c) Now, $c \in H_{ab} \Leftrightarrow \{a, c\} \subseteq H_{ab}$, by (a).
$$\Leftrightarrow H_{ac} \subseteq H_{ab}, \text{ by definition of } H_{ac}.$$
$$\Leftrightarrow g(H_{ac}) \le g(H_{ab}), \text{ the forwards implication holding by I1, and the reverse by remarking that } H_{ab} \cap H_{ac} \text{ is nonempty and then using H1 and I1.}$$

Thus $\cup\{H_{ab} : b \in S, g(H_{ab}) \leq \ell\} \subseteq H_a(\ell)$, by transitivity of $\leq$. The reverse inclusion holds by (a).

(d) By (a), $\{H_{aa} : a \in S\}$ cover S. By R4 and I1, $H_{aa} \subset H_{bb}$ is impossible. Hence, by H1, $\{H_{aa} : a \in S\}$ partition S. Using R4 and I1 again, this partition is clearly terminal and, by H1, a terminal partition is unique. Finally, H0 holds as $H \in \mathcal{H}$ and $H = \emptyset$ is impossible, since $\emptyset \subset H_{aa}$.

(e) is immediate from (a) and (c) using R4.

(f) R3 $\Rightarrow g(H_a(0)) = 0$ at once, while (e) implies $H_{aa} \subseteq H_a(0)$ and strict inclusion is impossible by I1.

(g) Suppose $g(H) = 0$ and consider the identity $H = \cup\{H \cap H_{aa} : a \in S, H \cap H_{aa} \neq \emptyset\}$ which holds by (d). It suffices to note that, by I1, there cannot be more than one member of the disjoint union in this identity. The reverse implication *is* condition R4. ∎

Proof of Proposition 5.6.2

Clearly $\tilde{H}1 \Rightarrow H1$. We establish the converse by induction on n. The case $n = 1$ of $\tilde{H}1$ holds trivially, while H1 implies (indeed, is equivalent to) the $n = 2$ version of $\tilde{H}1$. Suppose then that $\tilde{H}1$ holds for collections of n subsets of $\mathcal{H}$, and let $\{H_i : i = 1, \ldots, n+1\}$ be members of $\mathcal{H}$ with nonempty intersection. Let $H_{\underline{i}} = \cap\{H_i : i = 1, \ldots, n\}$ and $H_{\overline{i}} = \cup\{H_i : i = 1, \ldots, n\}$. Then:

$$\cap\{H_i : i = 1, \ldots, n+1\} = H_{\underline{i}} \cap H_{n+1} \in \{H_{\underline{i}}, H_{n+1}\} \text{ by H1}$$

while

$$\cup\{H_i : i = 1, \ldots, n+1\} = H_{\overline{i}} \cup H_{n+1} \in \{H_{\overline{i}}, H_{n+1}\} \text{ by H1.} \qquad \blacksquare$$

Proof of Proposition 5.6.3

It suffices to establish part (a). That R1 holds is immediate from Proposition 5.6.2. Inspection of the proof of Proposition 5.6.1 shows that its part (a) is always true, while its part (c) requires only GH2, H1, I1 and R1. We may therefore write

$$H_a(\ell) = \cup\{H_{ab} : b \in S, g(H_{ab}) \leq \ell\}$$

and note that $H_{aa} \subseteq H_{ab}$. Thus, $H_a(\ell)$ is a finite union of sets whose intersection contains H_{aa} and is therefore nonempty. Thus, we may use Proposition 5.6.2 to conclude that $H_a(\ell) = H_{ab}$ for some H_{ab} such that $g(H_{ab}) \leq \ell$. ∎

Proof of Proposition 5.6.4

(a) Suppose $(\mathcal{H}, g)$ is an IRGH. It suffices to show that R4 holds. But it is clear that H_{aa} is the member of the terminal partition which contains a. The result now follows by GI2. Conversely, suppose $(\mathcal{H}, g)$ is a Benzécri structure. It suffices to show H0, GH3 and GI2 hold. This is immediate from Proposition 5.6.1 parts (d) and (g).

(b) Suppose $(\mathcal{H}, g)$ is an indexed hierarchy. As S is finite it suffices, by Proposition 5.6.3, to show that GH2 and R4 hold. Clearly H2 $\Rightarrow$ GH2, while H3 $\Rightarrow H_{aa} = \{a\}$ and so I2 implies R4. Conversely, let $(\mathcal{H}, g)$ be a definite Benzécri structure. It suffices to show H0, H2, H3 and I2. Of these, H0 holds by Proposition 5.6.1(d), H2 is implied by GH2 and Proposition 5.6.2 as S is finite, H3 follows from R1 and the definition of definiteness of a Benzécri structure, while I2 then follows from Proposition 5.6.1(g). ∎

Proof of Proposition 5.6.5

(a)(i) By definition, $a \in H_{aa}$. Thus $H_{aa} \subseteq H$ trivially implies $a \in H$. Conversely if $a \in H$ then $H \cap H_{aa}$ is nonempty. Thus H1, and the fact that $\{H_{aa} : a \in S\}$ form a terminal partition, imply that $H_{aa} \subseteq H$.

(ii) This is clear from H1 and I1.

(b) (i) This follows at once from (a)(i), recalling $a \in H_{aa}$.

(ii) The first expression for H follows from (a)(i). The second is immediate from (a)(ii).

(iii) If $a \in H$, then $\forall b \in S$, $b \in H \Leftrightarrow H_{ab} \subseteq H$, by definition of H_{ab}.

$$\Leftrightarrow g(H_{ab}) \leq g(H), \text{ using (a)(ii)}.$$
$$\Leftrightarrow b \in H_a(g(H)), \text{ by definition of } H_a(\ell).$$

(c) By (b)(iii), $g(H) \in \{\tilde{\ell} : H = H_{\tilde{a}}(\tilde{\ell})$ for some $\tilde{a} \in S\}$ while, by R3, $H = H_{\tilde{a}}(\tilde{\ell}) \Rightarrow g(H) \leq \tilde{\ell}$. ∎

Proof of Lemma 5.6.1

For any nonempty subset K of L let $\underline{K}$ (respectively, $\overline{K}$) denote the (possibly empty) set of all lower (respectively, upper) bounds of K in L. The proof uses the following fact.

Fact B1. *Let K be any nonempty subset of an ordered set L.*

(i) *Suppose that $\underline{K}$ is nonempty. Then $\underline{K}$ is an ideal. Moreover,*

$$\vee \underline{K} \text{ exists} \Rightarrow [\vee \underline{K} = \max \underline{K} \equiv \wedge K \text{ and } \underline{K} = (\vee \underline{K}].]$$

(ii) *Dually, suppose instead that $\overline{K}$ is nonempty. Then $\overline{K}$ is a filter. Further,*

$$\wedge \overline{K} \text{ exists} \Rightarrow [\wedge \overline{K} = \min \overline{K} \equiv \vee K \text{ and } \overline{K} = [\wedge \overline{K}).]$$

Proof of Fact B1

Clearly $\underline{K}$ is an ideal. If $\vee \underline{K}$ exists, then it belongs to $\underline{K}$ since every member of K is an upper bound for $\underline{K}$. The rest is now immediate. ∎

(a) Suppose first LMIN and $\overline{L}$. Let K denote any nonempty subset of L. By LMIN, $0 \in \underline{K}$ while each member k of K is an upper bound for $\underline{K}$. Thus, by $\overline{L}$, $\vee \underline{K}$ exists and, by part (i) of Fact B1, is $\wedge K$. Conversely, suppose that L is a meet complete meet semi-lattice. Then $\min L = \wedge L$ exists. Let now K be any nonempty subset of L such that $\overline{K}$ is nonempty. Then $\wedge \overline{K}$ exists by hypothesis and, by part (ii) of Fact B1, is $\vee K$.

(b) Suppose first LTO, LMIN and $\overline{L}$. Let K be any filter in L. Then, as in the proof of (a) above, $\vee \underline{K}$ exists and equals $\wedge K$. If $\wedge K \in K$, then $K = [\wedge K)$, as K is a filter. Suppose then that $\wedge K \notin K$. Clearly, $\underline{K} \subseteq]\wedge K)$ while conversely:

$$\ell > \wedge K \Rightarrow \ell \notin \underline{K}, \text{ as } \underline{K} = (\wedge K] \text{ by Fact B1 (i)}$$
$$\Rightarrow \exists k \in K \text{ with } \ell > k, \text{ by LTO}$$
$$\Rightarrow \ell \in K, \text{ as } K \text{ is a filter.}$$

Hence $\underline{K} =]\wedge K)$. Conversely suppose LTO and LWF. Recall that LWF $\Rightarrow$ LMIN (see Proposition 4.6.1 (c)). Now let K be any nonempty subset of L such that $\overline{K}$ is nonempty, and hence a filter. Then, by LWF, $\overline{K}$ is of the form $[\ell)$ or $]\ell)$ for some $\ell \in L$. In the second case, $\ell \notin \overline{K}$ implies, as L is totally ordered, that $\ell < \overline{k}$ for some $\overline{k} \in K$. But then $\overline{K} =]\ell)$ implies $\overline{k} = \max K$ so that $\overline{K} = [\max K)$. Hence $\overline{K}$ is always of the form $[\ell)$ and so $\vee K = \ell$ always exists, as required.

(c) This is immediate from (b).

(d) Let K be any nonempty subset of L such that $\overline{K}$ is nonempty and hence a filter. Now it is clear that $\overline{K} = L \Leftrightarrow $ [LMIN and $K = \{0\}]$, in which case $\vee K = 0$. Otherwise $\overline{K}$ is a proper filter of L and hence, by LFILT, $\wedge \overline{K}$ exists. It follows that $\vee K = \wedge \overline{K}$ using Fact B1 (ii).

(e) Consider $L = (0, 1] \cup \{\tilde{\ell}\}$ where $(0, 1]$ inherits the usual order on R and $\forall \ell \in (0, 1], \ell \parallel \tilde{\ell}$. Clearly $\overline{L}$ holds. However, LFILT fails since $(0,1]$ is a proper filter with no infimum. Thus $\overline{L} \not\Rightarrow$ LFILT in general. Suppose now $\overline{L}$ and LTO. Let K be a proper filter of L. As K is proper, $\exists \ell \in L \backslash K$. As L is totally ordered and K is a filter:

$$\forall \ell \in L \backslash K, \forall k \in K, \ell < k.$$

Thus $\underline{K}$ is a nonempty set that is bounded above (by each element of K). Hence, using $\overline{L}$, $\vee \underline{K}$ exists and, using Fact B1 (i), equals $\wedge K$.

(f) Immediate from (d) and (e). ∎

Proof of Proposition 5.6.6

(a) Given Proposition 4.6.1 (c) and (g) and Theorem 4.8.1 and part (d) of the Lemma 5.6.1, the only implication here that is not immediate is $\overline{\text{LSU}} \Rightarrow $ LSU. Let $f \in \mathcal{F}_u$, $a \in S$ and $\ell \in L$ be given and let ϕ be the reflexive level foliation corresponding to f. We introduce the following three subsets of L. For both L_1 and L_2 the second, alternative, form follows at once from part 5(ii) of Theorem 5.5.1 which, we recall, does not require LTO (see Remark 5.5.2):

$$L_1 = \{\tilde{\ell} \in L : \phi(a, \ell) = \phi(\tilde{a}, \tilde{\ell}) \text{ for some } \tilde{a} \in S\} = \{\tilde{\ell} \in L : \phi(a, \ell) = \phi(a, \tilde{\ell})\}.$$
$$L_2 = \{\tilde{\ell} \in L : \phi(a, \ell) \subseteq \phi(\tilde{a}, \tilde{\ell}) \text{ for some } \tilde{a} \in S\} = \{\tilde{\ell} \in L : \phi(a, \ell) \subseteq \phi(a, \tilde{\ell})\}.$$
$$L_3 = \{f(a, b) : b \in S, f(a, b) \leq \ell\}.$$

We show first that:

(1) $L_2 = $ {all upper bounds of L_3}. Thus:

$$L_2 \text{ has a minimum } \Leftrightarrow L_3 \text{ has a least upper bound.}$$

(2) If L_2 has a minimum, then:
   (a) $\min L_2 = \vee L_3$,
   (b) $\phi(a, \ell) = \phi(a, \min L_2)$,

   and (c) L_1 has a minimum. Indeed, $\min L_1 = \min L_2$.

(3) Suppose that LTO holds. Then, L_1 has a minimum $\Rightarrow L_2$ has a minimum and $\min L_1 = \min L_2$.

Of these, (1) is immediate and implies 2(a). Suppose $\min L_2$ exists. Clearly, $\phi(a, \ell) \subseteq \phi(a, \min L_2)$. The reverse inclusion, and hence 2(b), follows on noting that $\ell \in L_2$. By 2(b), $\min L_2 \in L_1$. But $\min L_2$ is a lower bound for L_1, as $L_1 \subseteq L_2$. Thus, $\min L_2 = \min L_1$. It remains to show (3). Suppose then that $\min L_1$ exists and let $\tilde{\ell} \in L_2$. Clearly if $\tilde{\ell} \in L_1$, $\min L_1 \leq \tilde{\ell}$. If not, $\phi(a, \ell) \subset \phi(a, \tilde{\ell})$ and hence, using LTO, we conclude from part 6(ii) of Theorem 5.5.1 that $\ell < \tilde{\ell}$. But $\ell \in L_1$ and so $\min L_1 < \tilde{\ell}$. Hence, $\min L_1 = \min L_2$.

To prove $\overline{\text{LSU}} \Rightarrow \text{LSU}$ it suffices to show that $\min L_1$ exists. But $\overline{\text{LSU}}$ implies $\vee L_3$ exists, and the result now follows at once from (1) and (2) above.

(b) The only implication not immediate from parts (c) and (f) of the lemma, and part (a) here, is that LTO and LSU imply $\overline{\text{LSU}}$. This follows from (3) above. ∎

Proof of Theorem 5.7.1

Now (a) $\Rightarrow$ (b) at once using Corollary 5.2.1 while, given (a) and (b), (c) can be verified directly. It suffices then to show (a) which we establish by proving:

(1) Suppose LSU, so that α is defined. Then: LTO $\Rightarrow \alpha$ maps Φ_S^U into $\mathcal{B}$.

   Moreover, if $\phi \in \Phi_S^U$ and $(\mathcal{H}, g) = \alpha(\phi)$, then:

   (i) $H_{ab} = \phi(a, f(a, b))$, (ii) $g(H_{ab}) = f(a, b)$ and (iii) $H_a(\ell) = \phi(a, \ell)$.

(2) LTO $\Rightarrow \beta$ maps $\mathcal{B}$ into Φ_S^U.

(3) α and β are mutually inverse and both are isotone.

(1) Supposing LSU, let $\phi \in \Phi_S^U$ and let $(\mathcal{H}, g) = \alpha(\phi)$. We verify each of the axioms of a Benzécri structure in turn.

H1. Using LTO, this holds by Theorem 5.5.1.

GH2. This follows as, $\forall a \in S$, $\phi(a, \cdot)$ level $\Rightarrow \phi(a, \cdot)$ covers S while $\forall \ell \in L$, $a \in \phi(a, \ell)$.

We pause to note that, by definition of g,

$$g(\phi(a, \ell)) \leq \ell \qquad (*)$$

As $\phi(a, \cdot)$ is isotone, (*) implies $\phi(a, g(\phi(a, \ell))) \subseteq \phi(a, \ell)$. But using LTO we have, by (3) of the proof of Proposition 5.6.6, that the reverse inequality holds. Thus:

$$\phi(a, \ell) = \phi(a, g(\phi(a, \ell))). \qquad (**)$$

I1. Suppose $\phi(a, \ell) \subset \phi(a', \ell')$. Then by (**)

$$\phi(a, g(\phi(a, \ell))) \subset \phi(a', g(\phi(a', \ell'))).$$

Thus, by 6(ii) of Theorem 5.5.1, $g(\phi(a,\ell)) < g(\phi(a',\ell'))$, as required.

R1. Let $(a,b) \in S^2$. Then:

$$\begin{aligned}H_{ab} &= \cap\{\phi(c,\ell) : a \in \phi(c,\ell) \text{ and } b \in \phi(c,\ell)\}, \text{ by definition}\\ &= \cap\{\phi(a,\ell) : b \in \phi(a,\ell)\}, \text{ by (4) of Theorem 5.5.1.}\\ &= \phi(a, f(a,b)), \text{ as } \phi(a,\cdot) \text{ is isotone.}\\ &\in \mathcal{H}, \text{ by definition.}\end{aligned}$$

This also proves (i).

R2. $H_a(\ell) = \{b \in S : g(\phi(a, f(a,b))) \leq \ell\}$, using (i)
$= \{b \in S : f(a,b) \leq \ell\}$, as $g(\phi(a, f(a,b))) = f(a,b)$ using (*) and (**).
$= \phi(a,\ell) \in \mathcal{H}$, by definition.

This also proves (ii) and (iii).

R3. $g(H_a(\ell)) = g(\phi(a,\ell))$, by the form of $H_a(\ell)$ just found
$\leq \ell$, by (*).

R4. $H_{aa} = \phi(a,0)$ by the form of H_{ab} found above and the fact that $f(a,a) = 0$. Thus $g(H_{aa}) = 0$, using (*).

(2) Let $(\mathcal{H},g) \in \mathcal{B}$ and let $\phi = \beta(\mathcal{H},g)$. Then:

$$\forall (a,b) \in S^2, \ell \in L_{b,\phi(a,\cdot)} \Leftrightarrow b \in H_a(\ell) \Leftrightarrow g(H_{ab}) \leq \ell.$$

Thus $L_{b,\phi(a,\cdot)} = [g(H_{ab}))$. In particular, it is a principal filter and hence, using Theorem 4.5.2, ϕ is a level foliation. Further,

$$\begin{aligned}\phi \text{ is reflexive } &\Leftrightarrow \forall a \in S, \forall \ell \in L, (a,a) \in m(\ell), \text{ where } m \equiv \zeta_S^{-1}(\phi)\\ &\Leftrightarrow \forall a \in S, \forall \ell \in L, a \in H_a(\ell)\end{aligned}$$

which holds by Proposition 5.6.1(e). It only remains to show that $f \equiv (v \circ \zeta_S^{-1})\phi$ is ultrametric. Since LTO holds this happens, by Theorem 5.5.1, if and only if:

[HIER] $H_a(\ell) \cap H_{a'}(\ell') \in \{\emptyset, H_a(\ell), H_{a'}(\ell')\}$

and $b \in H_a(\ell) \Leftrightarrow a \in H_b(\ell)$.

The first of these holds by H1 and the second by Proposition 5.6.1(b).

(3) By (iii) of (1) above, $\beta \circ \alpha$ is the identity map on Φ_S^U. Now let $(\mathcal{H},g) \in \mathcal{B}$ and let $(\mathcal{H}',g') = (\alpha \circ \beta)(\mathcal{H},g)$. Then $\forall H \subseteq S$:

$H \in \mathcal{H}' \Leftrightarrow H = H_a(\ell)$ for some $a \in S, \ell \in L$
$\Leftrightarrow H \in \mathcal{H}$, the forwards implication holding by (iii) of (1) and the reverse by Proposition 5.6.5(b)(iii).

Thus $\mathcal{H}' = \mathcal{H}$. Proposition 5.6.5(c) now gives that $g' = g$. It therefore only remains to show that α and β are isotone. The first of these follows from (iii) of (1) above. The second is immediate. ∎

Proof of Theorem 5.8.1

(a) Clearly $\hat{f}_0 \in \hat{\mathcal{F}}_0$ obeys condition (i) of a subdominant, while (ii) follows by anti-symmetry of the order on L. Uniqueness is immediate.

(b) The forwards implication is immediate from the definition of $\hat{f}_0$, while the reverse implication again follows by anti-symmetry. ∎

Proof of Proposition 5.8.1

Let $\tilde{\mathcal{F}}_0$ be a nonempty bounded subset of $\mathcal{F}_{ot}$. Then $\tilde{f}_0 \equiv \vee \tilde{\mathcal{F}}_0$ is defined by $\overline{L}$. Clearly, $\tilde{f}_0$ is a dissimilarity. It suffices then to show transitivity. Suppose therefore that $\tilde{f}_0(a,b) \leq \ell$ and $\tilde{f}_0(b,c) \leq \ell$ for some $\ell \in L$. Then $\forall f \in \tilde{\mathcal{F}}_0$, $f(a,b) \leq \ell$ and $f(b,c) \leq \ell$, by transitivity of the order on L. But each f is ultrametric. Thus $\forall f \in \tilde{\mathcal{F}}_0$, $f(a,c) \leq \ell$ and hence, by definition, $\tilde{f}_0(a,c) \leq \ell$. ∎

References

Coppi, R., Bolasco, S. (1989), *Multiway Data Analysis*, North-Holland, Amsterdam.

Critchley, F. (1983), Ziggurats and dendrograms, Warwick Statistics Research Report n° 43, U.K.

Critchley, F., Van Cutsem, B. (1992), An order-theoretic unification and generalisation of certain fundamental bijections in mathematical classification – II, Joint Research Report, University of Warwick, U.K., and Université Joseph Fourier, Grenoble, France.

Critchley, F., Van Cutsem, B. (1994), An order-theoretic unification and generalisation of certain fundamental bijections in mathematical classification – I, In Van Cutsem, B., ed., *Classification and Dissimilarity Analysis*, Ch. 4, Lecture Notes in Statistics, Springer-Verlag, New York.

Jardine, N., Sibson, R. (1971), *Mathematical Taxonomy*, Wiley, London.

Van Cutsem, B. (1983a), Décomposition d'une ultramétrique : ultramétriques simples et semi-simples, Rapport de Recherche n° 388, Laboratoire IMAG, Grenoble, France.

Van Cutsem, B. (1983b), Ultramétriques, distances, ϕ-distances maximum dominées par une dissimilarité donnée, *Statist. Anal. Données*, 8, pp. 42–63.

Chapter 6.
The residuation model for the ordinal construction of dissimilarities and other valued objects*

Bruno Leclerc[†]

6.1. Introduction

The aim of this chapter is to present an ordinal model of valued objects, special cases of which appear in many contexts. The model lies on basic notions of ordered set theory: residuation or, equivalently, Galois connections; it is not new: explicitly proposed in fuzzy set theory by Achache (1982, 1988), it also underlies an order formalization of a Jardine and Sibson (1971) model given by Janowitz (1978; see also Barthélemy, Leclerc and Monjardet 1984a). Here, our main concern is to apply the model in order to obtain and study dissimilarities such as ultrametrics, Robinson or tree-compatible ones. Valued objects of other types, already considered in the literature, will be also given as examples: two types of valued non symmetric relations and two types of valued convex subsets. The chapter is neither a theoretical general presentation nor a detailed study of a few special cases. It is, tentatively, something between these extreme points of view. Some references are given to the reader interested to more details on a specific class of valued objects, or to more information about residuation (or Galois mappings) theory. In what follows, E will be a given finite set with n elements. Several families of combinatorial objects defined on E will be considered. In this introduction, we consider the basic (and well-known) example of ultrametrics, where these objects are the equivalence relations on the finite set E, that is all the binary relations $R \subseteq E \times E$ satisfying the following three properties, for all $a, b, c \in E$,

(E1) $(a, a) \in R$ (*reflexivity*)
(E2) $(a, b) \in R$ implies $(b, a) \in R$ (*symmetry*)
(E3) $(a, b) \in R$ and $(b, c) \in R$ imply $(a, c) \in R$ (*transitivity*)

The set $\mathcal{R}_E$ of all these equivalences is (partially) ordered by inclusion; the minimum equivalence is the *diagonal* $D = \{(a, a) : a \in E\}$ and the maximum one is $E \times E$. A dendrogram is a mapping $g : \mathbb{R}^+ \to \mathcal{R}_E$ satisfying the properties:

(D1) $\forall \lambda, \mu \in \mathbb{R}^+, \lambda \leq \mu$ implies $g(\lambda) \subseteq g(\mu)$ (*isotony*),
(D2) $g^{-1}(E \times E) \neq \emptyset$,
(D3) $\forall \lambda \in \mathbb{R}^+, \exists \epsilon > 0$ such that $g(\lambda) = g(\lambda + \epsilon)$.

---

* *In* Van Cutsem, B. (Ed.), (1994) *Classification and Dissimilarity Analysis*, Lecture Notes in Statistics, Springer-Verlag, New York.

[†] Centre d'Analyse et de Mathématiques Sociales, Ecole des Hautes Etudes en Sciences Sociales, Paris, France.

Notice that (D2) and (D3) imply that the set $\{\lambda \in \mathbb{R}^+ : (a,b) \in g(\lambda)\}$ has a minimum, for all $a, b \in E$. This minimum may be equal to 0 for some pairs of distinct elements a, b: we do not adopt here the usual convention $g(0) = D$.

The one-to-one correspondence between the set $\mathcal{D}_E$ of all the dendrograms on E and the set $\mathcal{U}_E$ of all the ultrametrics on E is well-known. An ultrametric is a mapping $u : E \times E \to \mathbb{R}^+$ such that, for all $a, b, c \in E$,

(U1) $u(a, a) = 0$
(U2) $u(a, b) = u(b, a)$
(U3) $u(a, c) \leq \max(u(a, b), u(b, c))$

The ultrametric associated with a dendrogram g is given by $u_g(a, b) = \min\{\lambda \in \mathbb{R}^+ : (a, b) \in g(\lambda)\}$. Conversely, let u be an ultrametric; for $\lambda \in \mathbb{R}^+$, the *threshold relation* $g_u(\lambda) = \{(a, b) \in E \times E : u(a, b) \leq \lambda\}$ satisfies (E1), (E2) and (E3) by, respectively, (U1), (U2) and (U3): the latter series of properties is in fact a generalization of the former. Then, the dendrogram associated with u is the mapping g_u defined above.

Let us examine the ingredients in this correspondence. The definition of dendrograms involves two ordered sets: an ordered set of objects (here, the set $\mathcal{R}_E$ of the equivalences on E partially ordered by inclusion) and an ordered set of values (here, $\mathbb{R}^+$, linearly ordered as usual). In the definition of ultrametrics, $\mathcal{R}_E$ does not directly appear, but only the set $E \times E$ of all the ordered pairs of E; properties (U1-U3) allow to assemble these pairs into threshold relations which are equivalences. Only order considerations appear in properties (D1-D3) and (U1-U3), as well as in the equalities defining u_g and g_u. As noticed above, the comparison of the series of properties (E1-E3) and (U1-U3) make ultrametrics, a special type of dissimilarity coefficients, look like a valued form of equivalence relations, a special type of binary relations. This remark is not original; several well-known properties of ultrametrics follow: for instance, the unique subdominant (maximum inferior) ultrametric generalizes the transitive closure of a reflexive and symmetric relation.

Similar properties have been recognized about other types of dissimilarities (Bertrand and Diday 1985, Diday 1986, Durand and Fichet 1988, Batbedat 1989), valued binary relations (Defays 1978, Doignon et al. 1986) or valued objects as those appearing in fuzzy set theory.

The paper organizes as follows: residuated and residual mappings are defined in Section 6.2 with minimal hypotheses, and the general properties of these mappings are recalled. It is assumed in Sections 6.3.1 and 6.3.2 that the sets P of objects and Q of values are endowed with lattice structures; several lattices of usual combinatorial objects are presented in Section 6.3.3 and, in Section 6.3.4, the general definition of the lattice Q lets multidimensional scales of values be possible. The first consequences of the hypotheses of lattice structures are investigated in Section 6.4.1; valued objects are precisely defined in Section 6.4.2, and several types of valued objects corresponding to the examples of Section 6.3.3 are constructed, with further results, sometimes new, in specific situations. The general lattice structures of the sets of valued objects is briefly presented in Section 6.5, with some references on applications. In the conclusion, we make some remarks on the background literature.

6.2. Residuated mappings and closure operators

6.2.1. Residuated and residual mappings

A *(partially) ordered set* P is a set endowed with a reflexive, antisymmetric and transitive binary relation, generally denoted by $\leq_P$, or simply by $\leq$ (by $\subseteq$ in the case of inclusion orders). Recall this means that,

$$\forall x, y, z \in P, \quad \begin{array}{lll} x \leq x & \text{holds}, & \\ x \leq y \text{ and } y \leq x & \text{imply} & x = y, \\ x \leq y \text{ and } y \leq z & \text{imply} & x \leq z. \end{array}$$

The order $\leq$ is *linear* if, moreover, either $x \leq y$ or $y \leq x$ for distinct x, y; in this case, P is also said to be a *chain*.

The duality that will appear in some places in the paper is nothing but order reversion. The *dual* P^d is the set P, now endowed with the reversed order $\leq^d$, where $x \leq^d y$ iff $y \leq x$.

A subset X of P is a *down-set* (resp. an *up-set*) if $x \in X$ and $y \leq x$ (resp. $y \geq x$) imply $y \in X$. The *principal ideal* generated by $x \in P$ is $\downarrow x = \{y \in P : y \leq x\}$, the down-set of all the elements of P inferior to x. The *principal filter* $\uparrow x$, an up-set, is dually defined.

Let $(P, \leq_P)$ and $(Q, \leq_Q)$ (or, simply, $(P, \leq)$ and $(Q, \leq)$) be two ordered sets. The set of all the mappings from P into Q is denoted, as usually, as Q^P. It is again an ordered set, with the pointwise order on mappings: for $f, f' \in Q^P$,

$$f \leq f' \iff \forall x \in P, \ f(x) \leq f'(x).$$

A mapping f from P into Q is said to be *isotone* if $x \leq y$ implies $f(x) \leq f(y)$. Given a subset Λ of Q and a mapping f from P into Q, the *converse image* $f^{-1}(\Lambda)$ is the subset $f^{-1}(\Lambda) = \{x \in P : f(x) \in \Lambda\}$ of P. It is not difficult to see that f is isotone iff it satisfies one of the following equivalent conditions (I) and (I'):

(I) $\quad \forall \lambda \in Q, \ f^{-1}(\downarrow \lambda)$ is a down-set of P,

(I') $\quad \forall \lambda \in Q, \ f^{-1}(\uparrow \lambda)$ is an up-set of P.

The set of all the isotone mappings from P into Q will be denoted as $\mathbf{Q}^P$, with bold letters.

In many, and various, situations, it is useful to consider a special class of isotone functions. In the following statement, Condition (R1), illustrated in Figure 6.2.1.a, is a strengthening of (I); Condition (R2) is the so-called *Pickert relation*; a mapping φ from P into P is said to be *extensive* if, with the pointwise order, $id_P \leq \varphi$, where id_P is the identity mapping onto P: in other terms, for all $x \in P$, one has $x \leq \varphi(x)$. Dually, φ is *anti-extensive* if $id_P \geq \varphi$.

Theorem 6.2.1. *The following three conditions are equivalent for a mapping $f: P \to Q$:*

(R1) for any $\lambda \in Q$, $f^{-1}(\downarrow \lambda)$ is a principal ideal of P,

(R2) The mapping f is isotone and there exists a mapping $g \in \mathbf{P}^Q$ such that, for any $\lambda \in Q$ and $x \in P$, $x \leq g(\lambda) \Leftrightarrow \lambda \geq f(x)$.

(R3) The mapping f is isotone and there exists a mapping $g \in \mathbf{P}^Q$ such that the composition mapping $\varphi = gf$ is extensive while $\psi = fg$ is anti-extensive.

Proof. Assume f satisfies (R1) and define g by $g(\lambda) = \max f^{-1}(\downarrow \lambda)$. Then, (R2) follows and, so, (R1) implies (R2). Starting from (R2), take successively $\lambda = f(x)$ and $x = g(\lambda)$ in order to obtain (R3). Assume (R3) is satisfied; let λ be an element of Q and $x \in f^{-1}(\downarrow \lambda)$, that is to say x satisfies $f(x) \leq \lambda$. Since gf is extensive and g is isotone, one has $x \leq gf(x) \leq g(\lambda)$; so, $f^{-1}(\downarrow \lambda) \subseteq \downarrow g(\lambda)$. For $y \in P$ such that $y \leq g(\lambda)$, one has $f(y) \leq fg(\lambda) \leq \lambda$, which completes the proof. ∎

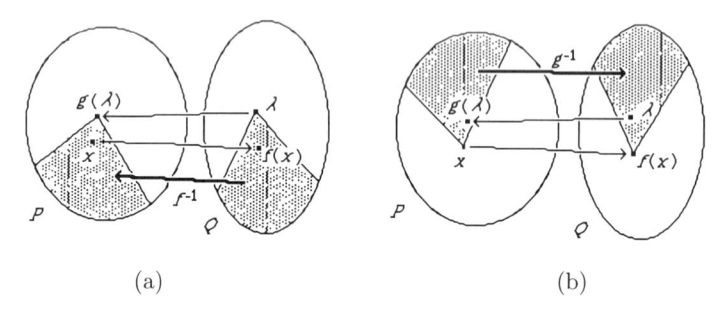

(a)   (b)

Figure 6.2.1

A mapping satisfying the conditions of Theorem 6.2.1 is said to be *residuated*. The mapping g associated to f by Conditions (R2) and (R3) is *residual*, that is it satisfies the following condition (R'1), a strengthening of condition (I') illustrated in Figure 6.2.1.b:

(R'1) for any $x \in P$, $g^{-1}(\uparrow x)$ is a principal filter of Q.

The set of all the residuated (residual) mappings from P to Q will be denoted as $\mathbf{R}(P,Q)$ ($\mathbf{R'}(P,Q)$), or simply as $\mathbf{R}$ ($\mathbf{R'}$). The correspondence between the sets $\mathbf{R}(P,Q)$ and $\mathbf{R'}(Q,P)$ defined above is one-to-one.

6.2.2. Closure and anticlosure operators

For the mappings f and g of the previous Section 6.2.1, the following inequalities also hold: $gf(g(\lambda)) \geq g(\lambda)$ by extensivity of gf, $fg(\lambda) \leq \lambda$ by anti-extensivity of fg and $g(fg(\lambda)) \leq g(\lambda)$ by isotony of g; so, $gfg = g$. Similarly, $fgf = f$.

These last equalities have important consequences. The composition mapping $gf = \varphi$ is isotone, extensive, and idempotent: it is a *closure operator* on P; on the other hand, $\psi = fg$, which is isotone, anti-extensive and idempotent, is an *anticlosure operator* on Q. The sets $f(P) = \psi(Q) = \Psi$ and $g(Q) = \varphi(P) = \Phi$ are order isomorphic by the restrictions of f and g. The elements of Φ (resp. of Ψ) will be said to be closed (resp. open). Given $x \in P$ (resp. $\lambda \in Q$), the closed (resp. open) element $\varphi(x)$ (resp. $\psi(\lambda)$) will be sometimes denoted as x' (resp. λ'). In fact, any closure operator φ on P is associated with a residuated-residual pair: take $\varphi(P)$ as ordered set Q, φ as f and $id_{\varphi(P)}$ as g.

Example. With $P = \mathcal{R}_E$ and $Q = \mathbb{R}^+$, conditions (D2) and (D3) of Section 6.1 are equivalent with a special instance of condition (R'1) above: set $f(a,b) = \min\{\lambda \in \mathbb{R}^+ : (a,b) \in g(\lambda)\}$. Then, for $R \in \mathcal{R}_E$, $g^{-1}(\uparrow R) = \uparrow \max\{f(a,b) : (a,b) \in R\}$. So, $\mathcal{D}_E = \mathbf{R}'(\mathbb{R}^+, \mathcal{R}_E)$: the dendrograms are exactly the residual mappings from $\mathbb{R}^+$ into $\mathcal{R}_E$. The residuated mapping associated with g is defined by $f(R) = \max\{f(a,b) : (a,b) \in R\}$. The "closed" equivalences are those appearing at the levels of the dendrogram while the "open" real numbers are the level values of the nodes of the dendrogram. This example makes it also appear that the correspondence between residual and residuated mappings is not yet the appropriate frame for ultrametrics: it must interfere a restriction of the residuated mappings; we will see that further hypotheses on P and Q (satisfied in the case $P = \mathcal{R}_E$ and $Q = \mathbb{R}^+$) are needed.

Many classical examples of closure operators may be found in the literature: topological, algebraic, convex closures; transitive closure of binary relations. An important result states that *the unicity of a maximal inferior element in a prescribed subset is characteristic of the existence of a closure operator*, and, similarly, *the unicity of a minimal superior element is characteristic of the existence of an anticlosure*. Formally:

Proposition 6.2.1. *Let φ be a closure operator on P and $\Phi = \varphi(P)$ the corresponding set of closed elements; then, for any $x \in P$, $\varphi(x)$ is the minimum of $(\uparrow x) \cap \Phi$. Conversely, if Φ is a subset of P such that, for all $x \in P$, $(\uparrow x) \cap \Phi$ has the minimum denoted $\varphi(x)$, the mapping φ is a closure operator on P, with Φ as set of closed elements.*

Proof. If φ is a closure operator, then, for all $x \in P$ and $y' \in (\uparrow x) \cap \Phi$, one has $x \leq y'$ and, by the isotony of φ, $\varphi(x) \leq \varphi(y') = y'$. For the converse, it is obvious that the mapping φ defined in such a way is idempotent, isotone and extensive. ∎

6.3. Lattices of objects and lattices of values

6.3.1. Lattices

A lower (upper) bound of a subset X of P is an element $y \in P$ such that $y \leq x$ ($x \leq y$) for all $x \in X$. A *lattice* $(P, \vee, \wedge, \leq)$ is an ordered set $(P, \leq)$ such that any pair x, y of elements of P has both a minimum upper bound (join), denoted by $x \vee y$ and a maximum lower bound (meet), denoted by $x \wedge y$. As binary operations, $\vee$ and $\wedge$ are associative, commutative and idempotent ($x \vee x = x \wedge x = x$). Moreover, they are linked by the *absorption laws*: $x \wedge (y \vee x) = x \vee (y \wedge x) = x$, for all $x, y \in P$. If exists, the minimum element of P, denoted $\underline{0}$ (resp. the maximum element of P, denoted $\underline{1}$) is neutral while $\underline{1}$ (resp. $\underline{0}$) is absorbant for the join (resp. for the meet) binary operation. For any $x \in P$, the sets $\uparrow x$ and $\downarrow x$ are sublattices of P. P is a *complete lattice* if, moreover, any subset X of P has a join and a meet, respectively denoted by $\vee X$ and $\wedge X$. In this case, elements $\underline{1}$ and $\underline{0}$ exist, with $\vee P = \wedge \emptyset = \underline{1}$ and $\vee \emptyset = \wedge P = \underline{0}$. All the lattices considered in the sequel are assumed to be complete, as it is the case for finite lattices or for, say, the closed real interval $[0, 1]$.

An element j of P is *join irreducible* if, for any subset X of P, $j = \vee X$ implies $j \in X$; when P is finite, an equivalent property is that j covers exactly one element of P; recall x covers y, what is denoted $y \prec x$, means $y \leq z < x$ implies $y = z$. Irreducible elements may be often thought of as elementary objects; they will have an important role throughtout the paper. Let $J(P)$ (or simply J) be the set of all the join irreducible elements of P (notice $\underline{0} \notin J$), and set $J[x] = \{j \in J(P) : j \leq x\}$ for all $x \in P$. When P is finite, the equality $x = \vee J[x]$ always holds. Then, J is the unique minimal subset generating the entire lattice by the join operation. One has $J[x \wedge y] = J[x] \cap J[y]$, but only $J[x] \cup J[y]) \subseteq J[x \vee y]$; the following property may be obtained by routine lattice considerations; it will be used in Section 6.4.1 below : $j \in J[x \vee y]$ iff $j \leq \vee (J[x] \cup J[y])$.

6.3.2. Distributivity

The classical definition of a distributive lattice is: a lattice P is *distributive* if it satisfies the distributivity laws: for all $x, y, z \in P, (x \vee y) \wedge z = (x \wedge z) \vee (y \wedge z)$ or, equivalently, $(x \wedge y) \vee z = (x \vee z) \wedge (y \vee z)$. To give the most known classes of distributive suffices for emphazising their importance: chains (linear orders) with the minimum and maximum as meet and join operations; direct product of chains with the pointwise minimum and maximum; and boolean lattices (sets of all the subsets of a given set) with the intersection and union.

In the case where P is finite, a useful characterization is based on a property of the join-irreducible elements (see for instance Monjardet 1990):

Proposition 6.3.1. *A finite lattice P is distributive if and only if the following condition (DL) holds :*

(DL) For all $j \in J$ and $X \subseteq P$, the inequality $j \leq \vee X$ implies that there exists some $x \in X$ such that $j \leq x$.

Proof. The inequality $j \leq \vee X$ and the distributivity hypothesis imply $j = j \wedge (\vee X) = \vee \{j \wedge x : x \in X\}$. Since j is join-irreducible, there is some x in X such that $j = j \wedge x$, that is $j \leq x$.

Conversely, let $x, y \in P$ such that $x \vee y$ exists, and $j \in J(x \vee y)$. By (DL), $j \leq x$ or $j \leq y$ holds. So, $J(x \vee y) \subseteq J(x) \cup J(y)$, that is $J(x \vee y) = J(x) \cup J(y)$. Then, the map $x \mapsto J(x)$ is an isomorphism between the lattice P and a sublattice of $\mathcal{P}(J)$. So, P is a distributive lattice. ∎

6.3.3. Lattices of objects: ten examples

In the model developed here, there is first a set P of objects of a given interesting kind. It is assumed now that P is finite and endowed with a lattice structure. Although this last condition is a quite strong one, it is satisfied in many important cases. We briefly present several examples of lattices of objects, not all binary relations.

Our simplest example (1) is the boolean lattice $(\mathcal{P}(E), \subseteq, \cup, \cap)$ of all the subsets of E, with $\underline{0} = \emptyset$ and $J = E$. Example (2) is the boolean lattice $(\mathcal{P}(E \times E), \subseteq, \cup, \cap)$ of all the binary relations on E, with two interesting variants leading to our examples (3) and (4).

(3) in the latter, the principal filter $\uparrow D$ is the boolean lattice of all the reflexive binary relations on E; its minimum element is the diagonal D and its join irreducible elements are the relations with the form $D \cup \{(a, b)\}$, a and b being distinct elements of E.

(4) the set of all the reflexive and symmetric relations on E is also a boolean lattice with D as minimum element. Its join irreducibles are the relations with the form $D \cup \{(a, b), (b, a)\}$, a and b being distinct elements of E; they are identifiable with the unordered pairs ab. So, the set of all the reflexive and symmetric relations will be identified with $\mathcal{P}(E^{(2)})$, where $E^{(2)}$ is the set of all the unordered pairs of distinct elements of E.

(5) The archetypal untrivial example, already used in Section 6.1, is the case where P is the set $\mathcal{R}_E$ of all the equivalence relations on E, with the inclusion order. It is well-known that $\mathcal{R}_E$ is a lattice: the meet of two equivalences R and R' is their intersection $R \cap R'$ and their join $R \vee R'$ is the transitive closure of their union; for the lattice $\mathcal{R}_E$, $D = \underline{0}$ and $E \times E = \underline{1}$. The join-irreducible elements are the same as in example (4) above, again identified with the elements of $E^{(2)}$.

(6) Delete the symmetry condition ($E2$) in the previous example. One obtains the set $\mathcal{P}_E$ of all the *preorders* (sometimes called *quasi-orders*) on E, with the same definition of meet and join operations, and the same minimum $\underline{0}$ and maximum $\underline{1}$. Equivalences are the symmetric preorders; for instance, there are five different equivalences on the set a, b, c among twenty-nine preorders. The join-irreducible elements of the lattice $\mathcal{P}_E$ are those of example (3) above.

(7) Assume E is linearly ordered. For instance, it is the chain $C_n = \{1 < 2 < 3 < \ldots < n\}$. An interval is then a subset I of E such that $a, c \in I$ and $a \leq b \leq c$ imply $b \in I$ (so, the empty set is an interval). The set I_E of all the intervals of E, endowed with the inclusion order, is a lattice: the meet of two intervals I and I' is their intersection $I \cap I'$ and their join $I \vee I'$ is the minimum interval including their union (the convex closure of their union); one has $\emptyset = \underline{0}$ and $E = \underline{1}$. The join-irreducible elements of I_E are the elements of E. Figure 6.3.1 shows the order diagram of this lattice for $n = 4$.

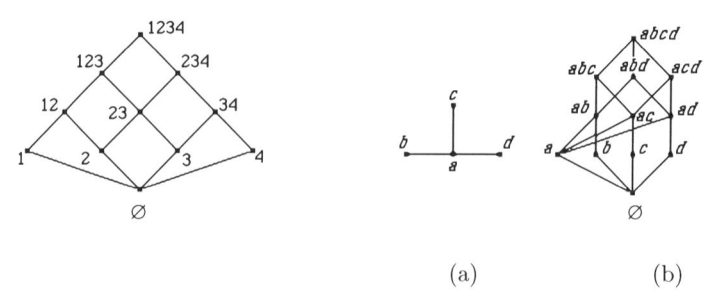

Figure 6.3.1   **Figure 6.3.2**

(8) A slight generalization of the previous case is when, instead of the chain C_n, E is the set of the vertices (or points) of a tree T, in the graph theory sense, that is an undirected, connected and acyclic graph (Figure 6.3.2.a). The set T_T of all the subtrees of T, that is of all subsets of E generating a connected and acyclic subgraph, is another example of a lattice of convex subsets. The join, meet, and join-irreducible elements are obtained on the same way. Figure 6.3.2.b shows the diagram of this lattice for the subtrees of the tree of Figure 6.3.2.a.

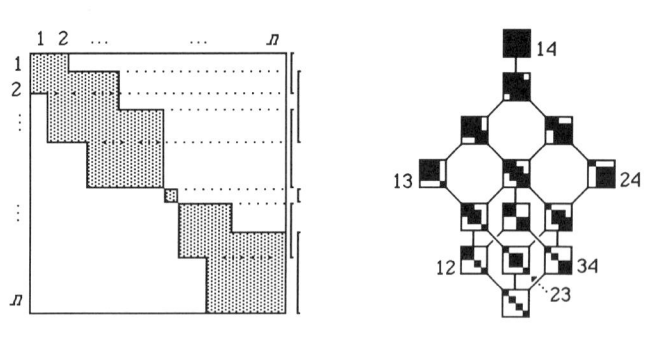

Figure 6.3.3   **Figure 6.3.4**

(9) Set $E = C_n$ as in example (7) above. We now consider binary relations on E. Here, S_n is the set of all these relations S satisfying conditions (E1) and (E2) of reflexivity and symmetry and the following condition (S3) of compatibility with the chain order on E.

(S3) $(a,c) \in S$ and $a \leq b \leq c$ imply $(a,b) \in S$ and $(b,c) \in S$.

Figure 6.3.3 gives the general form of the array representation of such a relation, which may be called, according to graph theory terminology (see e.g. Golumbic 1980), a *unit interval relation*. A set of clusters of elements of E, pairwise not included but possibly overlapping, is provided by the maximal intervals (indicated on the right of the

VI. RESIDUATION MODEL FOR THE ORDINAL CONSTRUCTION

array) I of C_n such that $(a,b) \in S$ for any $a,b \in I$. Since it is stable under union and intersection, the set S_n is a sublattice of $\mathcal{P}^{E(2)}$. Figure 6.3.4. gives its order diagram for $E = C_4 = \{1 < 2 < 3 < 4\}$. The join-irreducible elements do not all cover the 0, as in the previous examples. They are related with the pairs ab, written with the convention $a < b$: the relation S is join irreducible iff there exists a pair ab such that S is the minimum unit interval relation including ab. These irreducibles are indicated on Figure 6.3.4 by the corresponding pairs; notice that the restriction of the lattice order to the set of join-irreducibles is given by $ab \leq a'b'$ iff $a' \leq a < b \leq b'$ (e.g., one has $12 \leq 13$ in Figure 6.3.4).

The lattice of relations compatible with a tree A may be defined on the same way.

(10) Consider two positive integers m, p and the product order $E = C_m \times C_p$; let $\mathcal{B}_{m,p}$ be the set of all the subsets B of E (relations from C_m to C_p) satisfying the following property:

(B) $(a,b) \in B$, $a' \leq a$ and $b' \leq b$ imply $(a',b') \in B$.

Such relations are called *Ferrer relations*, or *biorders*, or *Guttmann scales*, compatible with C_m and C_p, in the literature. The situation is analog to example (9): the set $\mathcal{B}_{m,p}$ is stable under union and intersection and, so, it is a sublattice of $\mathcal{P}(C_m \times C_p)$. Figure 6.3.5 gives the diagram of this lattice for $m = 3$ and $p = 2$. Join-irreducible elements are again related with pairs; they are ordered according to the product order $C_m \times C_p$.

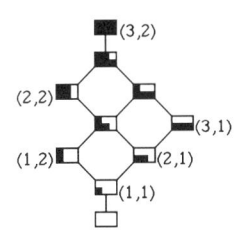

Figure 6.3.5

The above examples may be combined in order to obtain lattices such as, for instance, the lattice of all the partitions compatible to a given chain order ...

It is important to notice that, in Examples (7) to (10), the lattice structures vanish when the chain orders (or the tree structures) are no longer specified, as it is the case in many of the problems considered in the literature. The lattices in Examples (1) to (3) and (9) and (10) are distributive, as boolean or sublattices of boolean ones.

6.3.4. Lattices of values

In order to define valued objects, we also consider a second lattice $(Q, \vee, \wedge, \leq)$ of *values*, with a minimum 0 and a maximum ω; meet and join symbols, and some others, will be common to Q and P, but we will use greek letters for the elements or subsets of Q. It is not assumed that Q is finite, but only that it is complete. Within this general situation, two special cases will be considered, given here in order of decreasing strongness of the hypotheses:

(I) Q is a chain: in this case, the join (meet) is the supremum (infimum) operation, the maximum (minimum) for a finite number of elements. For instance, Q may be the set C_n considered above, or a closed real interval like $[0,1]$; most often, Q is the set $\overline{\mathbb{R}^+}$, that is $\mathbb{R}^+$ completed with a maximum infinite element ω.

(II) Q is a *distributive lattice*, that is to say it satisfies the distributivity laws; generally the ℓ-fold of a chain: C_n^ℓ, $[0,1]^\ell$ or $(\mathbb{R}^+)^\ell$.

In fact, we make the further hypothesis that Q is completely distributive, that is the distributivity laws remain valid for infinite families of elements of Q (see Birkhoff 1967, p. 119, for a precise definition); it is not a too strong hypothesis, since it is satisfied by the lattice $(\overline{\mathbb{R}^+})^\ell$.

Situation (II) occurs, for instance, when the values take the form of a k-tuple of real numbers. Another case is when they correspond with the presence or absence of several binary characters: if F is the set of these characters, then the values are elements of the boolean (distributive) lattice $\mathcal{P}(F)$.

A way of obtaining various types of lattices Q of values may be found in Symbolic Data Analysis. In this field, the variables are used for the description of classes of objects as well as single objects, and the obtained descriptions are directly used, without numerical transformations (see, e.g., Diday 1988). The range of a variable i is a set Q_i, ordered according to the generality of the description. Such an order is frequently a lattice, and the complete description of an object or a class is an element of a product lattice $Q = \times_i Q_i$. Then, residuated and residual mappings (or Galois connections: see the conclusion below) are used in order to construct particular classes of objects, associated with their description, and to study the structure of the set of these classes. Especially, with $Q = \{0,1\}^\ell$, this approach, which goes back to Barbut and Monjardet (1971), has been intensively developed by Wille and others (see, e.g., Wille 1982, Ganter et al. 1986, Duquenne 1987). Other examples of lattices of symbolic descriptions may be found in Daniel-Vatonne and de La Higuera (1993), Brito (1994).

6.4. Valued objects

6.4.1. Consequences of the lattice structures hypothesis

A first consequence of the hypotheses formulated above is that residuated and residual mappings admit now a simple fourth characterization:

Theorem 6.4.1. *Let f be a mapping from a finite lattice P to a complete lattice Q. The mapping f is residuated (residual) if and only if it satisfies the following condition R4 (R'4):*

(R4) $f(\underline{0}) = 0$ and, for all $x, y \in P$, $f(x \vee y) = f(x) \vee f(y)$.

(R'4) $f(\underline{1}) = \omega$ and, for all $x, y \in P$, $f(x \wedge y) = f(x) \wedge f(y)$.

Proof. Assume f is residuated. First, $\underline{0} \in f^{-1}(\downarrow 0) = f^{-1}(\{0\})$ and, so, $f(\underline{0}) = 0$. If x and y are two elements of P, then, by isotony of f, $f(x \vee y) \geq f(x)$ and $f(x \vee y) \geq f(y)$; so, $f(x \vee y) \geq f(x) \vee f(y)$. We will show that $f(x \vee y) \leq \lambda$ for any $\lambda \geq f(x) \vee f(y)$. From the isotony of g and the extensivity of gf, $\lambda \geq f(x)$ implies $g(\lambda) \geq gf(x) \geq x$; with the same inequalities for y, one obtains $g(\lambda) \geq x \vee y$ and $\lambda \geq fg(\lambda) \geq f(x \vee y)$. The equality $f(x \vee y) = f(x) \vee f(y)$ follows. Conversely, let $\lambda \in Q$; if (R4) is satisfied, the set $X = \{x \in P : f(x) \leq \lambda\}$ is unempty and $f^{-1}(\downarrow \lambda) = \downarrow \vee X$. So, f is residuated. The proof in the residual case is similar. ∎

When P is assumed to be complete, but not necessarily finite, Theorem 6.4.1 remains true with, instead of, say, (R4), the following condition: for all $X \subseteq P$, $f(\vee X) = \vee\{f(x) : x \in X\}$.

Now we consider the sets of closed or open elements.

Proposition 6.4.1. *If φ is a closure operator on the lattice P, then the set $\Phi = \varphi(P)$ of the closed elements contains $\underline{1}$ and is stable under the meet operation. Conversely, any subset ϕ of P containing $\underline{1}$ and stable under the meet is a set of closed elements for some closure operator on P.*

If ψ is an anticlosure operator on the lattice Q, then the set $\Psi = \psi(Q)$ of the open elements contains 0 and is stable under the join operation. Conversely, any subset Ψ of Q containing 0 and stable under the join is a set of open elements for some anticlosure operator on Q.

Proof. Let φ be a closure operator on P. One has $\varphi(\underline{1}) = \underline{1}$ by extensivity of φ. Let x', y' be two closed elements. The following inequalities hold: $x' \wedge y' \leq \varphi(x' \wedge y')$, by extensivity of φ; $\varphi(x' \wedge y') \leq \varphi(x')$ and $\varphi(x' \wedge y') \leq \varphi(y')$ by isotony and, so, $\varphi(x' \wedge y') \leq \varphi(x') \wedge \varphi(y') = x' \wedge y'$, this equality because of the idempotence. Finally, $\varphi(x') \wedge \varphi(y') = \varphi(x' \wedge y')$.

The converse follows from Proposition 6.2.1: the operator $x \mapsto \wedge\{y' \in \Phi : x \leq y'\}$ associates to $x \in P$ a unique minimal superior element of Φ.

The second part is similar, with, in the converse, the operator $\lambda \mapsto \vee\{\mu' \in \Psi : \lambda \leq \mu'\}$ as anticlosure operator. ∎

Section 6.3.3 provides several examples for the previous result. The sets $\mathcal{R}_E$ and $\mathcal{P}_E$ contain $E \times E$ and are stable under intersection: the corresponding closures in $\mathcal{P}(E \times E)$ are, respectively, the reflexive, symmetric and transitive one and the reflexive and transitive one; the sets $\mathcal{I}_E$ and $\mathcal{T}_T$ are defined by convex closures. Both a closure and an anticlosure may be associated with the sets S_n and $B_{m,p}$ which are stable under union and intersection.

When P is a lattice, the restriction to Φ of the order of P is again a lattice order, with the same meet as P and the join $\overline{\vee}$ defined by $x' \overline{\vee} y' = \wedge \{z' \in \Phi : x' \vee y' \leq z'\} = \varphi(x' \vee y')$; notice $x' \vee y' \leq x' \overline{\vee} y'$. Similarly, a set Ψ of open elements in Q is a lattice, with the same join as Q and the meet $\overline{\wedge}$ defined by $\lambda' \overline{\wedge} \mu' = \vee \{\nu' \in \Psi : \nu' \leq \lambda' \wedge \mu'\} = \psi(\lambda' \wedge \mu')$, with $\lambda' \overline{\wedge} \mu' \leq \lambda' \wedge \mu'$. In the residuation model of the previous section, the two lattices Φ and Ψ are isomorphic.

6.4.2. Valued objects: definition and examples

A consequence of the characterization (R4) of the residuated mappings is that a residuated mapping f on P is completely defined by its restriction to a subset X, provided X generates $P \setminus \{\underline{0}\}$ by the join operation. The minimum such generating subset is J, and the question then arises of what mappings $v : J \mapsto Q$ are just the restrictions to J of the residuated mappings? The following result gives three characterizations of such mappings. After proving it, we will comment properties (V2) and (V4), using the example of ultrametrics, and give several other examples.

Theorem 6.4.2. *The following four conditions are equivalent for a mapping $v : J \mapsto Q$:*

(V1) There is a residuated mapping $f : P \mapsto Q$ such that v is the restriction of f to J.

(V2) For any $j \in J$, $K \subseteq J$, $j \leq \vee K$ implies $v(j) \leq \vee \{v(k) : k \in K\}$.

(V3) For any $j \in J$, $v(j) = \wedge \{\vee_{k \in K} v(k) : K \subseteq J, j \leq \vee K\}$.

(V4) For any $\lambda \in Q$, there exists $x \in P$ such that $\{j \in J : v(j) \leq \lambda\} = J[x]$.

Proof. Let f be a residuated mapping from P into Q and v its restriction to J. Since $k \leq \vee K$ implies $v(k) = f(k) \leq f(\vee K) = \vee \{f(j) : j \in K\} = \vee \{v(j) : j \in K\}$, (V2) is satisfied by v.

Assume v satisfies (V2). Define f by $f(\underline{0}) = 0$ and $f(x) = \vee \{v(j) : j \in J[x]\}$; one has first $f(\underline{0}) = \vee \emptyset = 0$. Let $x, y \in P$ and $k \in J$ such that $k \leq \vee (J[x] \cup J[y])$; by (V2), one has $v(k) \leq \vee \{v(j) : j \in J[x] \cup J[y]\}$. Thus, $f(x \vee y) = \vee \{v(j) : j \in J[x \vee y]\} = \vee \{v(j) : j \leq \vee (J[x] \cup J[y])\} = \vee \{v(j) : j \in J[x] \cup J[y]\} = f(x) \vee f(y)$, and, by Theorem 6.4.1, f is a residuated function.

The equivalence of (V2) and (V3) is immediate (notice that $\{j\}$ is one of the sets K such that $j \leq \vee K$). Assume that, for some $\lambda \in Q$, there is no $x \in P$ such that $\{j \in J : v(j) \leq \lambda\} = J[x]$ and set $y = \vee \{j \in J : v(j) \leq \lambda\}$; such hypotheses imply that there is $k \leq y$ such that $v(k) \not\leq \lambda$, a contradiction with (V2); so, (V2) implies (V4). Assume v satisfies (V4) and let $k \in K, K \subseteq J$ such that $k \leq \vee K = x$. Set $\mu = \vee \{v(k) : k \in K\}$ and $y = \vee \{j \in J : v(j) \leq \mu\}$; then, $k \leq x \leq y$ and $k \in J[y]$, which is equivalent to $v(k) \leq \mu$. Thus, $v(k) \leq \mu$, which proves that (V2) is satisfied. ∎

VI. RESIDUATION MODEL FOR THE ORDINAL CONSTRUCTION

The mappings v satisfying conditions (V1-V4) will be called here *valued objects*. They are isotone: let $j, k \in J$ such that $j \leq k$; taking $K = \{k\}$ in (V2), one obtains $v(j) \leq v(k)$. The set of all the valued objects, a subset of $\boldsymbol{Q}^J$, will be denoted as $\boldsymbol{V}(P, Q)$.

Let us examine properties (V2) and (V4) in the case $P = \mathcal{R}_E$ and $Q = \mathbb{R}^+$ of Section 6.1. As seen in Section 6.3.3 (Example 3), the elements of J correspond to the unordered pairs ab of distinct elements of E. The formal condition $ab \leq \vee K$ means now that the pair ab belongs to the transitive closure of the pairs in K, that is to say there exists in K a sequence of pairs $aa_1, a_1a_2, a_2a_3, \ldots, a_{\ell-1}b$. Keeping in mind the fact that, in a chain like $\mathbb{R}^+$, the join of a finite subset is its maximum, condition (V2) becomes: for distinct elements $a_0 = a, a_1, a_2, a_3, \ldots, a_{\ell-1}, b = a_\ell \in E$, $v(ab) \leq \max_{0 \leq i \leq \ell-1} v(a_i a_{i+1})$; but it is well-known that this general inequality is implied by its restriction to the case $\ell = 2$, and we finally get the following condition, immediately equivalent with the ultrametric inequality (U3): for all distinct $a, b, c \in E, v(ac) \leq \max(v(ab), v(bc))$. So, up to some adaptations, one has $\boldsymbol{V}(\mathcal{R}_E, \mathbb{R}^+) = \mathcal{U}_E$. Under duality, the elements of $\boldsymbol{V}(\mathcal{R}_E, [0, 1])$ are also sometimes called "fuzzy equivalence relations" (with the "max-min transitivity"; see Kaufman 1973, Dubois and Prade 1980). Condition (V4) means that, for any $\lambda \in \mathbb{R}^+$, the threshold set of pairs $\{ab : v(ab) \leq \lambda\}$ is the set of all the pairs in R, for some equivalence R: another characteristic property of ultrametrics. More generally, a *threshold set* $T(v, \lambda) = \{j \in J : v(j) \leq \lambda\}$ is associated with any mapping $v \in \boldsymbol{Q}^J$ and with any value $\lambda \in Q$; for any $v, w \in \boldsymbol{Q}^J, \lambda, \mu \in Q$, the following inequalities hold:

$$v \leq w \Rightarrow T(v, \lambda) \supseteq T(w, \lambda) \text{ and } \lambda \leq \mu \Rightarrow T(v, \lambda) \subseteq T(v, \mu).$$

The mapping v is a valued object when all its threshold sets correspond to elements of P, that is to say when v satisfies Condition (V4). So, Theorem 6.4.2 completes the general theoretical scheme of Section 6.2: valued objects (like ultrametrics) satisfying Conditions (V2-V4) are the restricted forms of residuated mappings; they are also in one-to-one correspondence with residual mappings (like dendrograms). The role of Condition (V3) will appear in Section 6.5.

A second classical example is obtained with the set $\mathcal{P}_E$ instead of $\mathcal{R}_E$ (Example 6 in Section 6.3.3). The elements of $\boldsymbol{V}(\mathcal{P}_E, \mathbb{R}^+)$ are those mappings from $E \times E$ into $\mathbb{R}^+$ that satisfy conditions (U1) and (U3) of Section 6.1, but not always the condition (U2) of symmetry; these valued preorders, or, under duality, fuzzy preorders, are considered in the literature on preference modelling and aggregation (Defays 1978; see other references in Leclerc 1984a, 1991).

Examples 7 and 8 of Section 6.3.3 lead to valued intervals of a chain and subtrees of a tree. Roughly, for three join irreducibles a, b, c, the inequality $b \leq a \vee c$ means in these examples that b is between a and c. Condition (V2) then becomes $v(b) \leq v(a) \vee v(c)$ and, in the case where Q is a chain, $v(b) \leq \max(v(a), v(c))$. With the additional condition that $v^{-1}(0) \neq \emptyset$, valued intervals are sometimes called, under duality, "fuzzy numbers" (cf. Roubens 1986, Roubens and Vincke 1988).

The case where P is a distributive lattice deserves a particular study; the valued objects have then a very simple characterization:

Proposition 6.4.2. *If the lattice P of objects is distributive, then $\boldsymbol{V}(P,Q)$ is the set $\boldsymbol{Q}^J$ of all the isotone mappings from J into Q.*

Proof. It was already noticed that $\boldsymbol{V}(P,Q) \subseteq \boldsymbol{Q}^J$. For the converse inclusion, we use the property (DL) of distributive lattices recalled in Section 6.3.2 (Proposition 6.3.1): in condition (V2) of Theorem 6.4.2, $j \leq \vee K$ now means there exists $k \in K$ such that $j \leq k$; by isotony of v, one has then $v(j) \leq v(k) \leq \vee \{v(k) : k \in K\}$. ∎

This result applies to examples 1-4, 9 and 10 of Section 6.3.3. For boolean lattices like $\mathcal{P}(E)$, the set J is an antichain: the relation $j \leq k$ never holds for distinct $j, k \in J$. So, all the mappings from J into Q are isotone, independently of the type of Q; for instance, $\boldsymbol{V}(\mathcal{P}(E), Q) = Q^E$. The case $\boldsymbol{V}(\mathcal{P}(E^{(2)}), Q) = Q^{E^{(2)}}$ accounts for the set of all the dissimilarities on E.

In the case of S_n (Example 9 in Section 6.3.3), it was noticed that the elements of J may be identified with the pairs ab, now ordered by $ab \leq a'b'$ iff $a' \leq a < b \leq b'$; the valued compatible relations are the mappings r on $C_n \times C_n$ satisfying, for all $a, b \in C_n$, $r(a, a) = 0$, $r(a, b) = r(b, a)$ and, for $a' \leq a < b \leq b'$, $r(a, b) \leq r(a', b')$. In the case $Q = \mathbb{R}^+$, they are the dissimilarities *compatible with the (linear) order C_n*, in Diday's (1986) sense. The corresponding arrays are said to be *Robinson*.

Table 6.1 below gives a Robinson array with $E = C_5$ and $Q = \{0 < 1 < 2 < 3 < 4\}$; Figure 6.4.1 shows the threshold relations $R(r, \lambda)$, for $\lambda = 1, \ldots, 3$; one has also $R(r, 0) = D$ and $R(r, 4) = C_5 \times C_5$.

	1	2	3	4	5
1	0	1	2	3	4
2	1	0	2	3	3
3	2	2	0	2	3
4	3	3	2	0	1
5	4	3	3	1	0

Table 6.1

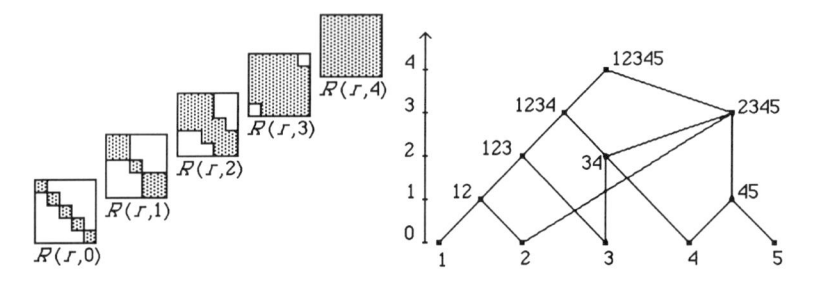

Figure 6.4.1

Figure 6.4.2

Figure 6.4.2 shows the (scaled according to r) diagram of the inclusion order of all the clusters (maximal intervals) appearing in these relations. Note that, since the set of intervals is not stable under intersection, there are crossings of lines: here, the subsets $234 = 1234 \cap 2345$ and $23 = 123 \cap 2345$ are not directly obtained as clusters. In order to avoid such crossings, one may add intersections and obtain a weakly indexed *pyramid* in Diday's sense (1986; see also Bertrand and Diday 1985). One may also consider only the *strictly Robinson* arrays, a wide special class with good properties (Durand and Fichet 1988).

Let us present also an example where Q is not a chain, but a boolean lattice. Table 6.2 gives a generalized Robinson array associated with an element of $\mathbf{V}(C_5, \mathcal{P}(\{\alpha, \beta, \gamma\}))$; Figure 6.4.3 shows the arrays of the corresponding untrivial threshold relations.

	1	2	3	4	5
1	$\emptyset$	β	$\alpha\beta$	$\alpha\beta\gamma$	$\alpha\beta\gamma$
2	β	$\emptyset$	α	$\alpha\gamma$	$\alpha\gamma$
3	$\alpha\beta$	α	$\emptyset$	γ	γ
4	$\alpha\beta\gamma$	$\alpha\gamma$	γ	$\emptyset$	γ
5	$\alpha\beta\gamma$	$\alpha\gamma$	γ	γ	$\emptyset$

Table 6.2

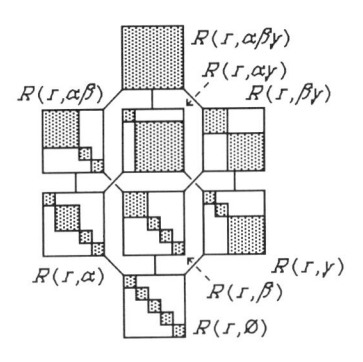

Figure 6.4.3

The dissimilarities compatible, on the same way, with a tree graph were studied first by Degenne and Vergès (1973; indeed they considered similarity indices) and more extensively recently by Batbedat (1989, 1990). The case where $P = \mathcal{B}_{m,p}$ (Example 10 of Section 6.3.3) is similar: one obtains the valued biorders compatible with the chains

C_m and C_p. Valued biorders have been mainly considered in mathematical psychology, in the context of "probabilistic consistency" (Doignon et al. 1986, Monjardet 1988).

6.5. Lattices of valued objects

When both P and Q are complete lattices, the pointwise order on the set Q^P of all the mappings from P on Q is again a complete lattice order with the pointwise meet and join operations. For $f, f' \in Q^P$, their join $f \vee f'$ and meet $f \wedge f'$ are defined by:

for any $x \in P, (f \vee f')(x) = f(x) \vee f'(x)$ and $(f \wedge f')(x) = f(x) \wedge f'(x)$;

by the completeness hypothesis, this definition extends to $\vee F$ and $\wedge F$, for all $F \subseteq Q^P$.

It is well-known that $\boldsymbol{Q^P}$, the set of all the isotone mappings, is stable for both operations; it is a sublattice of Q^P. What about residuated and residual mappings? The general answer is that $\boldsymbol{R}(P, Q)$ is again a complete lattice (Shmuely 1974). More precisely, $\boldsymbol{R}(P, Q)$ is stable for the pointwise join, but not for the meet; the situation is the same for $\boldsymbol{V}(P, Q)$ in Q^J. Similarly, $\boldsymbol{R}'(Q, P)$ is stable in P^Q for the meet, but not for the join.

Blyth and Janowitz (1972, p.32) propose as an exercise to show that the stability holds for both operations in the case where $\boldsymbol{P}$ and $\boldsymbol{Q}$ are chains. Barthélemy, Leclerc and Monjardet (1984a) remark it remains for, say, $\boldsymbol{R}'(Q, P)$ when Q is a chain and P is finite: the situation (I) of Section 6.3.4. An example is the stability for both join and meet of the set $D_E = \boldsymbol{R}'(\mathbb{R}^+, \mathcal{R}_E)$ of all the dendrograms on E (Boorman and Olivier 1973). Notice that, in this case, the corresponding sets of residuated mappings and valued objects are generally stable only for the join, as it is illustrated by the ultrametrics on E. Another case of double stability of $\boldsymbol{R}(P, Q)$ is pointed out below.

Indeed, the stability of $\boldsymbol{V}(P, Q)$ for the join follows also directly from the characterization (V2) of valued objects: for $j \in J$ and $K \subseteq J$ such that $j \leq \vee K$ and $v, v' \in \boldsymbol{V}(P, Q), v(j) \leq \vee\{v(k) : k \in K\}$ and $v'(j) \leq \vee\{v'(k) : k \in K\}$ imply:

$$\begin{aligned}(v \vee v')(j) &= v(j) \vee v'(j) \leq (\vee\{v(k) : k \in K\}) \vee (\vee\{v'(k) : k \in K\}) \\ &= \vee\{v(k) \vee v'(k) : k \in K\} \\ &= \vee\{(v \vee v')(k) : k \in K\}.\end{aligned}$$

Let $\underline{0}$ be the the minimum element of Q^J (that is the constant mapping $j \mapsto 0$ for all $j \in J$); it is again a valued object since it obviously satisfies Condition (V2). So, applying Proposition 6.4.1, and keeping in mind that any anticlosure operator in a complete lattice maps an element into its unique maximal inferior open one, we obtain the following generalization of the subdominant ultrametric:

Proposition 6.5.1. *There exists an anticlosure operator $\bar{\omega}$ in Q^J which maps any element w of Q^J into a unique maximal valued object inferior to w.*

This ordinal fitting of the "subdominant" valued object $\bar{\omega}(w)$ to any given $w \in Q^J$ is always possible, but not always obtained in a simple way. A general theoretical procedure is related to the characterization (V3) of the valued objects (see Polat and Flament 1980). Notice first that the inequality $w(j) \geq \wedge\{\vee_{k \in K} w(k) : K \subseteq J, j \leq \vee K\}$ always holds for all $j \in J$ and $w \in Q^J$; define $\bar{\omega}_1 : Q^J \to Q^J$ by $(\bar{\omega}_1(w))(j) = \wedge\{\vee_{k \in K} w(k) : K \subseteq J, j \leq \vee K\}$ (in fact, only the minimal subsets K such that $j \leq \vee K$ are to be taken in account); of course, $v \in \mathbf{V}(P,Q)$ implies $\bar{\omega}_1(v) = v$ and conversely. We omit here the proof of the following result, which emphasizes the role of the distributivity hypotheses in situations (I) and (II) of Section 6.3.4:

Proposition 6.5.2. *If Q is distributive, then $\bar{\omega}_1(w) = \bar{\omega}(w)$. If not, $\bar{\omega}_1(w)$ is isotone but not always an element of $\mathbf{V}(P,Q)$. Then, $\bar{\omega}(w)$ is obtained as $(\bar{\omega}_1)^\ell(w)$ for some finite integer ℓ.*

Let us consider again the case of ultrametrics ($J = E^{(2)}$ and $Q = \mathbb{R}^+$); an element d of $(\mathbb{R}^+)^{E^{(2)}}$ is a dissimilarity index. For two distinct elements a and b of E, the previous procedure consists of taking the maximum value of d on each path between a and b, and, then, taking the minimum of all these maximums as $u(a,b)$; see Gondran (1976) for an algebraic formalization of this procedure. In the much more efficient Johnson (1967) and Roux (1968) algorithm of obtention of the subdominant ultrametric by changes on triangles, the procedure is restrained to paths with only two edges: it then needs generally several iterations. Now, the most efficient algorithms construct first a minimum spanning tree of d (since Gower and Ross 1969). This example is a good illustration of the fact that the actual obtention of the subdominant valued object is related with the particular combinatorial or algebraic properties of P and Q: it would be unreasonable to propose any algorithm at the level of generality of this paper.

The case where P is distributive is again particularly interesting. By Proposition 6.4.2, $\mathbf{V}(P,Q)$ is then equal to the set $\mathbf{Q}^P$ of all the isotone mappings from Q on P, which is stable not only for the meet, but also for the join (this stability extends to $\mathbf{R}(P,Q)$ in situations I and II of Section 6.3.4). So, when P is distributive, there exists also a closure operator ξ in Q^J which maps any element w of Q^J into a unique minimal valued object superior to w. For instance, Durand and Fichet (1988) pointed out that a dissimilarity d on the chain C_n has a unique minimal superior dissimilarity compatible with the order on C_n (and also, as in the general case, a unique subdominant one). Of course, the situation is similar for valued biorders.

According to the classical results recalled in Section 6.4.1, the pointwise order on $\mathbf{V}(P,Q)$ is in all cases a lattice order with the meet $\bar{\wedge}$ defined by $v\bar{\wedge}v' = \bar{\omega}(v \wedge v')$ ($v\bar{\wedge}v' = v \wedge v'$ when P is distributive). The particular properties of the lattice $\mathbf{V}(P,Q)$ depend of those of P and Q; for instance, the lattice of ultrametrics, studied by Leclerc (1979, 1981; see also Barthélemy, Leclerc and Monjardet 1984a) inherits many properties from the lattice of equivalence relations. Especially, $\mathbf{V}(P,Q)$ has many dual copies of P as sublattices. The proof of the following result, based on routine considerations of lattice theory, is again omitted:

Proposition 6.5.3. Let $\alpha, \beta \in Q$ such that $\alpha < \beta$ and $\mathbf{V}_{\alpha,\beta}$ be the subset of those elements v of $\mathbf{V}(P,Q)$ such that $v(j) \in \{\alpha, \beta\}$, for all $j \in J$. Then, $\mathbf{V}_{\alpha,\beta}$ is dually isomorphic to P by $x \mapsto v_x$, where $v_x(j) = \beta$ if $j \leq x$ and $v_x(j) = \alpha$ if not.

Figure 6.5.1 shows a principal ideal of the lattice of all the ultrametrics on $\{a, b, c, d\}$ with values in the chain $\{0 < 1 < 2 < 3\}$; we let the reader verify that the ultrametrics with values 0 or 1 form a dual partition lattice. According to a formula given in Barthélemy, Leclerc and Monjardet (1984a), the entire lattice has 118 elements.

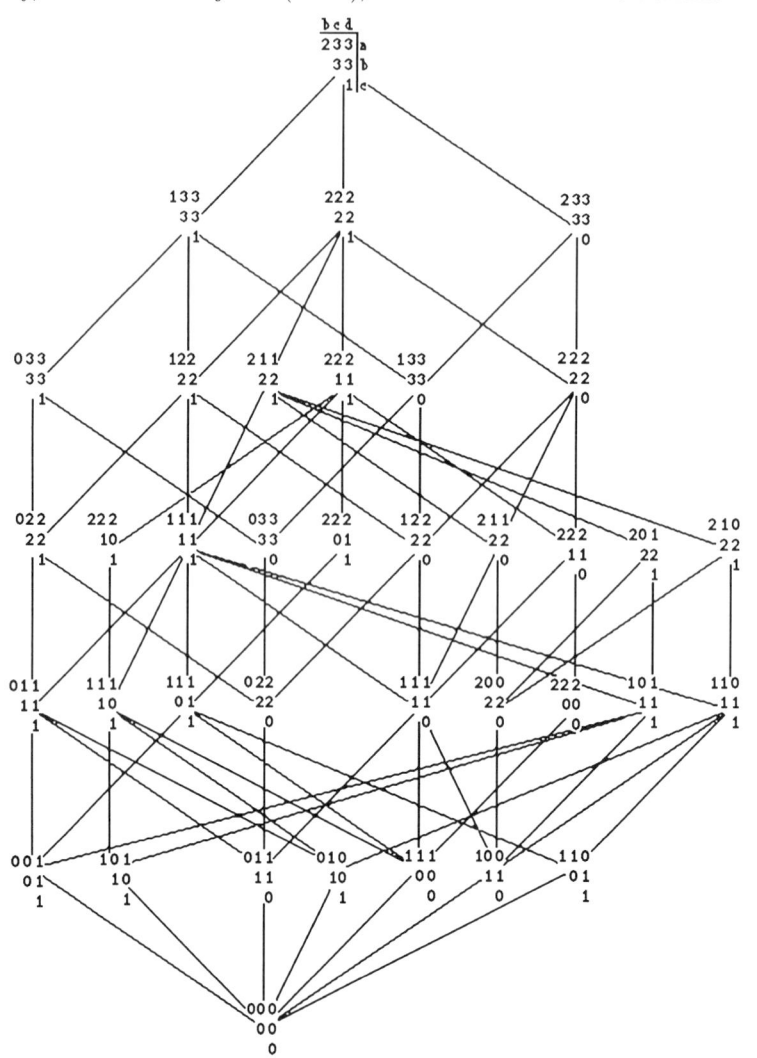

Figure 6.5.1

The lattice structure of $V(P,Q)$ is useful in the study of problems of comparison or aggregation (consensus) of valued objects. A first reference is the Boorman and Olivier (1973) paper where lattice metrics on equivalences are extended to dendrograms. Metrics directly defined on $\mathcal{U}_E$ are proposed in Leclerc (1981) and in Barthélemy, Leclerc and Monjardet (1986). The lattice structure on ultrametrics has an important role in the approaches of the consensus problem in Barthélemy, Leclerc and Monjardet (1984a,b, 1986). The general axiomatic results of Monjardet (1990) on the aggregation problem in lattices also apply to valued objects, and further specific axiomatic results on valued preorders and ultrametrics (Leclerc 1984a) extend to general valued objects (Leclerc 1991).

In the cases of Robinson arrays or valued biorders, the above results are essentially useful in modelling. When the orders on E are not fixed, as it is often the case in practice, the pointwise order on all Robinson arrays, or on all valued biorders, is no longer a lattice.

6.6. Notes and conclusions

It is a basic fact of ordered set theory that residuated/residual mappings and Galois connections are equivalent, under one order duality. Replace Q by its dual $R = Q^d$ in the theoretical considerations of Sections 6.2 and 6.4.1; then, in Theorems 6.2.1 and 6.4.1, conditions (R1-R4) take the following forms (G1-G4), where f and g get symmetric roles:

(G1) *for any $\lambda \in R$, $f^{-1}(\uparrow \lambda)$ is a principal ideal of P.*

(G2) *The mapping f is antitone ($x \leq y$ implies $f(x) \geq f(y)$) and there exists an antitone mapping g such that, for any $\lambda \in R$ and $x \in P$, $x \leq g(\lambda) \Leftrightarrow \lambda \leq f(x)$.*

(G3) *The mapping f is antitone and there exists an antitone mapping g such that the composition mappings $\varphi = gf$ and $\psi = fg$ are both extensive.*

(G4) $f(0) = \max R$ *and, for all $x, y \in P$, $f(x \vee y) = f(x) \wedge f(y)$.*

Recall that, in (R4), and now in (G4), P is assumed to be finite. The composition mappings fg and gf are both closure operators and the sets Ψ and Φ of closed elements are dually order isomorphic. Galois connections were first defined and studied by Öre (1944); an important case was previously investigated by Birkhoff (1967; 1st edition 1940). The term of residuation appears in a work of Dubreil and Croisot (1954; see also Croisot 1956). Other references on the origins of residuation theory may be found in the book of Blyth and Janowitz (1972), entirely devoted to this topic.

In this presentation, we had to choose between Galois and residuated mappings. From a theoretical point of view, residuated mappings have two good properties that Galois ones have not: the composition of two residuated mappings is again a residuated mapping (this follows immediately from property (R1)); residuated mappings are morphisms of complete join semilattices. On the other hand, in Galois formulations, one has only to consider one type of mappings, since g and f have symmetric roles. None of these properties is determinant for our subject and our main reason for the choice of residuation is just that it allows us to obtain dissimilarities and, so, to use directly the most known example of ultrametrics; the Galois model would lead to *similarity* coefficients. The cost is the duality appearing in Proposition 6.5.3, frequently awkward in intuitive - or formal - reasoning. In a work about general valued objects, the use of the Galois model often turns out to be easier: it leads to a lattice of valued objects which generalizes the lattice P without duality, and, more generally, properties of objects extend more directly to properties of the valued ones. An example is the transposition of Monjardet (1990) axioms in our 1991 paper on the aggregation of valued preferences.

Several types of mappings have been considered in this paper. Their relations are summarized hereunder, in the case where P and Q are assumed to be lattices with at least two elements and where, moreover, P is finite. A double arrow indicates a one-to-one correspondence; recall a distributive lattice where J is an antichain is boolean.

$$\begin{array}{cccccc}
\boldsymbol{V}(P,Q) & \subseteq & Q^J & \subseteq & Q^J \\
\updownarrow & = \text{if } P \text{ is distributive} & & = \text{if } J \text{ is an antichain} & \\
\boldsymbol{R}(P,Q) & \subseteq & Q^P & \subseteq & Q^P \\
\updownarrow & & & & \\
\boldsymbol{R}'(P,Q) & \subseteq & P^Q & \subseteq & P^Q
\end{array}$$

These relations are true under the hypotheses on the sets P and Q given in Section 6.3. The relaxation of some of these hypotheses may change the situation. Assuming that P and Q are ordered and J being a distinguished subset of P, it is natural, for a generalized definition of the valued objects, to start from a condition of the same type as (V4) in Theorem 6.4.2: the threshold subsets of J have a prescribed form. In several situations (Leclerc 1984b, Critchley and Van Cutsem 1994a,b), the obtained valued objects are in one-one correspondence with a class $\boldsymbol{R}''(P,Q)$ of mappings from P on Q which properly includes the residual ones.

A general presentation, like that of this paper, has several roles: to asemble models appearing in various fields (references are given along the paper and below) and to prepare the way for new ones of the same type; to precise the extent of some properties, like the existence of a subdominant dissimilarity of a given type; to give a general frame for problems such as comparison or aggregation ones (see the references at the end of Section 6.5). In all these directions, a unified presentation leads to unified, and frequently simplified, proofs.

As said in the introduction, the valued objects studied here have been formally defined, in the Galois form, by Achache (1982, 1988), in the general context of fuzzy sets theory. Instead of our pair (P, J), Achache considers a closure space, that is a set J, possibly infinite, endowed with a closure operator on $\mathcal{P}(J)$. In fuzzy sets, the set Q is generally the real interval $[0, 1]$, but to replace it by a complete lattice, as in Achache's work (and as in Section 6.3.4 above), is a classical generalization. Previously, the residuation model was described by Janowitz (1978), with the correspondence between dissimilarity coefficients and residuated mappings (or L-chains, with a chain L as ordered set Q). In his original theoretical approach of fuzziness, Negoita (1981) starts, in fact, from residual mappings, close to Janowitz's L-chains and, then, rediscovers Galois connections. Negoita also notices that the early model of fuzziness proposed by Gentilhomme (1968) is the particular case where $Q = \{0 < \frac{1}{2} < 1\}$. Real functions satisfying a condition close to (V4) are also studied, with applications in convex programming, by Volle (1985).

Acknowledgements

A first version of this paper appeared as a C.A.M.S. report (Leclerc (1990)). The author wishes to thank B. Monjardet and M.F. Janowitz for helpful discussions and advice in the last step of the preparation of this work.

References

Achache, A. (1982), Galois connection of a fuzzy subset, *Fuzzy Sets and Systems*, 8, pp. 215–218.

Achache, A. (1988), How to fuzzify a closure space, *J. Math.l Anal. Appl.*, 130, pp. 538–544.

Barbut, M., Monjardet, B. (1971), *Ordre et Classification, Algèbre et Combinatoire*, Hachette, Paris.

Barthélemy, J.P., Leclerc, B., Monjardet, B. (1984a), Ensembles ordonnés et taxonomie mathématique, In Pouzet, M., Richard, M., eds., *Orders, descriptions and roles, Annals of Discrete Mathematics*, 23, North-Holland, Amsterdam, pp. 523–548.

Barthélemy, J.P., Leclerc, B., Monjardet, B. (1984b), Quelques aspects du consensus en classification, in Diday, E., et al., eds., *Data Analysis and Informatics III*, North-Holland, Amsterdam, pp. 307-316.

Barthélemy, J.P., Leclerc, B., Monjardet, B. (1986), On the use of ordered sets in problems of comparison and consensus of classifications, *J. of Classification*, 3, pp. 187–224.

Batbedat, A. (1989), Les dissimilarités médas ou arbas, *Statist. Anal. Données*, 14, pp. 1–18.

Batbedat, A. (1990), *Les approches pyramidales dans la classification arborée*, Masson, Paris.

Bertrand, P., Diday, E. (1985), A visual representation of the compatibility between an order and a dissimilarity index: the pyramids, *Comput. Statist. Quart.*, 2, pp. 31–44.

Birkhoff, G. (1967), *Lattice Theory* (3rd ed.), Amer. Math. Soc., Providence.

Blyth, T.S., Janowitz, M.F. (1972), *Residuation Theory*, Pergamon Press, Oxford.

Boorman, S.A., Olivier, D.C. (1973), Metrics on spaces of finite trees, *J. Math. Psychol.*, 10, pp. 26–59.

Brito, P. (1994), Symbolic objects: order structure and pyramidal clustering, *IEEE Trans. Knowl. and Data Engin.*, to appear.

Critchley, F., Van Cutsem, B. (1994a), An order-theoretic unification and generalisation of certain fundamental bijections in mathematical classification - I, In Van Cutsem, B., ed., *Classification and Dissimilarity Analysis*, Ch. 4, Lecture Notes in Statistics, Springer-Verlag, New York.

Critchley, F., Van Cutsem, B. (1994b), An order-theoretic unification and generalisation of certain fundamental bijections in mathematical classification - II, In Van Cutsem, B., ed., *Classification and Dissimilarity Analysis*, Ch. 5, Lecture Notes in Statistics, Springer-Verlag, New York.

Croisot, R. (1956), Applications résiduées, *Ann. Sci. Ecole Norm. Sup. Paris*, 73, pp. 453–474.

Daniel-Vatonne, M.C., and de La Higuera, C. (1993), Les termes : un modèle de représentation et structuration de données symboliques, *Math. Inform. Sci. Humaines*, 122, pp. 41–63.

Defays, D. (1978), Analyse hiérarchique des préférences et généralisations de la transitivité, *Math. Sci. Humaines*, 61, pp. 5–27.

Degenne, A., and Vergès, P. (1973), Introduction à l'analyse de similitude, *Revue Française de Sociologie*, XIV, pp. 471–512.

Diday, E. (1986), Une représentation visuelle des classes empiétantes : les pyramides, *RAIRO Automat.-Prod. Inform. Ind.*, 20, pp. 475–526.

Diday, E. (1988), The symbolic approach in clustering and related methods of data analysis: the basic choices, In Bock, H.H., ed., *Classification and Related Methods of Data Analysis*, Amsterdam, North-Holland, pp. 673–683.

Doignon, J.-P., Monjardet, B., Roubens, M., Vincke, Ph. (1986), Biorder families, valued relations and preference modelling, *J. Math. Psych.* 30, pp. 435–480.

Dubois,D., Prade, H. (1980), *Fuzzy Sets and Systems: Theory and Applications*, Academic Press, New York.

Dubreil,P., Croisot, R. (1954), Propriétés générales de la résiduation en liaison avec les correspondances de Galois, *Collect. Math.*, 7, pp. 193–203.

Duquenne, V. (1987), Contextual implications between attributes and some representation properties for finite lattices, In Ganter, B., Wille, R., Wolf, K.E., ed., *Beiträge zur Begriffsanalyse*, Wissenchaftverlag, Mannheim, pp. 213–240.

Durand,C., Fichet, B. (1988), One-to-one correspondences in pyramidal representation: a unified approach, In Bock, H.H., ed., *Classification and Related Methods of Data Analysis*, North-Holland, Amsterdam, pp. 85–90.

Ganter, B., Rindfrey, K., Skorsky, M. (1986), Software for concept analysis, In Gaul, W., Schader, M., eds., *Classification as a Tool of Research*, North-Holland, Amsterdam, pp. 161–168.

Gentilhomme, Y. (1968), Les ensembles flous en linguistique, *Cahiers de Linguistique Théorique et Appliquée*, 5, pp. 47–65.

Golumbic, M.C. (1980), *Algorithmic Graph Theory and Perfect Graphs*, Academic Press, New York.

Gondran, M. (1976), La structure algébrique des classifications hiérarchiques, *Ann. INSEE*, 22-23, pp. 181–190.

Gower, J.C., Ross, G.J.S. (1969), Minimum spanning tree and single linkage analysis, *Applied Statistics*, 18, pp. 54–64.

Janowitz, M.F. (1978), An order theoretic model for cluster analysis, *SIAM J. Appl. Math.*, 37, pp. 55–72.

Jardine, N., Sibson, R. (1971), *Mathematical Taxonomy*, Wiley, London.

Johnson, S.C. (1967), Hierarchical clustering schemes, *Psychometrika*, 32, pp. 241–254.

Kaufman, A. (1973), *Introduction à la Théorie des Sous-ensembles Flous*, Masson, Paris. Translation (1975): *Introduction to the Theory of Fuzzy sets*, Academic Press, New York.

Leclerc, B. (1979), Semi-modularité des treillis d'ultramétriques, *C.R. Acad. Sci. Paris Sér. A*, 288, pp. 575–577.

Leclerc, B. (1981), Description combinatoire des ultramétriques, *Math. Sci. Humaines*, 73, pp. 5–31.

Leclerc, B. (1984a), Efficient and binary consensus functions on transitively valued relations, *Math. Social Sci.*, 8, pp. 45–61.

Leclerc, B. (1984b), Indices compatibles avec une structure de treillis et fermeture résiduelle, Rapport C.M.S. P.011, Centre d'Analyse et de Mathématiques Sociales, Paris, France.

Leclerc, B. (1990), The residuation model for ordinal construction of dissimilarities and other valued objects, Rapport C.M.S. P.063, Centre d'Analyse et de Mathématiques Sociales, Paris, France.

Leclerc, B. (1991), Aggregation of fuzzy preferences: a theoretic Arrow-like approach, *Fuzzy Sets and Systems*, 43, pp. 291–309.

Monjardet, B. (1988), A generalization of probabilistic consistency: linearity conditions for valued preference relations, In Kacprzyk, J., Roubens, M., eds., *Non-Conventional Preference Relations in Decision Making*, Lecture Notes in Econom. and Math. Syst., 301, Springer-Verlag, Berlin, pp. 36–53.

Monjardet, B. (1990), Arrowian characterization of latticial federation consensus functions, *Math. Soc. Sci.*, 20, pp. 51–71.

Negoita, C.V. (1981), *Fuzzy systems*, Abacus Press, Tunbridge Wells.

Öre, O. (1944), Galois connections, *Trans. Amer. Math. Soc.*, 55, pp. 494–513.

Polat, N., Flament, C. (1980), Applications galoisiennes proches d'une application entre treillis, *Math. Sci. Humaines*, 70, pp. 33–49.

Roubens, M. (1986), Comparison of flat fuzzy numbers, In Bandler, N., Kandel, A., eds., Proceedings of NAFIPS'86, pp. 462–476.

Roubens, M., Vincke, Ph. (1988), Fuzzy possibility graphs and their application to ranking fuzzy numbers, In Kacprzyk, J., Roubens, M., eds., *Non-conventional Preference Relations in Decision Making*, Springer-Verlag, Berlin, pp. 119–128.

Roux, M. (1968), Un algorithme pour construire une hiérarchie particulière, Thèse de 3ème cycle, Université Paris VI, France.

Shmuely, Z. (1974), The structure of Galois connections, *Pacific J. of Math.*, 54, pp. 209–225.

Volle, M. (1985), Conjugaison par tranches, *Ann. Mat. Pura Appl.*, 139, pp. 279–312.

Wille, R. (1982), Restructuring lattice theory : an approach based on hierarchies of concepts, In Rival, I., Reidel, D., eds., *Ordered Sets*, Dordrecht, pp. 445-470.

Chapter 7.
On exchangeability-based equivalence relations induced by strongly Robinson and, in particular, by quadripolar Robinson dissimilarity matrices[*]

Frank Critchley[†]

7.1. Overview

7.1.1. Preamble

Let I denote a totally ordered set of $n \geq 1$ elements. It is notationally convenient to identify I with the set of the first n integers or, on occasion, with the row vector $(1, \ldots, n)$. In the former case the order on I is embodied in the natural order on the reals, and in the latter case in the ordering amongst the elements of a vector.

A dissimilarity matrix (on I) is an $n \times n$ real matrix $\triangle \equiv (\delta_{ij})$ satisfying $\delta_{ij} = \delta_{ji} \geq \delta_{ii} = 0$ for all i and j in I. Thus dissimilarity matrices are considerably more general than distance matrices. The analysis of dissimilarity is a large and growing subject. In particular, it subsumes the distance geometry of finite point sets, multidimensional scaling and many methods of classification. Some key general references to these areas are Blumenthal (1953), de Leeuw and Heiser (1982) and Gordon (1987) respectively. Substantial contributions have appeared in Linear Algebra and Its Applications. These include Gower (1985), Mathar (1985), Critchley (1988a) and Hayden and Wells (1988). See also the books of collected works edited by de Leeuw et al. (1986), Diday et al. (1986) and Bock (1988), the recently founded Journal of Classification, and the other papers in this volume.

In this paper we consider two sub-classes of dissimilarity matrices: the quadripolar Robinson and the strongly Robinson matrices. These are defined and discussed in the following sub-section. The plan of the paper and its principal results are then outlined in Section 7.1.3. This paper is based ob Critchley (1988b).

Cognate work appears in Batbedat (1989, 1991, 1992) and Leclerc (1993) and in the references cited therein.

Attention is drawn to the fact that the notation and terminology of the present paper differ somewhat from that of the survey paper by Critchley and Fichet (1994) in this volume. A quadripolar dissimilarity here is called a tree dissimilarity there. The

---

[*] *In* Van Cutsem, B. (Ed.), (1994) *Classification and Dissimilarity Analysis*, Lecture Notes in Statistics, Springer-Verlag, New York.
[†] University of Birmingham, U.K.

essential difference is that I is here taken to be totally ordered and this order is then used to define the order of the rows and columns of a dissimilarity matrix $\triangle \equiv (\delta_{ij})$. It may then happen that, with respect to this order on I, the matrix $\triangle$ has the Robinsonian property. In contrast, in Critchley and Fichet (1994), the set I has no structure a priori. If such exists, a compatible order on I is then defined to be one such that, in it, the matrix $(\delta(i,j))$ is Robinsonian, in which case the dissimilarity itself is called Robinsonian.

7.1.2. Quadripolar, Robinson and strongly Robinson matrices

A dissimilarity matrix $\triangle$ (on I) is called a quadripolar matrix (on I) if it satisfies the four-point condition that:

$$\text{for all } i,j,k,l \text{ in } I, \quad \delta_{ik}^+ \leq \max\{\delta_{ij}^+, \delta_{jk}^+\} \tag{1}$$

where $\delta_{ik}^+ \equiv \delta_{ik} + \delta_{jl}$, $\delta_{ij}^+ \equiv \delta_{ij} + \delta_{kl}$ and $\delta_{jk}^+ \equiv \delta_{jk} + \delta_{il}$.

The importance of quadripolar matrices stems from the following characterisation. Consider a graph-theoretical tree whose edges have been assigned strictly positive lengths and suppose that each member of I is identified with a vertex of the tree (several members of I can be identified with the same vertex). Such a tree generates a dissimilarity matrix of path lengths between pairs of members of I. The dissimilarity matrices that can arise in this way are precisely the quadripolar matrices, and any quadripolar matrix identifies a unique such "additive tree" (Buneman, 1971). Quadripolar matrices have two consequent practical advantages. Firstly, the associated additive trees provide clear, readily assimilated visual displays of such dissimilarity data. And secondly, one may hope to identify the vertices that are not members of I as missing members of some evolutionary process about which only information on I remains. Indeed, this was part of Buneman's initial motivation in studying them.

A dissimilarity matrix $\triangle$ (on I) is called a Robinson matrix (on I) if:

$$\text{for all } i,j,k \text{ in } I, \quad i \leq j \leq k \text{ implies } \max\{\delta_{ij}, \delta_{jk}\} \leq \delta_{ik} \tag{2}$$

That is, if its elements do not decrease as you move away from the diagonal along any row or column. Thus the elements of a Robinson matrix respect the order on I in the sense that the further you move away from any element of I, in either direction, the greater the dissimilarity. Such matrices arise naturally in seriation and in other problems where there is an essentially one-dimensional structure in the data. For example, in the seminal paper by Robinson (1951), the elements of I were archaeological deposits for which it was reasonable to hypothesise that empirical measures of dissimilarity between them would, apart from error, satisfy the inequality in (2) whenever i occurred in time before j and j before k. The problem there was to then use this hypothesis to infer the time ordering of the deposits.

Now clearly (2) holds for a certain ordering on I if and only if it holds for the ordering which is the complete reverse of the original one. Such pairs of orderings are called dual. The idea of duality plays a key role in the study of Robinson matrices. Note that in Robinson's original practical problem the choice between two dual orderings can be made given the minimal proviso that one knows which of the two extreme deposits occurred first in time.

A dissimilarity matrix $\triangle$ (on I) is said to be a strongly Robinson matrix (on I) if it is Robinson and, moreover:

$$\text{for all } i \leq j \leq k \leq l \text{ in } I, \quad \delta_{ij} = \delta_{ik} \text{ implies } \delta_{hj} = \delta_{hk} \text{ for all } h \leq i \quad (3)$$
$$\text{and:} \quad \delta_{jl} = \delta_{kl} \text{ implies } \delta_{jm} = \delta_{km} \text{ for all } m \geq l \quad (3^*)$$

Both Robinson and strongly Robinson matrices arise naturally in the generalisation of hierarchical clustering to pyramids, also known as pseudo-hierarchies, introduced by Diday (1984) and Fichet (1984) and discussed in Bertrand and Diday (1985). A dissimilarity matrix $\triangle$ (on I) is called definite (or, in the French literature, proper) if $\delta_{ij} = 0$ implies $i = j$ and even (or semi-proper) if $\delta_{ij} = 0$ implies $\delta_{ik} = \delta_{jk}$ for all k in I. We have the following bijections:

(a) (Diday, 1986):
$$\{\text{all pyramids indexed in the broad sense}\} \xleftrightarrow{1-1} \{\text{all permutations of definite Robinson matrices}\},$$

(b) (Durand and Fichet, 1988):
$$\{\text{all indexed pseudo-hierarchies}\} \xleftrightarrow{1-1} \{\text{all permutations of strongly Robinson matrices}\},$$

(c) (Durand and Fichet, 1988):
$$\{\text{all weakly-indexed pseudo-hierarchies}\} \xleftrightarrow{1-1} \{\text{all permutations of even Robinson matrices}\},$$

and hence (d):
$$\{\text{all pyramids indexed in the strict sense}\} \xleftrightarrow{1-1} \{\text{all permutations of definite, strongly Robinson matrices}\}$$

in each of which "permutation" refers to a row-and-column permutation $\triangle \to P^T \triangle P$.

These results generalise the classical bijection (Johnson, 1967) between the ultrametric dissimilarity matrices and the dendrograms on I. The essence of these generalisations is to introduce the possibility of overlapping clusters within a hierarchical scheme by allowing nodes to have up to two predecessors. At the same time, it is required for clarity and for practical convenience that in the associated visual display there are no crossings between horizontal and vertical lines. It is essentially this requirement which corresponds to the property of being permutable to Robinson form. And it is this requirement which entails that all the clusters in the visual display are connected subsets with respect to the order on I.

7.1.3. Plan and principal results.

The plan of this paper is as follows. Some necessary preliminaries are dealt with in Section 7.2. In particular, quadripolar Robinsonianity and strong Robinsonianity are shown to be ordered restriction invariant properties. Following Durand (1988), the key concept of duality is also studied there. It is denoted by an asterisk throughout.

Section 7.3 characterises all quadripolar Robinson matrices of order four. Every order four principal sub-matrix of a quadripolar Robinson matrix is shown to be of one of three known types. This result yields, as a simple corollary, that the strongly Robinson matrices contain the quadripolar Robinson matrices. This important fact was proved first by Durand and Fichet (personal communication).

Having established that the quadripolar Robinson matrices are a special case, we discuss strongly Robinson matrices in generality in Section 7.4. The key idea of exchangeability is introduced here. Two members of I are said to be exchangeable with respect to a dissimilarity matrix if they have the same dissimilarities as each other with respect to every member of I. A strongly Robinson matrix is shown to induce four equivalence relations upon I corresponding to four variations on the theme of exchangeability. These four equivalence relations are then studied. An equivalence relation on I is said to be connected if its equivalence classes are necessarily connected with respect to the order on I. Thus, a connected equivalence relation respects the order on I in the same natural sense as do the clusters. Again, an equivalence relation on I is called internally even if it has the property that the nontrivial dissimilarities between members of any given equivalence class are all equal. Thus, an internally even equivalence relation has the convenient property that the restriction of a dissimilarity matrix to any of its equivalence classes can be summarised in a single number. It is shown in Section 7.4 that three of the four exchangeability-based equivalences are connected and that a different trio are internally even. The logical relations amongst these four equivalence relations are also reported there.

One motivation for studying these exchangeability relations is to obtain a natural global summary of a strongly Robinson matrix $\triangle$ in terms of:

(I) the restriction of $\triangle$ to each equivalence class

and

(II) the dissimilarity matrix induced by $\triangle$ upon the equivalence classes, whenever this is well-defined.

Note that internally even equivalence relations are particularly well-suited to part (I) of this summarisation process.

The matrix in (II) is called the reduced form of $\triangle$ with respect to the equivalence relation. It is well-defined if and only if the equivalence relation has the property that:

$$i \not\sim j \text{ and } j \sim k \text{ implies } \delta_{ij} = \delta_{ik},$$

which we term external evenness with respect to $\triangle$. It is shown in Section 7.5 that the three internally even exchangeability relations are externally even with respect to every strongly Robinson matrix. And that all four relations are externally even with respect to all quadripolar Robinson matrices.

It is also shown there that the properties of strong Robinsonianity and of quadripolar Robinsonianity are inherited by reduced forms. Thus the process of taking the reduced form of either sort of matrix can be iterated. As $|I|$ is finite, this process must converge in finitely many steps to an irreducible matrix.

Two final questions remain:

(i) what is the class of matrices that can arise as these limiting irreducible forms?

and

(ii) what are the logical relations between the limiting forms corresponding to the four different exchangeability-based equivalence relations?

These questions are answered for strongly Robinson matrices in Section 7.6 and for quad- ripolar Robinson matrices in Section 7.7. The key concept here is called regularity.

Section 7.7 highlights several properties that are peculiar to quadripolar Robinson matrices in the sense that they are not true for general strongly Robinson matrices. For example, we show there that a quadripolar Robinson matrix is regular if and only if each of its order four principal submatrices has type 1 according to the nomenclature established in Section 7.3. And that if a quadripolar Robinson matrix is regular then it is wholly determined by its first two superdiagonals. Thus, its $0(n^2)$ information can be perfectly encoded in $0(n)$ quantities. This greatly increases the power of stage (II) of the summarisation process in the case of quadripolar Robinson matrices. The properties of quadripolar Robinson matrices are developed further in Critchley (1988c, 1994). In particular, their irreducible forms are seen to have particularly simple visual representations which adds further to the power of the summarisation process.

7.2. Preliminaries

In this section we establish some necessary terminology and some useful general results. In particular, following Durand (1988), we explore the key idea of duality for dissimilarity matrices.

We define first two more types of dissimilarity matrix. A dissimilarity matrix $\triangle \equiv (\delta_{ij})$ on I is called:

(i) metric if $\delta_{ik} \leq \delta_{ij} + \delta_{jk}$ for all i, j, k in I.

(ii) ultrametric if $\delta_{ik} \leq \max(\delta_{ij}, \delta_{jk})$ for all i, j, k in I.

We omit the straightforward proof of the following result.

Proposition 7.2.1. Let $\triangle \equiv (\delta_{ij})$ be a dissimilarity matrix on I.

(i) $\triangle$ is even if it is either definite or metric or strongly Robinson.

(ii) $\triangle$ defines a metric on I via $\delta(i,j) = \delta_{ij}$ if and only if $\triangle$ is metric and definite.

(iii) $\triangle$ is quadripolar if and only if $\triangle$ is metric and the inequality in (1) above holds for all quadruples i, j, k, l of **distinct** elements of I.

(iv) $\triangle$ ultrametric implies $\triangle$ quadripolar.

(v) $\triangle$ is ultrametric and Robinson if and only if $\delta_{ik} = \max(\delta_{ij}, \delta_{jk})$ for all $i \leq j \leq k$ in I.

Let $\triangle \equiv (\delta_{ij})$ be a dissimilarity matrix on I and let $J \equiv (j_1, \ldots, j_N)$ be a row vector of $N \geq 1$ distinct elements of I. We say that J is ordered if $j_1 < \ldots < j_N$. The restriction of $\triangle$ to J, denoted $\triangle(J)$, is the dissimilarity matrix $\triangle' \equiv (\delta'_{gh})$ on $I' \overset{1-1}{\longleftrightarrow} \{1, \ldots, N\}$ defined by:

$$\text{for all } g, h = 1, \ldots, N, \quad \delta'_{gh} = \delta_{j_g j_h}.$$

We say that $\triangle(J)$ is an ordered restriction (of $\triangle$) if J is ordered. A property of dissimilarity matrices is said to be (ordered) restriction invariant if $\triangle(J)$ has that property whenever $\triangle$ does and $\triangle(J)$ is an (ordered) restriction of $\triangle$.

Proposition 7.2.2.

(i) Every restriction invariant property of dissimilarity matrices is ordered restriction invariant.

(ii) Definiteness, evenness, metricity, ultrametricity and quadripolarity are restriction invariant properties.

(iii) Robinsonianity and strong Robinsonianity are ordered restriction invariant properties. They are not restriction invariant.

Proof. Immediate. ∎

Let $\triangle \equiv (\delta_{ij})$ be a dissimilarity matrix on I. The dual of $\triangle$ is the dissimilarity matrix $\triangle^* \equiv (\delta^*_{ij})$ on I defined by:

$$\text{for all } i, j, \text{ in } I, \quad \delta^*_{ij} = \delta_{i^* j^*} \text{ where } i^* = (n+1) - i.$$

In other words, $\triangle^*$ contains the same elements as $\triangle$ but written out with the order of the rows and columns completely reversed. A property of dissimilarity matrices is said to be dual if $\triangle^*$ has that property whenever $\triangle$ does. We have the following result whose straightforward proof we omit.

Proposition 7.2.3.

(i) For any dissimilarity matrix $\triangle$ on I, $\triangle^{**} = \triangle$.

(ii) Every restriction invariant property of dissimilarity matrices is a dual property.

(iii) Robinsonianity and strong Robinsonianity are dual properties.

Corollary 7.2.1. *In particular, $\triangle$ is quadripolar and Robinson if and only if $\triangle^*$ is so.*

The fact that being quadripolar and Robinson is a dual property is both informative and convenient. For example, given any result about a quadripolar Robinson matrix $\triangle$, we can essentially write down at once a logically equivalent dual result about $\triangle$ by using the following simple procedure: apply the primal result to $\triangle^*$ and then translate this back into a result about $\triangle$ using $\triangle^{**} = \triangle$. Clearly the dual of the dual result is the primal. Pairs of such mutually dual results are stated in $\{(a), (a^*)\}$ form. This both gives added insight into the structure of the reported results and has the practical advantage that, of course, only one of them need be proved. Similar remarks apply of course in the strongly Robinson case.

7.3. Quadripolar Robinson matrices of order four

In this section, we characterise all quadripolar Robinson matrices of order four. This is sufficient to establish the important result that a quadripolar Robinson matrix of any order is strongly Robinson.

The following key definitions are used in the main result of this section, Theorem 7.3.1. Let $n \geq 4$, let $\triangle$ be a dissimilarity matrix on I, and let Q be an ordered quadruple (i, j, k, l) of distinct elements of I. That is, $1 \leq i < j < k < l \leq n$.

We say that Q has type 1 (with respect to $\triangle$) if:

$$\triangle(Q) = \begin{pmatrix} 0 & \varepsilon_i & \eta_i & \eta_i + \eta_j - \varepsilon_j \\ & 0 & \varepsilon_j & \eta_j \\ & & 0 & \varepsilon_k \\ & & & 0 \end{pmatrix}$$

for some $\varepsilon_i, \varepsilon_j, \varepsilon_k, \eta_i, \eta_j$ such that:

(i) $\min(\varepsilon_i, \varepsilon_j, \varepsilon_k) \geq 0$
(ii) $\max(\varepsilon_i, \varepsilon_j) \leq \eta_i \leq \varepsilon_i + \varepsilon_j$
(iii) $\max(\varepsilon_j, \varepsilon_k) \leq \eta_j \leq \varepsilon_j + \varepsilon_k$

and

(iv) $(\eta_i - \varepsilon_i) + (\eta_j - \varepsilon_k) > 0$,

the below diagonal elements being given by symmetry.

We say that Q has type 2 (with respect to $\triangle$) if:

$$\triangle(Q) = \begin{pmatrix} 0 & \beta & \beta & \delta \\ & 0 & \alpha & \gamma \\ & & 0 & \gamma \\ & & & 0 \end{pmatrix}$$

for some $\alpha, \beta, \gamma, \delta$ such that:

(i) $0 \leq \alpha \leq \min(\beta, \gamma)$

(ii) $\max(\beta, \gamma) \leq \delta$

and

(iii) $\delta < \beta + \gamma - \alpha$.

Finally, we say that Q has type 3 (with respect to $\triangle$) if:

$$\triangle(Q) = \begin{pmatrix} 0 & \beta & \beta & \beta+\gamma-\alpha \\ & 0 & \alpha & \gamma \\ & & 0 & \gamma \\ & & & 0 \end{pmatrix}$$

for some α, β, γ satisfying:

$$0 \leq \alpha \leq \min(\beta, \gamma).$$

For each $t = 1, 2, 3$, if Q has type t with respect to $\triangle$ we also say that $\triangle(Q)$ has type t and write $T_\triangle(Q) = t$, or just $T(Q) = t$ if there is no danger of confusion.

Theorem 7.3.1. *Let $\triangle$ be a quadripolar Robinson matrix on I. Then every ordered quadruple of distinct elements of I has one of the three mutually exclusive types 1,2 and 3 with respect to $\triangle$.*

Proof. If $n \leq 3$, there is nothing to prove. Suppose then that $n \geq 4$ and consider a quadruple $Q = (i, j, k, l)$ with $i < j < k < l$. As $\triangle$ is quadripolar, there are four mutually exclusive possibilities. In the notation of (1) above:

either [1] $\delta_{ij}^+ < \delta_{ik}^+ = \delta_{jk}^+$

or [2] $\delta_{jk}^+ < \delta_{ij}^+ = \delta_{ik}^+$

or [3] $\delta_{ik}^+ = \delta_{jk}^+ = \delta_{ij}^+$

or [4] $\delta_{ik}^+ < \delta_{jk}^+ = \delta_{ij}^+$.

But, as $\triangle$ is Robinson, $\delta_{ij}^+ \leq \delta_{ik}^+$. Therefore [4] is impossible. Again, as $\triangle$ is Robinson, $\delta_{ij}^+ = \delta_{ik}^+$ implies $\delta_{ij} = \delta_{ik}$ and $\delta_{jl} = \delta_{kl}$. Thus the cases [1], [2] and [3] are mutually exclusive and exhaustive and, moreover, $T_\triangle(Q) = t$ if and only if case $[t]$ occurs. ∎

For any row vector $J \equiv (j_1, \ldots, j_N)$ of $N \geq 1$ distinct elements of I, we define J^* to be the row vector $(j_N^*, \ldots, j_1^*)$ where, we recall, $i^* = (n+1) - i$ for all i in I. Thus, J^* is ordered if and only if J is. The following corollary is now immediate. It asserts that the type of a quadruple is invariant under duality.

Corollary 7.3.1. *Let $\triangle$ be a quadripolar Robinson matrix on I. Then, for every ordered quadruple Q of distinct elements of I,*

$$T_\triangle(Q) = T_{\triangle^*}(Q^*).$$

Theorem 7.3.1 also yields the following important result.

Corollary 7.3.2. (Durand and Fichet, personal communication) *For quadripolar matrices, Robinson $\Leftrightarrow$ strongly Robinson.*

Proof. The sufficiency of strong Robinsonianity is obvious. For necessity it suffices, by duality, to prove that a quadripolar Robinson matrix $\triangle$ satisfies (3). Suppose then that $h \leq i \leq j \leq k$ are in I and that $\delta_{ij} = \delta_{ik}$. We have to show that $\delta_{hj} = \delta_{hk}$. If $h = i$ or if $j = k$ this is trivial, while if $i = j$ it follows from $\delta_{ik} = 0$ and the evenness of $\triangle$ (Proposition 7.2.1). Suppose then that $h < i < j < k$. Now $T(h, i, j, k) = 2$ and $\delta_{ij} = \delta_{ik}$ imply $\delta_{hk} < \delta_{hj}$ contrary to the Robinsonianity of $\triangle$. Thus $T(h, i, j, k) \neq 2$. The result now follows from Theorem 7.3.1. ∎

Finally in this section, we note a third corollary of Theorem 7.3.1.

Corollary 7.3.3. *Let $\triangle$ be a quadripolar Robinson matrix on I. Then, for every ordered quadruple $Q \equiv (i, j, k, l)$ of distinct elements of I that is not of type 1 with respect to $\triangle$*

(a) $\delta_{hj} = \delta_{hj'} = \delta_{hk}$ and (a^*) $\delta_{jm} = \delta_{j'm} = \delta_{km}$

for all $j \leq j' \leq k$, for all $h \leq i$ and for all $m \geq l$.

Proof. By Theorem 7.3.1, $\delta_{ij} = \delta_{ik}$ and so, as $\triangle$ is Robinson, $\delta_{ij} = \delta_{ij'} = \delta_{ik}$ which implies (a) by Corollary 7.3.2. Part (a^*) follows by duality, using Corollary 7.3.1. ∎

7.4. Equivalence relations induced by strongly Robinson matrices

7.4.1. Exchangeability and connectedness

In this section we define and study four equivalence relations which a strongly Robinson matrix $\triangle$ induces upon I. These relations are based on a natural concept called exchangeability. Loosely speaking, two members of I are called exchangeable if, in some degree at least, they have the same dissimilarities.

An ordered row vector J of distinct elements of I is called **connected** if $k \in J$ whenever $j \leq k \leq l$ and $j, l \in J$. An equivalence relation on I is called connected if all of its equivalence classes are necessarily connected. Such relations are of particular interest since they respect the order on I in an obvious sense. Three of the following relations enjoy this property.

Let $\triangle$ be a strongly Robinson matrix on I and let $j, k \in I$. Then we say that, **with respect to** $\triangle$:

(i) j and k are **exchangeable**, and write $j \stackrel{1}{\sim} k$, if:

$$\delta_{ij} = \delta_{ik} \text{ for all } i \text{ in } I.$$

(ii) j and k are **eventually exchangeable**, and write $j \stackrel{2}{\sim} k$, if:

(a) $\exists i \leq \min(j, k)$ such that $\delta_{hj} = \delta_{hk}$ for all $h \leq i$

and

(b) $\exists l \geq \max(j, k)$ such that $\delta_{jm} = \delta_{km}$ for all $m \geq l$.

(iii) j and k are **externally exchangeable**, and write $j \overset{3}{\sim} k$, if:

$$\delta_{ij} = \delta_{ik} \text{ for all } i \text{ in } I \text{ different from } j \text{ and } k.$$

We give below a counterexample showing that $\overset{3}{\sim}$ is not connected. We therefore introduce what will be shown to be a stronger version of the same idea and say that:

(iv) j and k are **connectedly externally exchangeable**, and write $j \overset{4}{\sim} k$, if:

$$l \overset{3}{\sim} (l+1) \text{ for all } \min(j,k) \leq l < l+1 \leq \max(j,k).$$

We prove below that each $\overset{r}{\sim}$ ($r = 1, 2, 3, 4$) **is** an equivalence relation. We note the following two results first.

Proposition 7.4.1. *Let $\triangle$ be a strongly Robinson matrix on I and let $j, k \in I$. Then, for each $r = 1, 2, 3, 4$:*

$$j \overset{r}{\sim} k \text{ with respect to } \triangle \Leftrightarrow j^* \overset{r}{\sim} k^* \text{ with respect to } \triangle^*.$$

Proof. The result is immediate for $r = 1, 3$ and 4. For $r = 2$, we observe that condition (a) of the definition holds for $\triangle$ if and only if (b) holds for $\triangle^*$. Thus, by duality, (b) holds for $\triangle$ if and only if (a) holds for $\triangle^*$. ∎

Lemma 7.4.1. *Let $\triangle$ be a strongly Robinson matrix on I and let $j, k \in I$. Then:*

(i) $j \overset{1}{\sim} k$ if and only if $\delta_{jk} = 0$.

(ii) $j \overset{2}{\sim} k$ if and only if:

(a) $\exists i \leq \min(j,k)$ such that $\delta_{ij} = \delta_{ik}$

and

(b) $\exists l \geq \max(j,k)$ such that $\delta_{jl} = \delta_{kl}$

(iii) $j \overset{2}{\sim} k$ if and only if either $j \overset{1}{\sim} k$ or

(a$^+$) $\exists i < \min(j,k)$ such that $\delta_{ij} = \delta_{ik}$

and

(b$^+$) $\exists l > \max(j,k)$ such that $\delta_{jl} = \delta_{kl}$.

(iv) $j \overset{2}{\sim} k \Leftrightarrow$ (a) $\delta_{1j} = \delta_{1k}$ and (b) $\delta_{jn} = \delta_{kn}$

(v) if $\triangle$ is quadripolar Robinson and if $1 < j < k < n$, then:

$$j \overset{2}{\sim} k \Leftrightarrow T(1, j, k, n) \neq 1.$$

Proof.

(i) $j \overset{1}{\sim} k \Rightarrow \delta_{jk} = \delta_{jj} = 0$. Conversely $\delta_{jk} = 0$ implies $j \overset{1}{\sim} k$ as $\triangle$ is even.

(ii) Immediate from the definition and the strongly Robinson property.

(iii) Using (i), condition (a) holds if and only if $j \overset{1}{\sim} k$ or (a^+). Dually, (b) holds if and only if $j \overset{1}{\sim} k$ or (b^+).

(iv) Immediate from (ii) and the strongly Robinson property.

(v) Immediate from (iv) and Theorem 7.3.1. ∎

Theorem 7.4.1. *Let $\triangle$ be a strongly Robinson matrix on I. Then:*

(i) $\overset{1}{\sim}, \overset{2}{\sim}$ *and* $\overset{4}{\sim}$ *are connected equivalence relations upon I*

(ii) $\overset{3}{\sim}$ *is an equivalence relation on I. It is not connected, even for quadripolar Robinson matrices.*

Proof. It is clear from the definitions that each $\overset{r}{\sim}$ is reflexive and symmetric ($r = 1, 2, 3, 4$). Suppose now that $j \overset{r}{\sim} k$ and $k \overset{r}{\sim} l$ with j, k, and l distinct and $j < l$. To establish equivalence it suffices to show that $j \overset{r}{\sim} l$. We do this for each value of r in turn.

$r = 1$: For any i in I,
$$\delta_{ij} = \delta_{ik} \text{ as } j \overset{1}{\sim} k$$
$$= \delta_{il} \text{ as } k \overset{1}{\sim} l.$$

$r = 2$: Immediate from Lemma 7.4.1(iv).

$r = 3$: Let i be in I, i not equal to j or l. We have to show that $\delta_{ij} = \delta_{il}$. Suppose first that $i \neq k$. Then:
$$\delta_{ij} = \delta_{ik} \text{ as } j \overset{3}{\sim} k \text{ and } i \neq (j \text{ or } k)$$
$$= \delta_{il} \text{ as } k \overset{3}{\sim} l \text{ and } i \neq (k \text{ or } l).$$

Whereas if $i = k$, then:
$$\delta_{kj} = \delta_{lj} \text{ as } k \overset{3}{\sim} l \text{ and } j \neq (k \text{ or } l)$$
$$= \delta_{kl} \text{ as } j \overset{3}{\sim} k \text{ and } l \neq (j \text{ or } k).$$

$r = 4$: Of the three possible cases:

$$(1)\, j < k < l \text{ and } (2)\, k < j < l \text{ and } (2^*)\, j < l < k$$

it suffices, by duality, to consider (1) and (2).

It is immediate from the definition of $\overset{4}{\sim}$ that the required result holds in case (1) and that $[k < j < l$ and $k \overset{4}{\sim} l]$ implies $j \overset{4}{\sim} l$. This latter deals with case (2) and establishes that the $\overset{4}{\sim}$ classes are connected.

It only remains to show that

$$\left[k < j < l \text{ and } k \stackrel{r}{\sim} l\right] \text{ implies } j \stackrel{r}{\sim} l$$

for $r = 1$ and 2, but not for $r = 3$ even if $\triangle$ is quadripolar Robinson.

$\underline{r = 1}$: $\quad k \stackrel{1}{\sim} l \;\Rightarrow \delta_{kl} = 0,\;$ by Lemma 7.4.1(i)
$\qquad\qquad\qquad\quad\Rightarrow \delta_{jl} = 0,\;$ as $\triangle$ is Robinson
$\qquad\qquad\qquad\quad\Rightarrow j \stackrel{1}{\sim} l,\;$ by Lemma 7.4.1(i) again.

$\underline{r = 2}$: Immediate from Lemma 7.4.1(iv) and Robinsonianity.

$\underline{r = 3}$: Consider the case $n = 3$, $\delta_{12} = \delta_{23} = 1$ and $\delta_{13} = 2$. Then $1 \stackrel{3}{\sim} 3$ but neither $1 \stackrel{3}{\sim} 2$ nor $2 \stackrel{3}{\sim} 3$. ∎

For any strongly Robinson matrix $\triangle$ on I, and for each $r = 1, 2, 3, 4$, we denote the number of $\stackrel{r}{\sim}$ classes which $\triangle$ induces upon I by $m_\triangle(r)$, or just $m(r)$ when there is no danger of confusion. These classes are identified with ordered row vectors of distinct members of I and denoted $E_1^r, \ldots, E_{m(r)}^r$. When $r = 1, 2$, or 4, so that $\stackrel{r}{\sim}$ is connected, we label these classes such that for all $1 \leq i < j \leq m(r)$ the elements of E_i^r appear in I before the elements of E_j^r. When $r = 3$, we label such that the minimal element of E_i^3 appears in I before the minimal element of E_j^3 for all $1 \leq i < j \leq m(3)$. For each $r = 1, 2, 3, 4$ and for each $i = 1, \ldots, m(r)$, the number of elements of I in E_i^r is denoted by $n_i(r)$.

7.4.2. Internal evenness

We study next the restriction of a strongly Robinson matrix to a $\stackrel{r}{\sim}$ class and show that it takes a very simple form for $r = 1, 3$ and 4. An equivalence relation $\sim$ on I is called **internally even** with respect to a dissimilarity matrix $\triangle$ on I if:

for all i, j, k in I, $\quad [i, j, k$ distinct and $i \sim j \sim k]$ implies $\delta_{ij} = \delta_{jk} = \delta_{ik}$.

The discrete metric matrix of order n is the dissimilarity matrix $\triangle$ on I defined by $\delta_{ij} = 1$ for all $i \neq j$. It is denoted by S_n. We have the following results.

Lemma 7.4.2. *The restriction of a dissimilarity matrix to an internally even equivalence class is a nonnegative (possibly zero) multiple of the discrete metric matrix of the appropriate order.*

Proof. Immediate. ∎

Theorem 7.4.2. *Let $\triangle$ be a strongly Robinson matrix on I. Then:*

(i) *The relations $\stackrel{1}{\sim}, \stackrel{3}{\sim}$ and $\stackrel{4}{\sim}$ are internally even with respect to $\triangle$ and hence*

$$\triangle(E_i^r) \propto S_{n_i(r)}$$

for $r = 1, 3, 4$ and for $i = 1, \ldots, m(r)$.

(ii) The relation $\overset{2}{\sim}$ is not in general internally even with respect to $\triangle$, even if $\triangle$ is quadripolar Robinson.

Proof. (i) The second part of (i) follows from the first by Lemma 7.4.2. For the first part we consider each value of r separately. Let i, j, k be distinct and let $i \overset{r}{\sim} j \overset{r}{\sim} k$.

$\underline{r = 1}$: Immediate from Lemma 7.4.1(i).

$\underline{r = 3}$:
$$\begin{aligned}\delta_{ij} &= \delta_{jk} \text{ as } i \overset{3}{\sim} k \text{ and } j \neq (i \text{ or } k) \\ &= \delta_{ik} \text{ as } i \overset{3}{\sim} j \text{ and } k \neq (i \text{ or } j).\end{aligned}$$

$\underline{r = 4}$: Relabelling if necessary, we may suppose $i < j < k$. Now:
$$\begin{aligned} j \overset{4}{\sim} k &\Rightarrow l \overset{3}{\sim} (l+1) \text{ for all } j \leq l < l+1 \leq k \\ &\Rightarrow \delta_{il} = \delta_{i,l+1} \text{ for } l = j, \ldots, (k-1) \\ &\Rightarrow \delta_{ij} = \delta_{ik}.\end{aligned}$$

Thus, by duality, $i \overset{4}{\sim} j$ implies $\delta_{ik} = \delta_{jk}$.

(ii) Consider the case $n = 5$ and the quadripolar Robinson matrix $\triangle$ on I defined by:
$$\text{for all } i < j, \quad \delta_{ij} = \begin{cases} 1 & \text{if } i = 2, j = 3 \\ 2 & \text{otherwise.} \end{cases}$$
Then $E_1^2 = \{1\}, E_2^2 = \{2, 3, 4\}$ and $E_3^2 = \{5\}$ but $\triangle(E_2^2)$ is not a multiple of S_3. ∎

7.4.3. Logical relationships

We end this section by studying the logical relationships that exist among the four exchangeability-based equivalence relationships. We have the following result which, in particular, establishes that connectedly externally exchangeable members are indeed externally exchangeable.

Theorem 7.4.3. *Let $\triangle$ be a strongly Robinson matrix on I and let $j, k \in I$. Then:*

(i) $j \overset{2}{\sim} k \Leftarrow j \overset{1}{\sim} k \Rightarrow j \overset{4}{\sim} k \Rightarrow j \overset{3}{\sim} k$

(ii) *if $\min(j, k) = 1$ or $\max(j, k) = n$, then:* $j \overset{2}{\sim} k \Leftrightarrow j \overset{1}{\sim} k$.

(iii) *otherwise, if $\min(j, k) > 1$ and $\max(j, k) < n$, then:*

$$j \overset{3}{\sim} k \Rightarrow j \overset{2}{\sim} k$$
$$\Leftrightarrow \begin{cases} (a^+) \ \exists i < \min(j,k) \text{ such that } \delta_{ij} = \delta_{ik} \\ \text{and} \\ (b^+) \ \exists l > \max(j,k) \text{ such that } \delta_{jl} = \delta_{kl}. \end{cases}$$

Proof.

(i) The first two implications in (i) are immediate. For the third, it suffices to consider $j < k$. Suppose then that $j \overset{4}{\sim} k$. If $k = j + 1$, then $j \overset{3}{\sim} k$ is immediate. Suppose then that $k > j + 1$. Then by Theorem 7.4.2(i),

$$\delta_{jj'} = \delta_{jk} = \delta_{j'k} \text{ for all } j + 1 \leq j' \leq k - 1$$

and hence

$$\delta_{i,j+1} = \delta_{ik} \text{ for all } i < j \text{ and } \delta_{jl} = \delta_{k-1,l} \text{ for all } l > k$$

as $\triangle$ is strongly Robinson. Moreover,

$$j \overset{3}{\sim} (j+1) \Rightarrow \delta_{ij} = \delta_{i,j+1} \text{ for all } i < j,$$

while

$$(k-1) \overset{3}{\sim} k \Rightarrow \delta_{k-1,l} = \delta_{kl} \text{ for all } l > k.$$

Thus $j \overset{3}{\sim} k$ as required.

(ii) By Lemma 7.4.1(iii), $j \overset{2}{\sim} k$ if and only if either $j \overset{1}{\sim} k$ or both (a^+) and (b^+) above hold. But if $\min(j, k) = 1$, (a^+) can't hold; and if $\max(j, k) = n$, (b^+) is impossible.

(iii) In this case, $j \overset{1}{\sim} k$ implies both (a^+) and (b^+). The logical equivalence in part (iii) is now immediate from Lemma 7.4.1(iii). This equivalence makes it clear that, in this case, $j \overset{3}{\sim} k \Rightarrow j \overset{2}{\sim} k$. ∎

Example 1

The following example illustrates these four equivalence relations and how they differ. We take $n = 9$ and let $\triangle$ be the ultrametric Robinson matrix which, recalling Proposition 7.2.1, is completely defined by:

$$\delta_{12} = \delta_{45} = \delta_{56} = \delta_{89} = 3 \quad ; \quad \delta_{23} = \delta_{34} = \delta_{67} = 1 \quad \text{and} \quad \delta_{78} = 2$$

and is, in particular, quadripolar Robinson. The usual visual display of the associated dendrogram is shown in Figure 7.1.

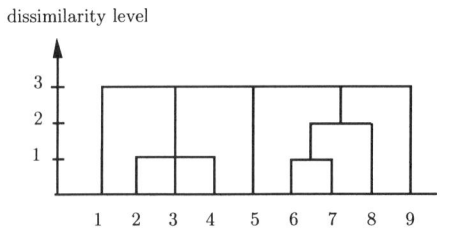

Figure 7.1. Visual display of the matrix of Example 1

The equivalence classes are given in Table 7.1 below.

r	$m(r)$	$E_1^r, E_2^r, \ldots, E_{m(r)}^r$
1	9	$\{i\}, i = 1, \ldots, 9.$
2	3	$\{1\}, \{2,3,\ldots,8\}, \{9\}.$
3	4	$\{1,5,9\}, \{2,3,4\}, \{6,7\}, \{8\}.$
4	6	$\{1\}, \{2,3,4\}, \{5\}, \{6,7\}, \{8\}, \{9\}.$

Table 7.1: Equivalence classes for the matrix of Example 1

Note that the $\overset{r}{\sim}$-classes are connected except when $r = 3$ (Theorem 7.4.1), and internally even except when $r = 2$ (Theorem 7.4.2). The logical relationships in Theorem 7.4.3 are also seen to hold. This example also illustrates Theorem 7.5.1 below. This theorem concerns a property called external evenness which we study next.

7.5. Reduced forms

7.5.1. External evenness

An equivalence relation $\sim$ on I is said to be externally even with respect to a dissimilarity matrix $\triangle$ on I if

$$i \not\sim j \text{ and } j \sim k \text{ imply } \delta_{ij} = \delta_{ik},$$

where $i \not\sim j$ denotes the negation of $i \sim j$. That is, if members of the same equivalence class have the same dissimilarities with respect to members of other classes. Let $\sim$ be such an equivalence relation on I with m equivalence classes labelled $E_1, \ldots, E_m$ and let $\tilde{I} \overset{1-1}{\longleftrightarrow} (1, \ldots, m)$. Then the $\sim$-reduced form of $\triangle$ is defined to be the dissimilarity matrix $\tilde{\triangle} \equiv (\tilde{\delta}_{ij})$ on $\tilde{I}$ given by:

$$\forall i, j \in \tilde{I}, \quad \tilde{\delta}_{ij} = \begin{cases} 0 & \text{if } i = j \\ \delta_{\tilde{i}\tilde{j}} & \text{if } i \neq j \end{cases}$$

where $\tilde{i} \in E_i$ and $\tilde{j} \in E_j$. Observe that $\tilde{\triangle}$ is well-defined precisely because $\sim$ is externally even.

In this section we study the $\overset{r}{\sim}$ − reduced forms of strongly Robinson matrices for each $r = 1, 2, 3, 4$. Consistent with our previous notation, the present symbols $m, E_i, \tilde{I}$ and $\tilde{\triangle}$ become $m(r), E_i^r, \tilde{I}^r$ and $\tilde{\triangle}^r$ respectively. Our first task is to check when $\tilde{\triangle}^r$ is well-defined.

Lemma 7.5.1. *Let $\triangle$ be a quadripolar Robinson matrix on I. Then, for every ordered quadruple $Q \equiv (i, j, k, l)$ of distinct elements of I,*

$$j \overset{2}{\not\sim} k \Rightarrow T_\triangle(Q) = 1.$$

Proof. As $j \overset{2}{\not\sim} k$, either $\delta_{ij} < \delta_{ik}$ or $\delta_{jl} > \delta_{kl}$ or both. The result now follows by Theorem 7.3.1. ∎

Theorem 7.5.1.
(i) *The identity equivalence relation is externally even with respect to any dissimilarity matrix.*
(ii) *For $r = 1, 3$, and 4, $\overset{r}{\sim}$ is externally even with respect to every strongly Robinson matrix on I. This is not so for $r = 2$.*
(iii) *$\overset{2}{\sim}$ is externally even with respect to every quadripolar Robinson matrix on I.*

Proof.
(i) Trivial.

(ii) Let $\triangle$ be a strongly Robinson matrix on I and suppose that $i \overset{r}{\not\sim} j \overset{r}{\sim} k$. For $r = 1$ or 3 it is immediate that $\delta_{ij} = \delta_{ik}$, as required. Consider now $r = 4$. If $j = k$, the result is trivial. Otherwise we may, by symmetry, suppose $j < k$ without loss of generality. As $\overset{4}{\sim}$ is connected (Theorem 7.4.1), $i \overset{4}{\not\sim} j \overset{4}{\sim} k$ implies $i < j$ or $i > k$. By duality, it suffices to consider the case $i < j$. As $\overset{4}{\sim}$ is internally even (Theorem 7.4.2), $\delta_{j,j+1} = \delta_{jk}$ and hence, as $\triangle$ is strongly Robinson, $\delta_{i,j+1} = \delta_{ik}$. But $j \overset{4}{\sim} k \Rightarrow j \overset{3}{\sim} (j+1) \Rightarrow \delta_{ij} = \delta_{i,j+1}$. Thus $\delta_{ij} = \delta_{ik}$ as required. Finally, consider $r = 2$. The dissimilarity matrix defined by $n = 5, \delta_{12} = \delta_{23} = \delta_{34} = 1$, all other above-diagonal elements equal 2 provides the counter-example we need. It is strongly Robinson and $2 \overset{2}{\not\sim} 3 \overset{2}{\sim} 4$ yet $\delta_{23} \neq \delta_{24}$.

(iii) Suppose that $\triangle$ is quadripolar Robinson. Recalling that $\overset{2}{\sim}$ is connected (Theorem 7.4.1), it suffices to consider the case $i < j < k$ by the same arguments as used above in the $r = 4$ case. Moreover, as $\triangle$ is strongly Robinson, it suffices to establish the result in the case where $(j, j+1, \ldots, k)$ is a complete $\overset{2}{\sim}$ class and where $i = j - 1$. As $j \overset{2}{\sim} k, \exists h \leq j$ such that $\delta_{gj} = \delta_{gk}$ for all $g \leq h$. If $h = j$ or $j - 1$, we are finished. If $h < j - 1$ then, as $(j-1) \overset{2}{\not\sim} j$, the quadruple $(h, j-1, j, k)$ of distinct elements of I has type 1 by Lemma 7.5.1. In particular, $\delta_{hk} + \delta_{j-1,j} = \delta_{hj} + \delta_{j-1,k}$ (see Theorem 7.3.1). But $\delta_{hj} = \delta_{hk}$ and so $\delta_{j-1,j} = \delta_{j-1,k}$ as required. ∎

Thus, for $r = 1, 3$ or 4, $\tilde{\triangle}^r$ is well-defined for any strongly Robinson matrix, while $\tilde{\triangle}^2$ is well-defined if either $\overset{2}{\sim}$ is the identity relation or if $\triangle$ is quadripolar Robinson.

7.5.2. Properties of reduced forms

The following result establishes the link between reduced forms and duality.

Proposition 7.5.1. *Let $\triangle$ be a dissimilarity matrix on I and let $\sim$ be an equivalence relation on I externally even with respect to $\triangle$. Then $\sim$ is externally even with respect to the dual matrix $\triangle^*$. Moreover, the $\sim$-reduced form of the dual is the dual of the $\sim$-reduced form. (We may therefore write $\tilde{\triangle}^*$ without confusion).*

Proof. Straightforward. ∎

It is natural to enquire about the properties of $\tilde{\triangle}^r$, particularly those inherited from $\triangle$.

Proposition 7.5.2.

(i) *Let $\triangle$ be a strongly Robinson matrix on I. Then, for $r = 1, 3$ and 4, $\tilde{\triangle}^r$ is strongly Robinson and definite.*

(ii) *Let $\triangle$ be a quadripolar Robinson matrix on I. Then, for each $r = 1, 2, 3, 4$, $\tilde{\triangle}^r$ is quadripolar Robinson and definite.*

Proof. Whenever it is well-defined, $\tilde{\triangle}^r$ is an ordered restriction of $\triangle$ for $r = 1, 2$ or 4 and can be taken to be so for $r = 3$. But, by Proposition 7.2.2, strong Robinsonianity and quadripolar Robinsonianity are ordered restriction invariant. Finally, by Lemma 7.4.1(i) and Theorem 7.4.3(i), $\delta_{ij} = 0 \Rightarrow i \stackrel{r}{\sim} j$ for each $r = 1, 2, 3, 4$. ∎

Let $\sim$ be an equivalence relation on I externally even with respect to a dissimilarity matrix $\triangle$ on I. Then the matrix $\triangle$ is said to be $\sim$-*irreducible* if $\triangle = \tilde{\triangle}$. Trivially, we have:

Proposition 7.5.3. *A dissimilarity matrix $\triangle$ is $\sim$-irreducible if and only if $\sim$ is the identity relation.*

Now Proposition 7.5.2 tells us that, for both strongly Robinson and quadripolar Robinson matrices separately, we may iterate the process of taking the $\stackrel{r}{\sim}$-reduced form. This process must converge to an $\stackrel{r}{\sim}$-irreducible matrix and can do so in at most one iteration. Theorem 7.5.2 and Theorem 7.5.3 spell out the details.

Theorem 7.5.2. *Let $\triangle$ be a strongly Robinson matrix on I and let $r = 1, 3$ or 4. Consider the sequence $(\triangle_t : t \geq 0)$ generated by $\triangle$ and r as follows:*

$$\triangle_0 = \triangle; \quad \triangle_{t+1} = \tilde{\triangle}_t^r \text{ for all } t \geq 0.$$

Then:

(i) *$\triangle_t$ is strongly Robinson and definite for all $t \geq 1$.*

(ii) *$\exists t$ such that $\triangle_u = \triangle_t$ for all $u \geq t$. Moreover, letting $\underline{t}$ denote the least such t, we have $\underline{t} \leq n - 1$.*

(iii) *If $r = 1$, $\underline{t} = 0$ or 1 with $\underline{t} = 0$ if and only if $\triangle$ is definite.*

Proof. (i) This is immediate from Proposition 7.5.2.

(ii) The $n = 1$ dissimilarity matrix is $\sim$-irreducible, for any $\sim$. The result now follows from observing that $\triangle_{t+1} \neq \triangle_t$ implies that $\triangle_{t+1}$ has strictly smaller order than $\triangle_t$.

(iii) By Lemma 7.4.1(i), a strongly Robinson dissimilarity matrix is $\overset{1}{\sim}$-irreducible if and only if it is definite. The result now follows, using part (i). ∎

Lemma 7.5.2. *Let $\triangle$ be a quadripolar Robinson matrix on I. Let j,k be in I with $j \in E_j^2$ and $k \in E_k^2$. Then*

$$j \overset{2}{\not\sim} k \text{ with respect to } \triangle \text{ implies } \tilde{j} \overset{2}{\not\sim} \tilde{k} \text{ with respect to } \tilde{\triangle}^2.$$

Proof. The result follows from the negation of the logical equivalence in Lemma 7.4.1(ii). ∎

Theorem 7.5.3. *Let $\triangle$ be a quadripolar Robinson matrix on I, let $r = 1, 2, 3$ or 4, and consider again the sequence defined by $\triangle_0 = \triangle; \triangle_{t+1} = \tilde{\triangle}_t^r$ for all $t \geq 0$. Then:*

(i) $\triangle_t$ *is quadripolar Robinson and definite for all $t \geq 1$.*

(ii) $\exists$ *a minimal $\underline{t}$ such that $\triangle_u = \triangle_{\underline{t}}$ for all $u \geq \underline{t}$. Moreover, $\underline{t} \leq n - 1$.*

(iii) *If $r = 1$, $\underline{t} = 0$ or 1 with $\underline{t} = 0$ if and only if $\triangle$ is definite.*

(iv) *If $r = 2, \underline{t} = 0$ or 1.*

Proof. (i) to (iii). Identical to the proof of Theorem 7.5.2, recalling that quadripolar Robinson implies strongly Robinson.

(iv) Lemma 7.5.2 implies that $\tilde{\tilde{\triangle}^2}^2 = \tilde{\triangle}^2$. ∎

In the notation of Theorem 7.5.2 and Theorem 7.5.3, the limiting matrix $\triangle_{\underline{t}}$ is called the limiting r-form of $\triangle$ and is denoted $\bar{\triangle}^r$.

Example 1 (continued). We use the ultrametric Robinson matrix of Example 1 to illustrate Theorem 7.5.3. Note that ultrametric Robinsonianity is an ordered restriction invariant property (Proposition 7.2.2). Thus, using Proposition 7.2.1 again, each reduced form is uniquely specified by the equivalence classes on which it is defined, these classes being ordered by their first elements, and the dissimilarities ε_i between the ith and $(i+1)^{\text{st}}$ of them. This latter list of dissimilarities is of course null whenever only a single equivalence class remains. The iterative process of taking the reduced form of the matrix of example 1 is summarised in Table 7.2 below. In this table all columns except the first two relate to *current* values, different iterations occupying different rows of the table.

The limiting r-forms of this matrix $\triangle$ are thus:

$$\bar{\triangle}^1 = \triangle, \bar{\triangle}^2 = \begin{pmatrix} 0 & 3 & 3 \\ & 0 & 3 \\ & & 0 \end{pmatrix}, \bar{\triangle}^3 = \bar{\triangle}^4 = (0).$$

r	t	m(r)	Equivalence classes	$\varepsilon_1,\ldots,\varepsilon_{m(r)-1}$
1	0			
2	1	3	$\{1\},\{2,\ldots,8\},\{9\}$	3, 3.
3		4	$\{1,5,9\},\{2,3,4\},\{6,7\},\{8\}$	3, 3, 2.
		2	$\{1,2,3,4,5,9\},\{6,7,8\}$	3
	3	1	I	
4		6	$\{1\},\{2,3,4\},\{5\},\{6,7\},\{8\},\{9\}$	3, 3, 3, 2, 3
		3	$\{1,2,3,4,5\},\{6,7,8\},\{9\}$	3, 3
	3	1	I	

Table 7.2: Reduced forms for the matrix of Example 1

7.6. Limiting r-forms of strongly Robinson matrices

It is of interest to ask in general what sort of matrices can arise as limiting r-forms. In one sense, the answer is very simple.

Proposition 7.6.1. $\triangle$ is the limiting r-form of a dissimilarity matrix if and only if $\triangle$ is $\overset{r}{\sim}$-irreducible.

Proof. Necessity is obvious. Conversely, if $\triangle$ is $\overset{r}{\sim}$-irreducible, then $\triangle = \bar{\triangle}^r$. ∎

It remains then to characterise the $\overset{r}{\sim}$-irreducible matrices. We do this here for the strongly Robinson case and in Section 7.7 for the quadripolar Robinson case. Let $\triangle$ be a strongly Robinson matrix on I. We say that $\triangle$ is regular if either $n \leq 3$ or $n \geq 4$ and the following condition holds:

for all $2 \leq j \leq n-2$, either $\delta_{j-1,j} < \delta_{j-1,j+1}$ or $\delta_{j,j+2} > \delta_{j+1,j+2}$.

The matrix $\triangle$ is called super-regular if either $n \leq 3$ or $n \geq 4$ and:

for all $2 \leq j \leq n-2$, either $\delta_{ij} < \delta_{i,j+1}$ for all $i < j$

or $\delta_{jk} > \delta_{j+1,k}$ for all $k > j+1$.

We say that $\triangle$ is end-regular if $n \geq 3$ and:

(a) $\delta_{13} > \delta_{23}$ and (a^*) $\delta_{n-2,n-1} < \delta_{n-2,n}$.

Proposition 7.6.2.

(i) $\triangle$ super-regular $\Rightarrow$ $\triangle$ regular $\Rightarrow$ $\delta_{j,j+1} > 0$ for all $2 \leq j \leq n-2$.

(ii) A regular matrix $\triangle$ is definite if and only if $n = 1$ or $n \geq 2$ and $\delta_{12} > 0$ and $\delta_{n-1,n} > 0$.

(iii) $\triangle$ end-regular $\Rightarrow$ $\delta_{12} > 0$ and $\delta_{n-1,n} > 0$.

(iv) $\triangle$ regular and end-regular $\Rightarrow$ $\triangle$ is definite.

Proof. Straightforward. ∎

We are now ready to characterise the limiting r-forms of strongly Robinson matrices.

Theorem 7.6.1. *Let $\triangle$ be a strongly Robinson matrix on I. Then:*

(i) $\triangle$ *is $\overset{1}{\sim}$-irreducible $\Leftrightarrow \triangle$ is definite.*

(ii) $\triangle$ *is $\overset{2}{\sim}$-irreducible $\Leftrightarrow \triangle$ is super-regular and definite.*

(iii) $\triangle$ *is $\overset{4}{\sim}$-irreducible $\Leftrightarrow n = 1$ or $n \geq 3$ and $\triangle$ is regular and end-regular.*

Proof. See Theorem 7.5.2(iii).

(ii) and (iii). The $n = 1$ dissimilarity matrix is $\sim$-irreducible for any $\sim$ and is definite and super-regular by definition. Suppose then that $n \geq 2$. Now

$$\triangle \text{ is } \overset{r}{\sim}\text{-irreducible} \Leftrightarrow j \overset{r}{\not\sim} k \text{ for all } j < k, \quad \text{by Proposition 7.5.3.}$$

$$\Leftrightarrow j \overset{r}{\not\sim} j+1 \text{ for all } j < n, \text{ as } \overset{2}{\sim} \text{ and } \overset{4}{\sim} \text{ are connected.}$$

Continuing first with $r = 2$, we have:

$\triangle$ is $\overset{2}{\sim}$-irreducible $\Leftrightarrow \delta_{12} > 0$ and $\delta_{n-1,n} > 0$ and, if $n \geq 4$, for all $2 \leq j \leq n - 2$:

either $\delta_{ij} < \delta_{i,j+1}$ for all $i \leq j$ or $\delta_{jk} > \delta_{j+1,k}$ for all $k \geq j + 1$,

using Theorem 7.4.3(ii) and Lemma 7.4.1(i),(ii),

$\Leftrightarrow \triangle$ is super-regular and definite,

using Proposition 7.6.2(i) and (ii).

Consider now $r = 4$. When $n = 2$, $1 \overset{4}{\sim} 2$ for any $\triangle$ and so $\triangle$ is not $\overset{4}{\sim}$-irreducible. Suppose then that $n \geq 3$. Continuing the above logical equivalences it is clear, from the definition of $\overset{4}{\sim}$ and the fact that $\triangle$ is strongly Robinson, that

$$\triangle \text{ is } \overset{4}{\sim}\text{-irreducible} \Leftrightarrow \triangle \text{ is regular and end-regular.}$$

This completes the proof. ∎

Proposition 7.5.3 gives a necessary and sufficient condition for a strongly Robinson matrix to be $\overset{3}{\sim}$-irreducible. Of course, this is hardly a neat condition and it appears that no such criterion exists in general. This reflects the fact that $\overset{3}{\sim}$ is not connected. Progress can however be made in particular cases. For instance, it is easy to show that:

Theorem 7.6.2. *Let $\triangle$ be a strongly Robinson matrix on I.*

(i) If $n = 1$, $\triangle$ is $\overset{3}{\sim}$-irreducible.

(ii) If $n = 2$, $\triangle$ is not $\overset{3}{\sim}$-irreducible.

(iii) If $n = 3$, $\triangle$ is $\overset{3}{\sim}$-irreducible if and only if δ_{12}, δ_{23} and δ_{13} are distinct.

Finally we note that, as is obvious directly from Theorem 7.4.3(i):

Theorem 7.6.3. *Let $\triangle$ be a strongly Robinson matrix on I. Then:*

$$\triangle \text{ is } \overset{2}{\sim}\text{-irreducible} \Rightarrow \triangle \text{ is } \overset{1}{\sim}\text{-irreducible}$$
$$\Uparrow$$
$$\triangle \text{ is } \overset{4}{\sim}\text{-irreducible} \Leftarrow \triangle \text{ is } \overset{3}{\sim}\text{-irreducible.}$$

7.7. Limiting r-forms of quadripolar Robinson matrices

In this section we develop the properties of regular quadripolar Robinson matrices and then use them to arrive at characterisations of their limiting r-forms.

When $n \geq 4$, Q_i denotes the consecutive ordered quadruple $(i, i+1, i+2, i+3)$ for $i = 1, \ldots, n-3$. The next result characterises regularity of quadripolar Robinson matrices in terms of the types of the Q_i.

Lemma 7.7.1. *Let $\triangle$ be a quadripolar Robinson matrix on I. Then:*

$$\triangle \text{ is regular} \Leftrightarrow Q_i \text{ has type 1 for all } 1 \leq i \leq n-3.$$

Proof. If $n \leq 3$, there is nothing to prove. If $n \geq 4$, the result is immediate from Theorem 7.3.1. ∎

If $n \geq 2$ and if $\triangle$ is a dissimilarity matrix on I, then for $t = 1, \ldots, n-1$ the tth superdiagonal of $\triangle$ is defined to be the row vector $(\delta_{1,1+t}, \delta_{2,2+t}, \ldots, \delta_{n-t,n})$ and is written δ_t. The next result makes clear a fundamental difference between strongly Robinson matrices in general and the particular case of quadripolar Robinson matrices. All the $n-1$ superdiagonals are in general needed in order to define a regular strongly Robinson matrix, whereas:

Theorem 7.7.1. *A regular quadripolar Robinson matrix $\triangle$ is completely determined by at most its first two superdiagonals. Specifically, when $n \geq 4$, we have the recurrence that for all $t = 3, \ldots, n-1$ and for all $i = 1, \ldots, n-t$:*

$$\delta_{i,i+t} = \delta_{i,i+t-1} + (\eta_{i+t-2} - \varepsilon_{i+t-2})$$
$$= (\eta_i + \eta_{i+1} + \ldots + \eta_{i+t-2}) - (\varepsilon_{i+1} + \ldots + \varepsilon_{i+t-2}),$$

where $\varepsilon \equiv (\varepsilon_1, \ldots, \varepsilon_{n-1}) = \delta_1$ and let $\eta \equiv (\eta_1, \ldots, \eta_{n-2}) = \delta_2$.

Proof. If $n \leq 3$, the result is trivial. Suppose then that $n \geq 4$. Lemma 7.7.1 establishes the recurrence for $t = 3$. Suppose now inductively that the recurrence holds for all $t' < t$ and consider, for any $i = 1, \ldots, n-t$, the quadruple $(i, i+t-2, i+t-1, i+t)$. Then

$$(\delta_{i,i+t-1} - \delta_{i,i+t-2}) + (\delta_{i+t-2,i+t} - \delta_{i+t-1,i+t}) = (\eta_{i+t-3} - \varepsilon_{i+t-3}) + (\eta_{i+t-2} - \varepsilon_{i+t-1})$$

which is strictly positive as Q_{i+t-3} has type 1. Thus $(i, i+t-2, i+t-1, i+t)$ has type 1 and so $\delta_{i,i+t} = \delta_{i,i+t-1} + (\eta_{i+t-2} - \varepsilon_{i+t-2})$ which in turn, using the inductive hypothesis again, is equal to $(\eta_i + \ldots + \eta_{i+t-3} + \eta_{i+t-2}) - (\varepsilon_{i+1} + \ldots + \varepsilon_{i+t-3} + \varepsilon_{i+t-2})$, as required. ∎

It is this theorem which entails that, for quadripolar Robinson matrices, regular implies super-regular. Indeed, we have more. Let $\triangle$ be a quadripolar Robinson matrix on I. Then $\triangle$ is said to be *totally of type 1* if either $n \leq 3$ or $n \geq 4$ and *every* ordered quadruple of distinct elements of I has type 1. We have:

Theorem 7.7.2. *Let $\triangle$ be a quadripolar Robinson matrix on I. Then the following are equivalent:*

(1) $\triangle$ *is regular*

(2) $\triangle$ *is super-regular*

(3) $\triangle$ *is totally of type 1.*

Proof. If $n < 3$, the result is trivial. Suppose then that $n \geq 4$. We prove $(1) \Rightarrow (3) \Rightarrow (2) \Rightarrow (1)$.

$(1) \Rightarrow (3)$:

For $i = 1, \ldots, n-3$ let $\mu_i = (\eta_i - \varepsilon_i) + (\eta_{i+i} - \varepsilon_{i+2})$ in the notation of Theorem 7.7.1. As $\triangle$ is regular, each $\mu_i > 0$. Let now $i < j < k < l$ belong to I. Using the recurrence in Theorem 7.7.1 and simplifying, we find

$$(\delta_{ik} - \delta_{ij}) + (\delta_{jl} - \delta_{kl}) = \sum_{s=j-1}^{k-2} \mu_s.$$

As this is strictly positive (i, j, k, l) has type 1, by Theorem 7.3.1.

$(3) \Rightarrow (2)$:

Let $2 \leq j \leq n-2$. As Q_{j-1} has type 1,

either (a) $\delta_{j-1,j} < \delta_{j-1,j+1}$ or (b) $\delta_{j,j+2} > \delta_{j+1,j+2}$.

If (a) then for any $i < j - 1, \delta_{ij} < \delta_{i,j+1}$ since $(i, j-1, j, j+1)$ has type 1 implies that $\delta_{i,j+1} - \delta_{ij} = \delta_{j-1,j+1} - \delta_{j-1,j}$. The proof that (b) implies $\delta_{jk} > \delta_{j+1,k}$ for all $k > j + 1$ is similar.

$(2) \Rightarrow (1)$:

This is a special case of the first implication in Proposition 7.6.2(i). ∎

We are now ready to characterise the limiting r-forms of quadripolar Robinson matrices. The differences from the strongly Robinson case are clear from comparing the next theorem and its corollary with Theorems 7.6.1 to 7.6.3.

Theorem 7.7.3. *Let $\triangle$ be a quadripolar Robinson matrix on I. Then:*

(i) $\triangle$ *is $\overset{1}{\sim}$-irreducible $\Leftrightarrow \triangle$ is definite.*

(ii) $\triangle$ *is $\overset{2}{\sim}$-irreducible $\Leftrightarrow \triangle$ is regular and definite.*

(iii) $\triangle$ *is $\overset{3}{\sim}$-irreducible $\Leftrightarrow$* $\begin{cases} n = 1 \\ \text{or} \quad n = 3 \text{ and } \delta_{12}, \delta_{23} \text{ and } \delta_{13} \text{ are distinct} \\ \text{or} \quad n \geq 4 \text{ and } \triangle \text{ is } \overset{4}{\sim}\text{-irreducible} \end{cases}$

(iv) $\triangle$ *is $\overset{4}{\sim}$-irreducible $\Leftrightarrow n = 1$ or $n \geq 3$ and $\triangle$ is regular and end-regular.*

Proof. Apart from the $n \geq 4$ case of (iii), the theorem is immediate from Theorems 7.6.1, 7.6.2 and 7.7.2; and in that case the necessity of $\triangle$ being $\overset{4}{\sim}$-irreducible follows

VII. STRONGLY AND QUADRIPOLAR ROBINSON DISSIMILARITY MATRICES   195

from Theorem 7.6.3. Suppose then that $n \geq 4$ and that $\triangle$ is $\overset{4}{\sim}$-irreducible. We have to prove that $j \overset{3}{\not\sim} k$ for all $j < k$ (Proposition 7.5.3).

Now, $1 \overset{4}{\not\sim} 2 \Rightarrow \delta_{12} > 0$ and $\delta_{13} > \delta_{23}$
$\Rightarrow \delta_{1j} > \delta_{2j}$ for all $j \geq 2$, as $\triangle$ is totally of type 1 (Theorem 7.7.2)
$\Rightarrow \delta_{1j} > \delta_{ij}$ for all $2 \leq i \leq n$ and for all $j \geq i$, as $\triangle$ is Robinson
$\Rightarrow \delta_{1n} > \delta_{in}$ for all $2 \leq i \leq n$
$\Rightarrow 1 \overset{3}{\not\sim} i$ for all $2 \leq i \leq n-1$.

Dually, $(n-1) \overset{4}{\not\sim} n \Rightarrow i \overset{3}{\not\sim} n$ for all $2 \leq i \leq n-1$.
Consider now any $2 \leq j \leq n-2$. Then:

$j \overset{4}{\not\sim} (j+1) \Rightarrow$ either (a) $\delta_{j-1,j} < \delta_{j-1,j+1}$ or (b) $\delta_{j,j+2} > \delta_{j+1,j+2}$.

Now,
(a) $\Rightarrow \delta_{ij} < \delta_{i,j+1}$ for all $i \leq j$, as $\triangle$ is totally of type 1
$\Rightarrow \delta_{1j} < \delta_{1,j+1}$
$\Rightarrow \delta_{1k} < \delta_{1,j+1}$ for all $1 \leq k \leq j$, and $\delta_{1j} < \delta_{1l}$ for all $l \geq j+1$, as $\triangle$ is Robinson.
$\Rightarrow k \overset{3}{\not\sim} (j+1)$ for all $2 \leq k \leq j$ and $j \overset{3}{\not\sim} l$ for all $l \geq j+1$.

Dually,
(b) $\Rightarrow k \overset{3}{\not\sim} (j+1)$ for all $1 \leq k \leq j$ and $j \overset{3}{\not\sim} l$ for all $j+1 \leq l \leq n-1$.

Thus, for each $2 \leq j \leq n-2$:

$j \overset{4}{\not\sim} j+1 \Rightarrow j \overset{3}{\not\sim} l$ for all $j+1 \leq l \leq n-1$ and $k \not\sim (j+1)$ for all $2 \leq k \leq j$.

Finally, as $\triangle$ is Robinson:

$1 \overset{3}{\sim} n \Leftrightarrow \delta_{12} = \delta_{13} = \ldots = \delta_{1,n-1} = \delta_{2n} = \delta_{3n} = \ldots = \delta_{n-1,n}$.

But by the above, for any $2 \leq j \leq n-2$, either $\delta_{1j} < \delta_{1,j+1}$ or $\delta_{jn} > \delta_{j+1,n}$. Thus $1 \overset{3}{\not\sim} n$, as required. ∎

Corollary 7.7.1. Let $\triangle$ be a quadripolar Robinson matrix on I. Then:

$\triangle$ is $\overset{3}{\sim}$-irreducible $\Rightarrow \triangle$ is $\overset{4}{\sim}$-irreducible $\Rightarrow \triangle$ is $\overset{2}{\sim}$-irreducible $\Rightarrow \triangle$ is $\overset{1}{\sim}$-irreducible.

When $n \neq 2$ or 3,

$\triangle$ is $\overset{3}{\sim}$-irreducible $\Leftrightarrow \triangle$ is $\overset{4}{\sim}$-irreducible.

Proof. Immediate from Theorems 7.6.3 and 7.7.3. ∎

Example 2. We illustrate Theorem 7.7.3 by giving examples of quadripolar Robinson matrices that are $\overset{r}{\sim}$-irreducible. When $r = 1$ these are just the definite matrices. For $r > 1$, we consider first the special case of ultrametric Robinson matrices. Table 7.3 lists some $\overset{2}{\sim}$-irreducible ultrametric Robinson matrices which are again specified in terms of their first super-diagonal. The first of these is the limiting 2-form of the matrix of Example 1. Figure 7.2 shows the usual visual displays of the associated dendrograms.

Label	n	$\varepsilon_1, \ldots, \varepsilon_{n-1}$
A	3	3, 3
B	6	1, 1.5, 2, 2.5, 3
C	6	1, 2, 3, 2, 1

Table 7.3. Examples of $\overset{2}{\sim}$-irreducible ultrametric Robinson matrices

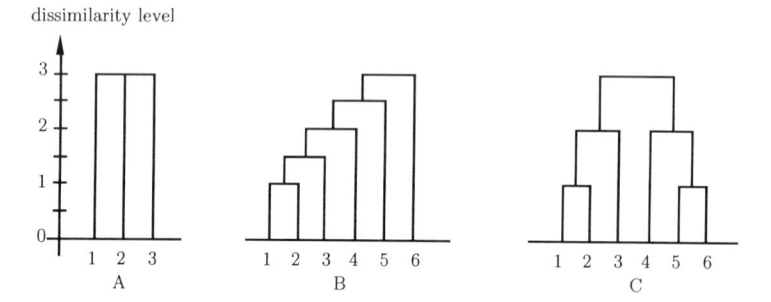

Figure 7.2. Visual displays of the 3 examples of Table 7.3

The ultrametric Robinson matrix of Example 1 has $\bar{\Delta}^3 = \bar{\Delta}^4 = (0)$. This is true generally.

Theorem 7.7.4. *Let Δ be an ultrametric Robinson matrix on I. Then:*

$$\Delta \text{ is } \overset{3}{\sim}-\text{irreducible} \Leftrightarrow \Delta \text{ is } \overset{4}{\sim}-\text{irreducible} \Leftrightarrow n = 1.$$

Proof. By Theorem 7.7.3 and Corollary 7.7.1 it suffices to prove that $n > 1$ implies Δ is not $\overset{4}{\sim}$-irreducible. Suppose then that $n \geq 2$ and let $g < h$ be such that:

$$\delta_{gh} = \min\{\delta_{ij} \mid i < j\}.$$

Then $\delta_{gi} = \delta_{ih} = \delta_{gh}$ for any $g < i < h$, as Δ is ultrametric Robinson. Thus $g \overset{4}{\sim} h$. ∎

The main thrust of this theorem is a positive one. Consider again the summarisation process described at the outset of this paper (see (I) and (II) of Section 7.1.3). When $r = 1$ or 2 we have seen that this process can terminate at a nontrivial matrix. However,

Theorem 7.7.4 tells us that that process goes the whole way if either $\overset{3}{\sim}$ or $\overset{4}{\sim}$ is used on an ultrametric Robinson matrix. Of these $\overset{4}{\sim}$ is to be preferred, as it is connected.

For nontrivial limiting 3 - or 4 - forms we need then to look more widely. Recalling Proposition 7.6.2 and Theorem 7.7.3 it will be enough, apart from details at the ends or when $n = 3$, to describe regular quadripolar Robinson matrices of order at least four. It follows from Theorem 7.3.1 and 7.7.2 that the topology of the tree structure underlying such a matrix is as shown in Figure 7.3.

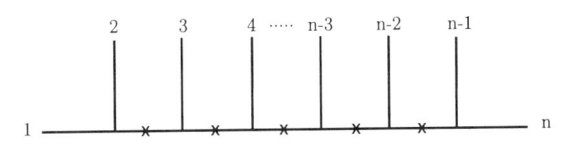

Figure 7.3. Topology of the tree structure of a regular quadripolar Robinson matrix of order at least 4
Note. The edges marked with a cross are those whose lengths cannot vanish

These examples complete our study of the exchangeability-based equivalence relations $\overset{r}{\sim}$, ($r = 1, 2, 3, 4$). Quadripolar Robinson matrices are studied further in Critchley (1988c, 1994) where both local and global characterisations are obtained.

References

Batbedat, A. (1989), Les dissimilarités médas ou arbas, *Statist. Anal. Données*, 14, pp. 1–18.

Batbedat, A. (1991), Phylogénie et dendrogrammes, Journées de Statistique, Strasbourg, France.

Batbedat, A. (1992), Les distances quadrangulaires qui ont une orientation, *RAIRO-Rech. Opér.*, 26, pp. 15–29.

Bertrand, P., Diday, E. (1985), A visual representation of the compatibility between an order and a dissimilarity index: the pyramids, *Comput. Statist. Quart.*, 2, pp. 31–42.

Blumenthal, L.M. (1953), *Theory and Applications of Distance Geometry*, Clarendon Press, Oxford.

Bock, H.H., ed. (1988), *Classification and Related Methods of Data Analysis*, North-Holland, Amsterdam.

Buneman, P. (1971), The recovery of trees from measures of dissimilarity, In Hodson, F.R., Kendall, D.G., Taŭtu, P., eds., *Mathematics in the Archaeological and Historical Sciences*, The University Press, Edinburgh, pp. 387–395.

Critchley, F. (1988a), On certain linear mappings between inner-product and squared-distance matrices, *Linear Algebra Appl.*, 150, pp. 91–107.

Critchley, F. (1988b), On exchangeability-based equivalence relations induced by strongly Robinson and, in particular, by quadripolar Robinson dissimilarity matrices, Warwick Statistics Research Report n° 152, U.K.

Critchley, F. (1988c), On quadripolar Robinson dissimilarity matrices, Warwick Statistics Research Report n° 153, U.K., To appear in Diday, E., et al., eds (1994), *Proc. IFCS93 meeting*, Springer-Verlag, New York.

Critchley, F., Fichet, B. (1994), The partial order by inclusion of the principal classes of dissimilarity on a finite set, and some of their basic properties, In Van Cutsem, B., ed., *Classification and Dissimilarity Analysis*, Lecture Notes in Statistics, Ch. 2, Springer-Verlag, New York.

de Leeuw, J., Heiser, W. (1982), Theory of multidimensional scaling, In Krishnaiah, P.R., ed., *Handbook of Statistics, Vol. 2*, North-Holland, Amsterdam, Ch. 13.

de Leeuw, J., Heiser, W., Meulman, J., Critchley, F., eds. (1986), *Multidimensional Data Analysis*, DSWO Press, Leiden.

Diday, E. (1984), Une représentation visuelle des classes empiétantes: les pyramides, Rapport de Recherche n° 291, INRIA, Rocquencourt, France.

Diday, E. (1986), Orders and overlapping clusters in pyramids, In de Leeuw, J., Heiser, W., Meulman, J., Critchley, F., eds., *Multidimensional Data Analysis*, DSWO Press, Leiden, pp. 201–234.

Diday, E., Escoufier, Y., Lebart, L., Pagès, J.P., Schektman, Y., Tomassone, R., eds. (1986), *Data Analysis and Informatics 4*, North-Holland, Amsterdam.

Durand, C. (1988), Une approximation de Robinson inférieure maximale, Rapport de Recherche Laboratoire de Mathématiques Appliquées et Informatique, n° 88-02, Université de Provence, Marseille, France.

Durand, C., Fichet, B. (1988), One-to-one correspondences in pyramidal representation: a unified approach, In Bock, H.H., ed., *Classification and Related Methods of Data Analysis*, North-Holland, Amsterdam, pp. 85–90.

Fichet, B. (1984), Sur une extension de la notion de hiérarchie et son équivalence avec certaines matrices de Robinson, Journées de Statistique, Montpellier, France.

Gordon, A.D. (1987), A review of hierarchical classification, *J. Roy. Statist. Soc. A*, 150, pp. 119–137.

Gower, J.C. (1985), Properties of Euclidean and non-Euclidean distance matrices, *Linear Algebra Appl.*, 67, pp. 81–97.

Hayden, T.L., Wells, J. (1988), Approximation by matrices positive semidefinite on a subspace, *Linear Algebra Appl.*, 109, pp. 115–130.

Johnson, S.C. (1967), Hierarchical clustering schemes, *Psychometrika*, 32, pp. 241–253.

Leclerc, B. (1993), Minimum spanning trees for tree metrics: abridgements and adjustments, Research Report C.M.S. P.084, Centre d'Analyse et de Mathématique Sociales, Paris, France.

Mathar, R. (1985), The best Euclidean fit to a given distance matrix in prescribed dimensions, *Linear Algebra Appl.*, 67, pp. 1–6.

Robinson, W.S. (1951), A method for chronological ordering of archaeological deposits, *American Antiquity*, 16, pp. 293–301.

Chapter 8.
Dimensionality problems in L_1-norm representations[*]

Bernard Fichet[†]

8.1. Introduction

The use of the L_1-norm via the least absolute deviations appeared early in the field of statistics. Generally people mention Boscovich (1757) and Laplace (1793), and for further historical aspects the reader will may consult Farebrother (1987). However, many periods of silence followed those pioneering works. There are many reasons for this lack of development. In particular, some computational difficulties arose in the city-block approaches. During the last two decades, we note a growing interest in statistical methods based upon the L_1-norm. This phenomenon concerns many areas such as robustness, nonparametric analysis, multidimensional scaling. In this last field, dimensionality problems are crucial. This chapter is devoted to theoretical and computational results on the dimension of an L_1-figure.

In Section 8.2, we give some preliminaries and some notations strongly connected with the notations of Critchley and Fichet (1994). In Section 8.3, we deal with general results on distances of L_p-type. We recall an upperbound for the dimensionality and prove that the cone of k-dimensional distances of L_p-type is closed. In addition, embeddability of an ultrametric space in a Euclidean space is extended to embeddability in an L_p-space. We focus on the dimension of L_1-figures in Section 8.4. An upperbound is established for the embedding of tree metric spaces. Furthermore we exhibit an n-point figure which necessitates more than n axes. Computational aspects are investigated in Section 8.5. In particular, we pay attention to some approximations of a data dissimilarity by a semi-distance of L_1-type.

Most of those results have been either published by the author (1986, 1987, 1988) or announced in different meetings such as the International Statistical Institute session held in Paris (1989) and the conference of the International Federation of Classification Societies held in Edinburgh (1991). Note that Ball (1990) discovered independently some of those results and gave a fundamental lower bound for the dimensionality, which deeply extends our above-mentioned example. References will be done throughout the chapter.

---

[*] *In* Van Cutsem, B. (Ed.), (1994) *Classification and Dissimilarity Analysis*, Lecture Notes in Statistics, Springer-Verlag, New York.
[†] Université d'Aix-Marseille II, France.

8.2. Preliminaries and notations

8.2.1. Dissimilarities

In the sequel, we will use some basic tools and some notations which have been introduced in Critchley and Fichet (1994) of this volume. So, we will be concerned with a finite set I of cardinality $n > 1$ and, for numerical investigations, I will be very often identified with $\{1, \ldots, n\}$. Denote by $\mathcal{D}$ the set of real functions d mapping I^2 into $\mathbb{R}$ that are symmetric and vanish on the diagonal. Clearly $\mathcal{D}$ is a real vector space of dimension $\frac{n(n-1)}{2}$.

A canonical basis is given by all $d^{\{i,j\}}$ ($i \neq j$) defined by

$$\forall (k,l) \in I^2,\ d^{\{i,j\}}(k,l) = \begin{cases} 1 & \text{if } \{k,l\} = \{i,j\} \\ 0 & \text{else.} \end{cases}$$

For topological arguments, $\mathcal{D}$ will be supposed to be endowed with any (given) norm.

A dissimilarity is a nonnegative element of $\mathcal{D}$ ($d(i,j) \geq 0,\ \forall\, i \neq j$). We denote by $\mathcal{D}_+$ the set of all dissimilarities.

8.2.2. Some notations

We recall here some usual definitions for particular dissimilarities and precise the notations for the sets of such dissimilarities.

A semi-distance is a dissimilarity obeying the well-known triangle inequality $\forall (i,j,k) \in I^3,\ d(i,j) \leq d(i,k) + d(k,j)$. The corresponding set is denoted by $\mathcal{D}_\infty$.

An ultrametric is a dissimilarity obeying the ultrametric inequality $\forall (i,j,k) \in I^3,\ d(i,j) \leq \max[d(i,k), d(j,k)]$. The corresponding set is denoted by $\mathcal{D}_u$.

A tree semi-distance d is such that (I, d) is isometrically embeddable in a weighted tree endowed with the usual induced distance. A particular case is given by chain semi-distances for which the tree has only two terminal vertices. The sets of tree and chain semi-distances are denoted $\mathcal{D}_t$ and $\mathcal{D}_{ch}$, respectively.

A semi-distance d is said to be of ℓ_p^N-type ($1 \leq p < \infty$) if and only if (I, d) is isometrically embeddable in ℓ_p^N, i.e. iff there exist real numbers x_{ik}, $i \in I$, $k = 1, \ldots, N$, such that

$$\forall\, (i,j) \in I^2,\ d^p(k,l) = \sum_{k=1}^{N} |x_{ik} - x_{jk}|^p.$$

It is said to be of L_p-type if it is of ℓ_p^N-type for some N. The sets of semi-distances of ℓ_p^N-type and L_p-type are denoted $\mathcal{D}_{p(N)}$ and $\mathcal{D}_p$, respectively. Thus $\mathcal{D}_{p(N)}$ is increasing in N and $\mathcal{D}_p = \cup_N \mathcal{D}_{p(N)}$.

Observe that we have not to investigate the semi-distances of L_∞-type. Indeed, according to a nice result of Frechet (1910), every finite metric space (I, d) is embeddable in ℓ_∞^{n-1}. See for example, Critchley and Fichet (1994). Note that this result justifies our notation $\mathcal{D}_\infty$ for the set of semi-distances.

For any $d \in \mathcal{D}_+$ and for any $\alpha > 0$, we define d^α by $d^\alpha(i,j) = (d(i,j))^\alpha$, with the convention $0^\alpha = 0$. Again, for any subset $\mathcal{D}_* \subseteq \mathcal{D}_+$, $\mathcal{D}_*^\alpha$ stands for $\{d^\alpha \mid d \in \mathcal{D}_*\}$. In particular, for every $p > 1$, we will take an interest in the set $\mathcal{D}_p^p$, set of dissimilarities with a pth-root of L_p-type. For $p = 2$, they correspond to quasi-hypermetrics explored in Critchley and Fichet (1994).

8.2.3. Some characterisations

General conditions for embeddability of a normed space or a metric space in some $L_p(E, \mu)$ have been established by Bretagnolle, Dacunha-Castelle and Krivine (1966). However, except for $p = 2$ or $p = 1$, to our knowledge there does not exist any numerical procedure to establish whether or not a finite metric space is of L_p-type.

Semi-distances of L_2-type are also called Euclidean and $\mathcal{D}_2$ is also denoted $\mathcal{D}_e$. Following the work of Schœnberg (1935), their characterisation is amply discussed in Critchley and Fichet (1994).

(Semi)-distances of L_1-type have been advocated under a variety of names, such as city-block or Hamming (semi)-distances. They have been explored in different areas by many authors. After the pioneering work of Deza (1960) in connection with coding theory and embeddability on a hypercube, one may mention Bretagnolle, Dacunha-Castelle and Krivine (1966), Assouad (1977), Avis (1977), Assouad and Deza (1982), Fichet (1986, 1987), Le Calvé (1987) until the paper of Ball (1990) previously cited. The main characterisation of $\mathcal{D}_1$ is expressed in terms of dichotomies. Recall that a dichotomy is associated with a nontrivial partition (J, J^c) of I. For every J ($\emptyset \subset J \subset I$), the dichotomy d_J is defined by

$$d_J(i,j) = \begin{cases} 1 & \text{if } i \in J, \ j \in J^c \text{ or } i \in J^c, \ j \in J, \\ 0 & \text{else.} \end{cases}$$

Then $\mathcal{D}_1$ appears to be the conic-convex hull of dichotomies. Moreover, each ray spanned by a dichotomy is an extreme ray of $\mathcal{D}_\infty$, hence of $\mathcal{D}_1$. See, for example, Critchley and Fichet (1994). Thus $\mathcal{D}_1$ is a closed convex polyhedral cone.

We end this section with the following remark. Let $d \in \mathcal{D}_1$. Then, there are positive coefficients α_k and dichotomies d_{J_k} such that $d = \sum_k \alpha_k d_{J_k}$. Putting $x_{ik} = 1$ (resp. 0) iff $i \in J_k$ (resp. $i \notin J_k$), we get an equivalent definition of semi-distances of L_1-type: $d \in \mathcal{D}_1$ iff there exist an integer K, positive real numbers α_k, $k = 1, \ldots, K$ and real numbers x_{ik}, $i \in I$, $k = 1, \ldots, K$ equalling 0 or 1 such that

$$\forall (i,j) \in I^2, \quad d(i,j) = \sum_k \alpha_k |x_{ik} - x_{jk}|.$$

Consequently, for every $p \geq 1$, $d^{1/p}$ is of L_p-type provide $d \in \mathcal{D}_1$, i.e. $\mathcal{D}_1 \subseteq \mathcal{D}_p^p$. In particular, $(I, \sqrt{d})$ is embeddable on the vertices of a rectangular parallelepiped.

8.3. Dimensionality for semi-distances of L_p-type

First we give two simple examples. They show that some usual dimensionality properties of Euclidean geometry are generally false with the L_p-norm, $p \neq 2$. Two isometrical figures may generate spaces of different dimensions. Consider the following L_1-norm figures in $\mathbb{R}$ and $\mathbb{R}^2$, respectively

$$x = 0,\ y = 1,\ z = 2 \quad ; \quad x' = (0,0),\ y' = (1,0),\ z' = (1,1).$$

A figure in $\mathbb{R}^N$ generating a subspace of dimension $N' < N$, does not admit necessarily an isometrical figure in $\mathbb{R}^{N'}$. We are indebted to Carole Durand for her helpful assistance in constructing the following data. In $\mathbb{R}^3$, endowed with the L_1-norm, let us consider the points

$$O = (0,0,0)\ ;\ A = (2,-2,-4)\ ;\ B = (4,2,-2)\ ;\ C = (3,5,2)\ ;\ D = (1,4,3).$$

They are in a plane of equation: $x - y + z = 0$. An easy calculation shows that for every triangle, the three points are not metrically aligned. Consequently there does not exist any isometrical figure in $\mathbb{R}^2$, as shown by the following lemma.

Lemma 8.3.1. *Among five points of $\mathbb{R}^2$ endowed with the L_1-norm, are three distinct points metrically aligned.*

Proof. Let A, B, C, D, E be five points in $\mathbb{R}^2$. Denote (x_M, y_M) the coordinates of a point M. Without loss of generality, we may suppose

$$x_A \leq x_B \leq x_C \leq x_D \leq x_E \quad ; \quad y_A \leq y_B.$$

Now, assume that for any triangle the vertices are not metrically aligned. Considering the triangles ABC, ABD and BCD successively, we have $y_C < y_D < y_B$. Consequently, from the triangle BDE we have $y_D < y_E$. But the triangle CDE gives $y_E < y_D$. Contradiction. ∎

The preceding remarks lead us to give the following definition.

Definition 8.3.1. For every $p \geq 1$ and for any $d \in \mathcal{D}_p$, the (L_p)-dimension of (I,d) or the (L_p)-rank of d is the smallest N such that $d \in \mathcal{D}_{p(N)}$. □

Clearly an L_2-dimension is at most $(n-1)$ and using Frechet's theorem the same upperbound holds for an L_∞-dimension. From the definition of a distance of L_1-type, we get an easy upperbound for an L_1-dimension. Indeed two or more axes with coordinates arranged in a common rank order (up to the converse order) may be contracted to a single axis. Thus an upperbound is $n!/2$. The characterisation of $\mathcal{D}_1$ in terms of dichotomies gives another upperbound $(2^{(n-1)} - 1)$. In fact, as proved by Fichet (1988) and Ball (1990), a smaller upperbound, valid for every p, may be given. Note that it is not immediate that an upperbound is finite for a fixed p. We have the following proposition.

Proposition 8.3.1. Let $p \geq 1$ and let $d \in \mathcal{D}_p$. Then the dimension of (I,d) is at most $r = n(n-1)/2$, so that $\mathcal{D}_p = \mathcal{D}_{p(r)}$.

Proof. Clearly $\mathcal{D}_p^p$ is the convex hull of $\mathcal{D}_{ch}^p$, i.e. $\mathcal{D}_p^p = \text{conv}(\mathcal{D}_{ch}^p)$. Then we may apply Caratheodory's theorem (see, for example, Grünbaum (1967, p 15)). Intersecting D_p^p with the affine hyperplane passing through the basic vectors gives the result. ∎

Now, we give some topological results. The cone $\mathcal{D}_p$ is closed as proved by Fichet (1986, 1988) and rediscovered by Ball (1990). The result is extended by Benayade and Fichet (1993) with a similar proof. This is the next proposition.

Proposition 8.3.2. For every $1 \leq p < \infty$ and every N, $\mathcal{D}_{p(N)}$ is closed.

Proof. Clearly $\mathcal{D}_{p(N)}^p$ is a cone whose vectors are the sum of at most N vectors of $\mathcal{D}_{ch}^p$. Let S be the unit sphere with the centre at the origin. Define the set A as the set of vectors which are a convex linear combination of at most N vectors of $\mathcal{D}_{ch}^p \cap S$

$$d \in A \iff d = \sum_{k=1}^{N} \lambda_k d_k \, , \, d_k \in \mathcal{D}_{ch}^p \cap S \, , \, 0 \leq \lambda_k \leq 1 \, , \, \sum_{k=1}^{N} \lambda_k = 1.$$

Some easy geometrical arguments show that $\mathcal{D}_{p(N)}^p$ is the conic hull of A, i.e. $\mathcal{D}_{p(N)}^p = C(A)$. Clearly, $\mathcal{D}_{ch}$ is closed in $\mathcal{D}$, hence in $\mathcal{D}_+$. See, for example Critchley and Fichet (1994). Consequently $\mathcal{D}_{ch}^p$ is closed in $\mathcal{D}_+$, hence in $\mathcal{D}$. Thus $\mathcal{D}_{ch}^p \cap S$ is compact. Then some easy arguments show that A is compact. Since the origin is not in A, some similar arguments show that $C(A)$ is closed in $\mathcal{D}$, hence in $\mathcal{D}_+$. It follows that $\mathcal{D}_{p(N)}$ is closed in $\mathcal{D}_+$, hence in $\mathcal{D}$. ∎

Proposition 8.3.1 and Proposition 8.3.2. give the following corollary.

Corollary 8.3.1. For every $p \geq 1$, $\mathcal{D}_p$ is closed.

A priori, embeddability in ℓ_p or in $L_p(\Omega, \mathcal{F}, P)$ defines an extension of semi-distances of L_p-type. For finite metric spaces, this is not true as observed by Le Calvé (1987), Fichet (1988) or Ball (1990). We have the following proposition.

Proposition 8.3.3. Let (I,d) be a finite metric space. Then for every $p \geq 1$ the following three properties are equivalent.

i) (I,d) is isometrically embeddable in ℓ_p^N for some N,

ii) (I,d) is isometrically embeddable in ℓ_p,

iii) (I,d) is isometrically embeddable in $L_p(\Omega, \mathcal{F}, P)$ for some probability space $(\Omega, \mathcal{F}, P)$.

Proof. i) $\Rightarrow$ ii) obvious.

ii) $\Rightarrow$ i) obvious, by Corollary 8.3.1.

i) $\Rightarrow$ iii) For every $i \in I$, let x_{ik}, $k = 1, \ldots, N$ be the coordinates of the corresponding point in $\mathbb{R}^N$. Choose $\Omega = (0, N)$ endowed with its Borel subsets. Define P as the uniform probability and for every $i \in I$, consider the random variable X_i such that $X_i(x) = N^{1/p} x_{ik}$ whenever $x \in (k-1, k)$.

iii) $\Rightarrow$ i) Denote X_i, $i \in I$ the random variables of $L_p(\Omega, \mathcal{F}, P)$, more simply noted L_p, such that $\forall\, i, j \in I$, $(\mathbb{E}|X_i - X_j|^p)^{1/p} = d(i,j)$. For every $i \in I$, we define $X_i^+ = \sup(X_i, 0)$; $X_i^- = \sup(-X_i, 0)$. Then for every $i \in I$, there are two increasing sequences of discrete nonnegative random variables X_{im}^+ and X_{im}^- converging to X_i^+ and X_i^-, respectively. Putting $X_{im} = X_{im}^+ - X_{im}^-$, the sequence X_{im} converges to X_i and $|X_{im}| \leq |X_i|$ for every $i \in I$ and every m. Since I is finite, for every m there exists a measurable partition $(A_1, \ldots, A_{L_m})$ of Ω such that, for every $i \in I$ and every k, X_{im} is constant on A_k, i.e. $X_{im} = x_{ik}$ (say) on A_k. Then we define the dissimilarity d_m by

$$\forall\, (i,j) \in I^2, \quad d_m(i,j) = \left(\sum_k \alpha_k |x_{ik} - x_{jk}|^p \right)^{1/p}, \text{ where } \alpha_k = P[A_k].$$

Clearly $d_m \in \mathcal{D}_p$ and

$$\forall\, (i,j) \in I^2, \quad d_m(i,j) = (\mathbb{E}|X_{im} - X_{jm}|^p)^{1/p}.$$

Since $|X_i| + |X_j|$ is in L_p and $|X_{im} - X_{jm}| \leq |X_i| + |X_j|$, by Fatou's lemma $d_m(i,j)$ tends to $d(i,j)$. With Corollary 8.3.1, the proof is complete. ∎

Note that the following less restrictive condition may be given in ii): the embedding $\{x_i,\ i \in I\}$ is such that $(x_i - x_j)$ is in ℓ_p for every $(i,j) \in I^2$. We cannot do that for iii).

Proposition 8.3.4.
$$\bigcup_{1 \leq p < \infty} \mathcal{D}_{p(n-1)} = \bigcup_{1 \leq p < \infty} \mathcal{D}_p = \mathcal{D}_\infty.$$

Proof. Let $d \in \mathcal{D}_\infty$. By Frechet's theorem, there are x_{ik}, $i \in I$, $k = 1, \ldots, (n-1)$ such that $\forall\, (i,j) \in I^2$, $d(i,j) = \max_k |x_{ik} - x_{jk}|$. Defining, for every $p \geq 1$, the semi-distance d_p by $d_p(i,j) = (\sum_k |x_{ik} - x_{jk}|^p)^{1/p}$ for every pair (i,j) yields the result by an obvious limit and Proposition 8.3.2. ∎

Remark 8.3.1. It is well-known that, for $n \leq 4$, every semi-distance is of L_1-type so that $\bigcup_{1 \leq p < \infty} \mathcal{D}_p = \mathcal{D}_\infty$. See for example, Critchley and Fichet (1994). But for $n \geq 5$, $\bigcup_{1 \leq p < \infty} \mathcal{D}_p \subset \mathcal{D}_\infty$. That proves in passing that $\bigcup_{1 \leq p < \infty} \mathcal{D}_p$ is not closed. The following example shows the strict inclusion. Consider the distance d defined on $I = \{1, 2, 3, 4, 5\}$ by

$$d(1,2) = d(2,3) = d(3,4) = d(4,1) = 1,$$
$$d(1,3) = d(2,4) = 2,$$
$$d(1,5) = d(2,5) = d(3,5) = 1,$$
$$d(4,5) = 2.$$

This is a distance introduced by Fichet and Le Calvé in order to prove the existence of a distance with a non-Euclidean square root. See Joly and Le Calvé (1986). Indeed, some geometrical arguments show that $\sqrt{d} \notin \mathcal{D}_e$, hence $d \notin \mathcal{D}_1$. But the restriction of d to $\{1,2,3,4\}$ is not of L_p-type for every $1 < p < \infty$. See Appendix 1. Thus $d \notin \bigcup_{1 \leq p < \infty} \mathcal{D}_p$.

Bretagnolle, Dacunha-Castelle and Krivine (1966) proved that every L_p-space embeds isometrically in an L_1-space, whenever $1 \leq p \leq 2$. However ℓ_p^3 does not embed isometrically in an L_1-space, for $p > 2$. See Dor (1976), following a result of Witsenhausen (1973) for $p > 2.7$. Note that Corollary 8.3.1, Proposition 8.3.4 and the previous distance yield an easy example showing that $\mathcal{D}_p \not\subseteq \mathcal{D}_1$ or even $\mathcal{D}_p \not\subseteq \mathcal{D}_q$ q fixed, for some p sufficiently large. □

A famous theorem due to Holman (1972) stipulates that every ultrametric is Euclidean. This result has been extended by Fichet (1986, 1988). This is the following proposition.

Proposition 8.3.5. *For any $p \geq 1$, every ultrametric is of L_p-type, i.e. $\mathcal{D}_u \subseteq \mathcal{D}_p$.*

Proof. It is well-known that $\mathcal{D}_u \subseteq \mathcal{D}_1$. See, for example, Critchley and Fichet (1994). Therefore: $\mathcal{D}_u \subseteq \mathcal{D}_p^p$. But for every ultrametric d, d^p is also ultrametric. Thus $d \in \mathcal{D}_p$. ■

Finally, we note that every metric triangle is embeddable in ℓ_p^2. See Appendix 2.

8.4. Dimensionality for semi-distances of L_1-type

Recall the geometrical upperbound given in the previous section for the dimensionality $\frac{n(n-1)}{2}$. For semi-distances of L_1-type, we have a smaller upperbound (note the improvement !).

Proposition 8.4.1. *Let $d \in \mathcal{D}_1$. Then the dimension of (I, d) is at most $\frac{n(n-1)}{2} - 1$.*

Proof. First, suppose that d is in the boundary of $\mathcal{D}_1$. Recall that $\mathcal{D}_1 = \text{conv}(\mathcal{D}_{ch})$. Then, intesecting $\mathcal{D}_1$ with the affine hyperplane passing through the basic vectors, an extension of Caratheodory's theorem shows that d is the sum of at most $\frac{n(n-1)}{2} - 1$ vectors of $\mathcal{D}_{ch}$. See, for example, Blackwell and Girshick (1966, p. 36). Another proof can be obtained by solving via the simplex method, the additive constant problem applied to $(d - c_0 d_1)$, where $c_0 > 0$ is fixed and $d_1(i,j) = 1$ whenever $i \neq j$. See the next section.

Now, let $d \in \overset{\circ}{\mathcal{D}}_1$. Let c_1 be the smallest nonnegative constant such that $(d - c_1 d_1) \in \mathcal{D}_1$. Since $\mathcal{D}_1$ is closed, c_1 exists, $c_1 > 0$ and $(d - c_1 d_1)$ is in the boundary of $\mathcal{D}_1$. Thus, d which is equal to $(d - c_1 d_1) + c_1 d_1$, can be written as

$$d = \sum_{k=1}^{K} \alpha_k d_{J_k} + \frac{c_1}{2} \sum_i d_{\{i\}}, \quad K \leq \frac{n(n-1)}{2} - 1, \; \alpha_k > 0, \; \forall\, k = 1, \ldots, K.$$

It is obvious that any 1-dichotomy $d_{\{i\}}$ can be associated with any dichotomy d_{J_k} to define one axis. Thus the proof is complete. ■

An ultrametric distance is both Euclidean and of L_1-type. Its L_2-rank is proved to be $(n-1)$. See, for example, Critchley and Fichet (1994). The L_1-rank is smaller. This is the following proposition.

Proposition 8.4.2. *Any tree semi-distance, henceforth any ultrametric, has an L_1-rank at most $\frac{(n+1)}{2}$ or $\frac{n}{2}$ according as n is odd or even.*

For the proof we shall need the following lemma.

Lemma 8.4.1. *Let $T = (X, E)$ be a tree with $2p$ terminal vertices. Then the edges are covered by p chains (paths).*

Proof. We proceed by induction. The result is trivially true for $p = 1$. Now, suppose $p > 1$. For every terminal vertex x, we note x' the first vertex whose valency (degree) is greater than two along the chain starting from x. Now, choose a vertex v of valency $r > 2$. If we suppress the edges associated with v, we obtain r connected components $C_1, \ldots, C_r$. Let x_1, x_2 be two terminal vertices of T which are in C_1 and C_2, respectively. If $r \geq 4$ or if $x_1' = x_2' = v$ does not hold, we obtain a subtree with $2(p-1)$ terminal vertices by suppressing the vertices (different from v) between x_1 and x_1' and between x_2 and x_2'. Moreover, the chain linking x_1 and x_2 covers the cancelled edges. By the inductive hypothesis, the result follows.

Now, suppose $r = 3$, $x_1' = x_2' = v$. Let x_3 be a terminal vertex of T in C_3. Since the number of terminal vertices of T is even, necessarily $x_3' \neq v$. Then, considering the chain between x_1 and x_3, we can proceed as previously. This completes the proof. ∎

Proof of Proposition 8.4.2. For a given element d in $\mathcal{D}_t$, let $\{x_i, i \in I\}$ be an embedding of (I, d) in some weighted tree (T, w), with $T = (X, E)$. Without loss of generality, we may suppose that every terminal vertex of T is one of the x_i, $i \in I$. Thus T has at most n terminal vertices. A dichotomy decomposition of d can be exhibited from the edges of T. Indeed, if we suppress an edge $e \in E$ in the tree, we obtain two connected components defining a bipartition of X and an induced bipartition (J_e, J_e^c) of I. From the condition imposed on the tree, we have $\emptyset \subset J_e \subset I$. Then d is shown to obey $d = \sum_{e \in E} w(e) d_{J_e}$. See, for example, Critchley and Fichet (1994). Furthermore, it is obvious that the dichotomies associated with the edges of a given chain, define a 1-dimensional semi-distance of L_1-type. Adding in case of necessity a new terminal vertex to the tree, we can apply the lemma and the proof is complete. ∎

The following lemma gives a lower bound for the dimensionality.

Lemma 8.4.2. *Assume that $d \in \mathcal{D}_1$ has a single dichotomy decomposition $d = \sum_{k=1}^K \alpha_k d_{J_k}$, $\alpha_k > 0$. Suppose that the first m dichotomies obey the following two properties.*

i) $J_1, \ldots, J_m$ *have the same cardinality,*

ii) $\exists i \in I : \forall k = 1, \ldots, m, \; i \in J_k$.

Then the dimension of (I, d) is at least m.

Proof. Consider an embedding of (I, d) in some ℓ_p^N. Each axis defines a chain semi-distance which is a positive linear combination of some d_{J_k}, $k = 1, \ldots, K$. Conversely, any d_{J_k} occurs at least on one axis. But it is immediate that two distinct dichotomies d_{J_k}, d_{J_h}, $k \leq m$, $h \leq m$, cannot occur on a same axis provide the conditions i) and ii) are fulfilled. ∎

Now we exhibit some results for small values of n. As given in Fichet (1989) or in Ball (1990), the dimension is always at most 2 for $n = 4$ and at most 3 for $n = 5$. See Appendix 3. In Appendix 4, we exhibit a distance $d^{(6)}$ for $n = 6$ and a distance $d^{(7)}$ for $n = 7$, which have a single dichotomy decomposition. They obey

$$d^{(6)} = (d_{126} + d_{146} + d_{156} + d_{236} + d_{346} + d_{356}) + 2(d_{14} + d_{34})$$
$$+ (d_2 + d_4 + d_5 + d_6)$$

$$d^{(7)} = (d_{124} + d_{145} + d_{146} + d_{147} + d_{234} + d_{345} + d_{346} + d_{347})$$
$$+ (d_{14} + d_{34}) + (d_2 + d_4 + d_5 + d_6 + d_7)$$

(where a dichotomy $d_{\{u\}}$, $d_{\{u,v\}}$, $d_{\{u,v,w\}}$ is more simply denoted d_u, d_{uv}, d_{uvw} respectively). Both $d^{(6)}$ and $d^{(7)}$ fulfil the conditions of Lemma 8.4.2. Thus the dimension of $(I, d^{(6)})$ is at least 6. It is exactly 6 by easy arguments. Ball (1990) asserts that, for $n = 6$, any dimension is shown to be at most 6 by ad hoc methods. For $n = 7$, the dimension of $(I, d^{(7)})$ is at least 8, and easy arguments show that it is exactly 8. Note the deep extensions established by Ball (1990): for every $n \geq 4$ there is a distance of L_1-type having a rank at least $\frac{(n-2)(n-3)}{2}$, and for every p $(1 < p < 2)$ and every $n \geq 3$, there is a distance of L_p-type having a rank at least $\frac{(n-1)(n-2)}{2}$.

Let us end this section by noting that to our knowledge, there does not exist any efficient procedure to exhibit the rank of any $d \in \mathcal{D}_1$. The following general procedure (but generally non-tractable !) may be proposed. First observe an obvious correspondence between the embeddings of (I, d) and the dichotomy decompositions of d. Consequently the smallest dimension is the smallest dimension over all the dichotomy decompositions. As proved by Benayade and Fichet in a forthcoming paper, any dichotomy decomposition is a convex combination of a fixed finite system of dichotomy decompositions, called minimal. Then clearly, the smallest dimension is the smallest dimension over all minimal dichotomy decompositions. Now fix a (minimal) dichotomy decomposition. It is obvious that any corresponding embedding depends only on the set S of dichotomies occurring in the decomposition. Moreover, each axis is associated with a subset of S, called below 1-dimensional. It is obvious that we can modify the isometrical embedding to another one in order to have a maximal 1-dimensional subset of S, say S_1. It is quite easy to define an algorithm for constructing the maximal 1-dimensional subsets of S, denoted by $S_1, \ldots, S_K$. Then the smallest dimension for S is one plus the smallest dimension over $k = 1, \ldots, K$ for $S \setminus S_1, \ldots, S \setminus S_K$. We can iterate the procedure.

8.5. Numerical characterisation of semi-distances of L_1-type

As noted by Avis (1977) and explored by Fichet (1987), the geometrical nature of $\mathcal{D}_1$ yields a numerical characterisation in terms of linear programming techniques.

8.5.1. Solving the general problem

We denote by $\{d_{J_1}, \ldots, d_{J_s}\}$, $s = 2^{(n-1)} - 1$, the whole set of dichotomies. From Section 8.2, it comes that $d \in \mathcal{D}_1$ iff there exist nonnegative coefficients α_k, $k = 1, \ldots, s$ such that

$$d = \sum_{k=1}^{s} \alpha_k d_{J_k}. \tag{8.1}$$

Let $\{d_1, \ldots, d_r\}$, $r = \frac{n(n-1)}{2}$, be any basis of $\mathcal{D}$. By projection, (8.1) is equivalent to the linear system

$$A\alpha = D, \tag{8.2}$$

where A is the $r \times s$ matrix of coordinates of dichotomies and D is the vector matrix of coordinates of d. For example, in the canonical basis, the terms of A are equal to 0 or 1 and the coordinates of D are all $d(i,j)$, $i < j$.

Thus $d \in \mathcal{D}_1$ iff there exists $\alpha \geq 0$ obeying (8.2). In other words, $d \in \mathcal{D}_1$ iff the domain defined by the constraints $A\alpha = D$, $\alpha \geq 0$, is nonempty.

Adding new variables, the problem is solved by the simplex method as follows. First, changing if necessary the sign of some rows in (8.2), we get an equivalent system

$$A_*\alpha = D_*, \quad D_* \geq 0. \tag{8.3}$$

Then we consider the following minimisation problem

$$\min F = \beta'\mathbf{1} \quad \text{under the constraints}$$

$$A_*\alpha + \mathbb{I}\beta = D_*, \quad \alpha \geq 0, \ \beta \geq 0, \tag{8.4}$$

where $\mathbb{I}$ and $\mathbf{1}$ denote the identity matrix and the vector matrix with terms equal to 1, respectively. Then, clearly the new domain is nonempty and $d \in \mathcal{D}_1$ iff the minimum is 0. It is well-known that the simplex method yields in a finite number of iterations a new and equivalent form for (8.4)

$$\min F = F_0 + \xi'C, \ C \geq 0, \quad \text{under the constraints}$$

$$B\xi + \mathbb{I}\eta = E, \quad E \geq 0, \tag{8.5}$$

where the variables defined by ξ and η are, up to a permutation, those of α and β. Thus $d \in \mathcal{D}_1$ iff $F_0 = 0$ and a dichotomy decomposition is given by the values E of the variables η and the corresponding dichotomies.

Now, supposing $d \in \mathcal{D}_1$, i.e. $F_0 = 0$, we study the uniqueness of the solution. First it is obvious that the set of solutions is obtained by vanishing in (8.5) the variables ξ_k of ξ equal to some β_l or such that the corresponding coefficient c_k of C is strictly positive. Thus the set of solutions is given by a subsystem of (8.5)

$$B^*\xi_* + \mathbb{I}\eta = E, \tag{8.6}$$

where B^* is a submatrix of B, possibly empty, obtained by deleting some columns of B. It is obvious that $B^* = \emptyset$ is a sufficient condition for uniqueness. That is a necessary and sufficient condition in case of nondegeneracy, i.e. if all the coefficients e_j in the right hand member E of (8.6) are strictly positive. Now we suppose degeneracy and $B^* \neq \emptyset$. Supposing d different from the null dissimilarity, we observe that, due to the simplex method, there is not any degeneracy on each row of (8.6). We prove that (8.6) may be reduced when for a row j, $e_j = 0$ and η_j is some β_l. Indeed, if all the coefficients B^*_{jk} of the jth row of B^* are nonpositive, that row j and possibly the columns k of B^* such that $B^*_{jk} < 0$ may be cancelled. In contrast, if one $B^*_{jk} > 0$, the variable ξ^*_k may replace η_j and the new matrix B^* has one column less. Thus, iteratively we get a new subsystem

$$\widetilde{B}\widetilde{\xi} + \mathbb{I}\widetilde{\eta} = \widetilde{E}, \qquad (8.7)$$

where all variables in $\widetilde{\eta}$ are some α_k. Necessary and sufficient conditions previously discussed for uniqueness stay valid for (8.7) and if $\widetilde{B} \neq \emptyset$ with degeneracy, uniqueness may be established or not by solving a new linear program

$$\max \sum \widetilde{\xi}_k \quad \text{under the constraints (8.7)}.$$

Finally, observe that before doing that, the system (8.7) may be possibly still iteratively reduced. Indeed, if there is degeneracy for some row j, all columns k of $\widetilde{B}$ such that $\widetilde{B}_{jk} > 0$ may be cancelled provide all the coefficients of the row are nonnegative.

8.5.2. Reducing the problem

Since the matrix A contains $(2^{(n-1)} - 1)$ columns, the basic system (8.1) has high complexity. Therefore, the general procedure based on the simplex method is impracticable, except for small values of n, $n \leq 12$, with a usual P.C. However, in particular cases depending either on the problem or on the data, the procedure may be considerably reduced. We give here some examples.

First note that the general procedure stays valid to establish whether or not a dissimilarity belongs to a convex subcone of $\mathcal{D}_1$ spanned by some dichotomies. Now, the following question illustrates some potential applications. Is a dissimilarity d a tree semi-distance, with a fixed configuration of units on the vertices of a fixed support tree ? It is well-known that a tree semi-distance is of L_1-type. Cancelling one edge e_k of the tree yields two connected components, hence a bipartition $\{J_k, J^c_k\}$ of I. Then a dichotomy decomposition of the tree semi-distance is $\sum \alpha_k d_{J_k}$, where α_k is the weight assigned to e_k. See for example Critchley and Fichet (1994). Thus answering the question is proving that d belongs to the polyhedral subcone spanned by the given dichotomies. Note that in a similar way, we may answer the following more general question. Is a dissimilarity the sum of h tree semi-distances of the previous type ? A fixed chain is a particular case of a fixed support tree and, as noted in Benayade and Fichet (1993), the question may be equivalently expressed as follows: does a dissimilarity d belong to $D_{1(h)}$, with fixed within-dimensional ranks for units ? That is exactly the question formulated by Eisler (1973) in his pioneering work for introducing the L_1-norm in multidimensional scaling. Eisler gave a solution in a rather complicated way. In fact, we have a pure linear programming problem, with $h \times (n-1)$ columns in the matrix A of coefficients.

Even in the general case of semi-distances of L_1-type, the system (8.1) may be reduced. We give two examples. First, suppose that the given dissimilarity d contains some triples of the type (i,j,k) metrically aligned, i.e. such that $d(i,j) = d(i,k)+d(j,k)$. Then clearly any dichotomy occurring in any dichotomy decomposition must obey the same equalities. Thus many dichotomies are not concerned with the decomposition. More precisely, those are of the type d_J such that i and j are in J and k in J^c, or vice-versa.

Secondly, suppose that d is constructed inductively and assume that the restriction d' of d to $I' = I\setminus\{n\}$ is of L_1-type and has a single dichotomy decomposition $d' = \sum \alpha_k d'_{J_k}$. Putting $L_k = J_k \cup \{n\}$ for every k, a decomposition of d is necessarily in the form

$$d = \sum_k \gamma_k d_{J_k} + \sum_k (\alpha_k - \gamma_k) d_{L_k} + \gamma_0 d_{\{n\}}, \quad 0 \leq \gamma_k \leq \alpha_k, \; \gamma_0 \geq 0.$$

For every k and every $i = 1, \ldots, (n-1)$, putting x_{ik} equal to 1 or 0 according as $i \in J_k$ or $i \notin J_k$ respectively, we get the simple system

$$\begin{cases} \forall i = 1, \ldots, (n-1), & \sum_k (2x_{ik} - 1)\gamma_k + \gamma_0 = d(i,n) - \sum_k (1 - x_{ik})\alpha_k, \\ \forall k, & \gamma_k + \gamma'_k = \alpha_k, \end{cases}$$

with the constraints $\forall k$, $\gamma_k \geq 0$, $\gamma'_k \geq 0$; $\gamma_0 \geq 0$. Note, that only $(n-1)$ new artificial variables must be added. Fichet (1992) uses such a procedure for computing a "spherical extension" of a (semi)-metric space of L_1-type.

8.5.3. Approximations

In this section we deal with the problem of approximating a data dissimilarity by a semi-distance of L_1-type. Three types of approximations are discussed.

8.5.3.1. Least absolute deviations approximations

First we observe that the minimisation problem (8.4) yields an L_1-norm approximation. Indeed, choosing the canonical basis, the coefficients of β may be relabelled as β_{ij}, $i < j$, so that they may be considered as the coordinates of a dissimilarity d_β. Putting $d' = \sum_k \alpha_k d_{J_k}$, we have $d_\beta = d - d'$ so that (8.4) is equivalent to

$$\min\{||d - d'||_1 \;:\; d' \in \mathcal{D}_1, \, d' \leq d\},$$

where $||.||_1$ stands of the L_1-norm on $\mathcal{D}$, with respect to the canonical basis. Solving (8.4) we get a lower L_1-norm approximation. Note that any solution given by (8.5) is equipped with a dichotomy decomposition. The reader will may compare that approximation to the one by a (squared) Euclidean semi-distance obtained by cancelling the negative eigenvalues of the Torgerson's matrix, which a posteriori appears to be an L_2-norm (upper) approximation.

Similarly, an upper L_1-norm approximation solves the problem

$$\min\{||d - d'||_1 \;:\; d' \in \mathcal{D}_1, \, d' \geq d\}.$$

A projection on the canonical basis gives the minimisation problem

$$\min \sum_{i<j} \beta_{ij} \quad \text{under the constraints } A\alpha - \mathbb{I}\beta = D, \; \alpha \geq 0, \; \beta \geq 0.$$

However, in this case we need a feasible solution. It is obtained by adding new artificial variables $\gamma_{ij} \geq 0$, and by minimising $\sum \gamma_{ij}$. The existence of an upperbound of L_1-type implies the existence of a feasible solution.

Having a lower approximation $d_\wedge$ and an upper approximation $d^\wedge$, we may look for a compromise by solving

$$\min_t ||d - d_\wedge - t\left(d^\wedge - d_\wedge\right)||_1.$$

This last problem is closed to the L_1-norm regression problem, and is solved via the simplex method.

8.5.3.2. Least squares approximation

Suppose $\mathcal{D}$ endowed with an inner product $<.,.>$ and let $||.||_2$ denote the induced norm. Then, the least squares problem is

$$\min\left\{||d - d'||_2 \; : \; d' \in \mathcal{D}_1\right\}.$$

In particular, when the inner product is chosen to define the canonical basis to be orthonormal, we have

$$\min\left\{\sum_{i<j}\left(d(i,j) - \sum_k \alpha_k d_{J_k}(i,j)\right)^2 \; : \; \alpha_k \geq 0, \; \forall \, k = 1, \ldots, s\right\}.$$

Since $\mathcal{D}_1$ is closed and convex, there is a single solution. Geometrically, we have to project d on $\mathcal{D}_1$. Several algorithms of projection on a closed convex polyhedral cone have been proposed, as, for example, in Tenenhaus (1988), following the work of Lawson and Hanson (1974). However, the great number of extreme rays will be a limit for tractability.

8.5.3.3. The additive constants

The general problem is stated as follows. Fix $d_* \in \mathcal{D}_1$. Then, for any $d \in \mathcal{D}_+$, the general additive constant problem is

$$\min\left\{c \geq 0 \; : \; (d + cd_*) \in \mathcal{D}_1\right\}. \tag{8.8}$$

Since $\mathcal{D}_1$ is closed, the problem has a solution c_0 provide the domain is non-empty. In particular that is true when d_* belongs to $\overset{\circ}{\mathcal{D}}_1$: for c sufficiently large, $(\frac{1}{c}d + d_*)$, hence $(d + cd_*)$ belongs to $\mathcal{D}_1$. Clearly (8.8) is equivalent to

$$\min F = c \quad \text{under the constraints}$$

$$\sum_{k=1}^{s} \alpha_k d_{J_k} - cd_* = d, \quad \alpha_k \geq 0, \; \forall \, k = 1, \ldots, s; \; c \geq 0.$$

A projection on any basis of $\mathcal{D}$ gives a new linear programming problem with $2^{(n-1)}$ variables. First of all we need a feasible solution which exists if and only if the domain is non-empty. Addition of new parameters solves this question. When a solution c_0 exists, we get a dichotomy decomposition of $(d + c_0 d_*)$.

Now, we give some examples. For a weight $w_i > 0$ assigned to each unit $i \in I$, we define the dissimilarity d_w by $d_w(i,j) = \frac{1}{2}(\frac{1}{w_i} + \frac{1}{w_j})$ whenever $i \neq j$. If $w_i = 1$ for every $i \in I$, d_w is d_1, the distance of the unit regular simplex. The distance d_w is a particular tree distance, so that $d_w \in \mathcal{D}_1$. See Critchley and Fichet (1994). Moreover, using symmetry arguments or some arguments given below, d_1 is in $\overset{\circ}{\mathcal{D}}_1$. Consequently d_w which can always be written as $d_w = (d_w - \varepsilon d_1) + \varepsilon d_1$ for some $\varepsilon > 0$, is also in $\overset{\circ}{\mathcal{D}}_1$. Then the weighted additive constant problem is (8.8) with $d_* = d_w$. The problem admits a solution.

Using a new basis alternative to the canonical basis, Fichet (1987) obtained directly a feasible solution when the weights are equal to 1. We give here a complete investigation. First note that the 2-dichotomies $d_{\{i,j\}}$, $i < j$, form a basis of $\mathcal{D}$ for $n > 4$. In this basis, any $d \in \mathcal{D}$ has coordinates

$$D_{ij} = -\frac{1}{2}\left[d(i,j) - \frac{1}{n-4}\left(d(i,.) + d(.,j)\right) + \frac{1}{(n-2)(n-4)} d(.,.)\right], i < j.$$

where a dot denotes summation over an omitted index. See for example Critchley and Fichet (1994). In particular d_1 has coordinates equal to $\frac{1}{2(n-2)}$. Consequently, d_1 is in $\overset{\circ}{\mathcal{D}}_1$. Relabelling the dichotomies as d_{J_k}, $k = 1, \ldots, s' = s - \frac{n(n-1)}{2}$ and $d_{\{i,j\}}$, $i < j$, $(d + c d_1)$ belongs to $\mathcal{D}_1$ iff it is written as

$$d + c d_1 = \sum_{k=1}^{s'} \alpha_k d_{J_k} + \sum_{i<j} \beta_{ij} d_{\{i,j\}}, \quad \alpha_k \geq 0, \beta_{ij} \geq 0.$$

By projection on the basis, we get the system

$$A\alpha + \mathbb{I}\beta - [c/(n-2)]\mathbf{1} = D. \tag{8.9}$$

The explicit values of the coefficients of A may be given. Indeed, a simple calculation shows that any dichotomy d_J has the following coordinate on the vector $d_{\{i,j\}}$

$$\begin{cases} \dfrac{(n-k)(n-k-2)}{(n-2)(n-4)} & \text{if } i \in J, j \in J, \\ \dfrac{k(k-2)}{(n-2)(n-4)} & \text{if } i \notin J, j \notin J, \\ \dfrac{-(k-2)(n-k-2)}{(n-2)(n-4)} & \text{if } i \in J, j \notin J \text{ or } i \notin J, j \in J, \end{cases}$$

where k stands for the cardinality of J. That proves in passing that except the basic vectors, none of the dichotomies is in the nonnegative orthant. Then, if in (8.9) $D \geq 0$, d is in the nonnegative orthant, hence in $\mathcal{D}_1$, so that $c_0 = 0$. In contrast, let $D_{i'j'}$ realize the minimum of D_{ij}, $i < j$. Subtraction of the row (i', j') from the other rows gives a feasible solution, with the basic variables β_{ij}, $(i,j) \neq (i',j')$ and c. The minimum constant c_0 is at most $-2(n-2)D_{i'j'}$.

Another example privileges a unit of I. Identifying here I with $\{0,\ldots,(n-1)\}$, we choose d_* in (8.8) equal to d'_1, where $d'_1(0,i) = \frac{1}{2}$, $d'_1(i,j) = 1$ for every $i,j = 1,\ldots,(n-1)$, $i \neq j$. The distance d'_1 is of L_1-type. However, d'_1 is in the boundary of $\mathcal{D}_\infty$, hence of $\mathcal{D}_1$. A priori the domain defined in (8.8) may be empty. Using the basis formed by the 1- dichotomies $d_{\{i\}}$, $i = 1,..,(n-1)$ and the 2-dichotomies $d_{\{i,j\}}$, $i,j = 1,..,(n-1)$, $i < j$, and using the explicit coordinates of a vector in this basis, Fichet (1992) proved that the additive constant problem has always a solution.

8.6. Appendices

Throughout this section, a 1-dichotomy $d_{\{u\}}$, a 2-dichotomy $d_{\{u,v\}}$ and a 3-dichotomy $d_{\{u,v,w\}}$ is more simply denoted d_u, d_{uv} and d_{uvw}, respectively.

8.6.1. Appendix 1

Lemma 8.6.1. *Let d be the distance on $\{1,2,3,4\}$ defined by*

$$d(1,2) = d(2,3) = d(3,4) = d(4,1) = 1\,;\ d(1,3) = d(2,4) = 2.$$

Then $d \notin \mathcal{D}_p$ for every $1 < p < \infty$.

Proof. We begin with the following remark. Let $\{d_m,\ m \in M\}$ be a finite family of distances defined on a set I. For every $1 \leq p < \infty$ define $d' = [\sum_m d_m^p]^{1/p}$. Then d' is a distance. Moreover, if $d'(i,j) = d'(i,k) + d'(k,j)$, then necessarily $d_m(i,j) = d_m(i,k) + d_m(k,j)$, for every m. Indeed, we have

$$\forall m \in M,\ \forall\ (i,j,k) \in I^3,\quad d_m(i,j) \leq d_m(i,k) + d_m(k,j) \tag{1}$$

and, by the Minkowski's inequality, we obtain

$$\left(\sum_m d_m^p(i,j)\right)^{1/p} \leq \left(\sum_m (d_m(i,k) + d_m(k,j))^p\right)^{1/p}$$
$$\leq \left(\sum_m d_m^p(i,k)\right)^{1/p} + \left(\sum_m d_m^p(k,j)\right)^{1/p}. \tag{2}$$

Moreover, if the extreme terms in (2) are equal, clearly the inequalities of (1) are equalities for every m. Now we apply this remark to ℓ_p^N. Suppose that three vectors x_i, x_j, x_k are metrically aligned, i.e. $||x_i - x_j|| = ||x_i - x_k|| + ||x_k - x_j||$. Then, their coordinates obey $x_{im} \leq x_{jm} \leq x_{km}$ (up to the converse order).

Now, we consider an embedding in some ℓ_p^N for the given distance d. Denote the vectors by $x_i = (x_{im})_{m=1,\ldots,N}$, $i = 1,\ldots,4$. Since for every $i \in I$ ($I = \{1,2,3,4\}$), there exist $j, k \in I$ obeying $d(j,k) = d(j,i) + d(i,k)$, every coordinate lies between the other coordinates on each axis. In other words, both the smallest coordinate and the greatest coordinate are equal to another coordinate. With the constraints given by the triangular equalities, we have, up to the converse order, either

$$x_{1m} = x_{4m} \leq x_{2m} = x_{3m} \tag{3}$$

or

$$x_{1m} = x_{2m} \leq x_{3m} = x_{4m}. \tag{4}$$

Let us denote by N_1 and N_2 the axes giving (3) and (4), respectively. Putting $a_m = |x_{1m}-x_{2m}|$, $\forall m \in N_1$ and $b_m = |x_{1m}-x_{3m}|$, $\forall m \in N_2$ and defining $a = (\sum_{m \in N_1} a_m^p)^{1/p}$, $b = (\sum_{m \in N_2} b_m^p)^{1/p}$, it is clear that all the axes of N_1 (resp. N_2) may be contracted in a single axis m_1 (resp. m_2) in the following way

$$x_{1m_1} = x_{4m_1} = 0 \; ; \; x_{2m_1} = x_{3m_1} = a,$$
$$x_{1m_2} = x_{2m_2} = 0 \; ; \; x_{3m_2} = x_{4m_2} = b.$$

The values of d give $a = 1$, $b = 1$ and $a^p + b^p = 2^p$. That achieves the proof. ■

8.6.2. Appendix 2

Lemma 8.6.2. *Every metric triangle is embeddable in ℓ_p^2 for $1 \leq p \leq \infty$.*

Proof. For $p = \infty$, the result is given by Frechet's theorem. Now suppose $p < \infty$. Let us denote by d the distance and by A, B, C the points. We put $d(A, B) = \alpha$, $d(A, C) = \beta$, $d(B, C) = \gamma$ with $\gamma \leq \beta \leq \alpha$. Then we consider the following points of $\mathbb{R}^2$: $x_A = (0, 0)$; $x_B = (\alpha, 0)$; $x_C = (u, v)$, $0 \leq u \leq \alpha$; $0 \leq v$. The triple (x_A, x_B, x_C) is an embedding in ℓ_p^2 iff

$$u^p + v^p = \beta^p \; ; \; (\alpha - u)^p + v^p = \gamma^p$$

or equivalently

$$u^p + v^p = \beta^p \; ; \; u^p - (\alpha - u)^p = \beta^p - \gamma^p.$$

An easy calculation shows that the equation $\beta^p - \gamma^p - u^p + (\alpha - u)^p = 0$, $0 \leq u \leq \alpha$, has a single solution $u_o \leq \beta$. The proof is complete. ■

8.6.3. Appendix 3

We begin with the following two lemmas. Their proof is very easy.

Lemma 8.6.3. *For $n \geq 3$,*

$$\forall i \in I, \; \sum_{j \neq i} d_{ij} = \sum_k d_k + (n - 4) \, d_i.$$

Lemma 8.6.4. *For $n = 5$, up to all the permutations,*

$$d_{12} + d_{23} + d_{13} = d_{45} + d_1 + d_2 + d_3.$$

Then we have the following lemmas.

Lemma 8.6.5. For $n = 4$, any L_1-dimension is at most 2.

Proof. Let $d \in \mathcal{D}_1$. Then d may be written as

$$d = \sum_{i=1}^{4} \alpha_i d_i + \sum_{i=2}^{4} \beta_i d_{1i}, \quad \alpha_i \geq 0, \ \beta_i \geq 0. \tag{1}$$

Let $a = \min(\beta_2, \beta_3, \beta_4)$. By Lemma 8.6.3, $d = d + a \sum_{i=1}^{4} d_i - a \sum_{i=2}^{4} d_{1i}$. Thus, one 2-dichotomy may be omitted in (1). If for example $\beta_2 = 0$, we have

$$d = (\alpha_1 d_1 + \beta_3 d_{13} + \alpha_2 d_2) + (\alpha_4 d_4 + \beta_4 d_{14} + \alpha_3 d_3). \qquad \blacksquare$$

Lemma 8.6.6. For $n = 5$, any L_1-dimension is at most 3.

Proof. Let $d \in \mathcal{D}_1$. Clearly the dimension of (I, d) depends only on the dichotomies occuring in the dichotomy decompositions. In the sequel, we modify a subset S of the whole set of dichotomies, such that d may be written as a nonnegative linear combination of the elements of S. We start with the whole set. By Lemma 8.6.3 related to the unit 1, without loss of generality (w.l.o.g.) we may suppose $d_{12} \notin S$. Now we consider two cases: a) w.l.o.g. $d_{34} \notin S$ and b) $d_{34}, d_{35}, d_{45} \in S$.

a) Using Lemma 8.6.3 related to the unit 5, w.l.o.g. we may suppose $d_{15} \notin S$. Using Lemma 8.6.4, we may suppose that one of the dichotomies d_{24}, d_{25}, d_{45} is not in S. Then

If $d_{24} \notin S$, $S = \{(d_1, d_{13}, d_{25}, d_2), (d_3, d_{23}, d_{45}, d_4), (d_5, d_{35}, d_{14})\}$.

If $d_{25} \notin S$, $S = \{(d_1, d_{13}, d_{24}, d_2), (d_3, d_{23}, d_{45}, d_4), (d_5, d_{35}, d_{14})\}$.

If $d_{45} \notin S$, $S = \{(d_1, d_{13}, d_{25}, d_2), (d_3, d_{23}, d_{14}, d_4), (d_5, d_{35}, d_{24})\}$.

In each case, the parentheses indicate a 3-dimensional representation.

b) Applying Lemma 8.6.3 related to the units 3, 4 and 5 successively, we may suppose that for every $i = 3, 4, 5$ there exists $j_i \neq i$ such that $d_{ij_i} \notin S$. If one the units j_i, $i = 3, 4, 5$ is greater than 2, this is the case a). Consequently, we must only treat the following cases: b1) w.l.o.g. $d_{13}, d_{14} \notin S$; $d_{25} \notin S$ and b2) w.l.o.g. $d_{13}, d_{14}, d_{15} \notin S$.

b1) $S = \{(d_2, d_{23}, d_{45}, d_4), (d_3, d_{34}, d_{15}, d_1), (d_5, d_{35}, d_{24})\}$. The parentheses indicate a 3-dimensional representation.

b2) Using Lemma 8.6.4 related to the units 2, 3 and 4, and adding d_{15} in S in case of necessity, we may suppose that one of the dichotomies d_{23}, d_{24}, and d_{34} is not in S. When $d_{34} \notin S$, that is the case a). Thus we may suppose w.l.o.g. $d_{23} \notin S$. Using Lemma 8.6.3 related to the unit 5, we may suppose that one of the dichotomies d_{15}, d_{25}, d_{35} and d_{45} is not in S. When $d_{35} \notin S$, or $d_{45} \notin S$ that is the case a). When $d_{25} \notin S$, that is the case b1). Finally, when $d_{15} \notin S$, $S = \{(d_2, d_{25}, d_{34}, d_3), (d_4, d_{24}, d_{35}, d_5), (d_1, d_{45})\}$. The parentheses indicate a 3-dimensional representation. The proof is complete. $\blacksquare$

8.6.4. Appendix 4

For $I = \{1, \ldots, n\}$, $n \geq 5$, let $d^{(n)}$ be the distance on I, defined by

$$d^{(n)}(1,2) = d^{(n)}(2,3) = d^{(n)}(3,4) = d^{(n)}(4,1) = 6$$
$$d^{(n)}(1,3) = d^{(n)}(2,4) = 10.$$
$$\forall\, j, k \geq 5,\ j \neq k,\ d^{(n)}(1,j) = d^{(n)}(2,j) = d^{(n)}(3,j) = 6$$
$$d^{(n)}(4,j) = 10\,;\ d^{(n)}(j,k) = 6.$$

Lemma 8.6.7. *The distance $d^{(5)}$ is of L_1-type and admits the following single dichotomy decomposition*

$$d^{(5)} = (d_{12} + d_{15} + d_{23} + d_{35}) + 3(d_{14} + d_{34}) + (d_2 + d_4 + d_5).$$

Proof. First we observe that the given dichotomy decomposition is right. Now, for the uniqueness, we write that $d^{(5)}$ obeys

$$d^{(5)} = \sum_{i<j} \alpha_{ij} d_{ij} + \sum_i \gamma_i d_i,\quad \alpha_{ij} \geq 0,\ \gamma_i \geq 0,\ \forall\, i, j = 1, \ldots, 5,\ i < j. \qquad (1)$$

Clearly

$$6 \sum_{i<j} \alpha_{ij} + 4 \sum_i \gamma_i = \sum_{i<j} d(i,j) = 72. \qquad (2)$$

$d^{(5)}(1,3) + d^{(5)}(2,4) + d^{(5)}(2,5) + d^{(5)}(4,5)$ gives

$$3\alpha_{12} + 3\alpha_{14} + 3\alpha_{15} + 3\alpha_{23} + 2\alpha_{24} + 2\alpha_{25} + 3\alpha_{34} + 3\alpha_{35} + 2\alpha_{45}$$
$$+\gamma_1 + 2\gamma_2 + \gamma_3 + 2\gamma_4 + 2\gamma_5 = 36. \qquad (3)$$

Subtracting twice (3) from (2) gives

$$6\alpha_{13} + 2\alpha_{24} + 2\alpha_{25} + 2\alpha_{45} + 2\gamma_1 + 2\gamma_3 = 0,$$

so that necessarily

$$\alpha_{13} = \alpha_{24} = \alpha_{25} = \alpha_{45} = \gamma_1 = \gamma_3 = 0. \qquad (4)$$

Using (4) gives

$$d^{(5)}(1,2) + d^{(5)}(2,3) - d^{(5)}(1,3) = 2\gamma_2,$$
$$d^{(5)}(1,4) + d^{(5)}(3,4) - d^{(5)}(1,3) = 2\gamma_4,$$
$$d^{(5)}(1,5) + d^{(5)}(3,5) - d^{(5)}(1,3) = 2\gamma_5,$$
$$d^{(5)}(3,4) + d^{(5)}(3,5) - d^{(5)}(4,5) = 2\alpha_{23},$$
$$d^{(5)}(2,3) + d^{(5)}(3,4) - d^{(5)}(2,4) = 2\alpha_{35}.$$

Thus necessarily

$$\gamma_2 = \gamma_4 = \gamma_5 = \alpha_{23} = \alpha_{35} = 1. \qquad (5)$$

Now, using (4) and (5) gives

$$d^{(5)}(3,4) = \alpha_{14} + 3,$$
$$d^{(5)}(1,5) = \alpha_{12} + \alpha_{14} + 2,$$
$$d^{(5)}(2,3) = \alpha_{12} + \alpha_{34} + 2,$$
$$d^{(5)}(1,2) = \alpha_{14} + \alpha_{15} + 2,$$

so that necessarily

$$\alpha_{14} = 3, \ \alpha_{12} = 1, \ \alpha_{34} = 3, \ \alpha_{15} = 1. \tag{6}$$

Then the uniqueness is proved. ∎

Lemma 8.6.8. *The distance $d^{(6)}$ is of L_1-type and admits the following single dichotomy decomposition*

$$d^{(6)} = (d_{124} + d_{125} + d_{126} + d_{145} + d_{146} + d_{156})$$
$$+ 2(d_{14} + d_{34}) + (d_2 + d_4 + d_5 + d_6).$$

Proof. First we observe that the given dichotomy decomposition takes the values of $d^{(6)}$ on the pairs $(i, 6)$, $i = 1, \ldots, 5$. Furthermore, its restriction to the set $\{1, \ldots, 5\}$ is the dichotomy decomposition established in the previous lemma. Thus, the given dichotomy decomposition is right. In order to prove the uniqueness, we use the previous lemma. Necessarily $d^{(6)}$ must be written as

$$d^{(6)} = [\alpha_{12}d_{12} + (1 - \alpha_{12})d_{126}] + [\alpha_{15}d_{15} + (1 - \alpha_{15})d_{156}]$$
$$+ [\alpha_{23}d_{23} + (1 - \alpha_{23})d_{236}] + [\alpha_{35}d_{35} + (1 - \alpha_{35})d_{356}]$$
$$+ [\alpha_{14}d_{14} + (3 - \alpha_{14})d_{146}] + [\alpha_{34}d_{34} + (3 - \alpha_{34})d_{346}]$$
$$+ [\gamma_2 d_2 + (1 - \gamma_2)d_{26}] + [\gamma_4 d_4 + (1 - \gamma_4)d_{46}]$$
$$+ [\gamma_5 d_5 + (1 - \gamma_5)d_{56}] + \gamma_6 d_6,$$

with the constraints

$$\alpha_{12}, \alpha_{15}, \alpha_{23}, \alpha_{35} \in [0,1]; \ \alpha_{14}, \alpha_{34} \in [0,3]; \ \gamma_2, \gamma_4, \gamma_5 \in [0,1]; \ \gamma_6 \geq 0.$$

We have

$$d^{(6)}(1,6) + d^{(6)}(3,6) = -2\gamma_2 - 2\gamma_4 - 2\gamma_5 + 2\gamma_6 + 16, \tag{1}$$
$$d^{(6)}(4,6) + d^{(6)}(5,6) = -2\alpha_{12} - 2\alpha_{23} - 2\gamma_2 + 2\gamma_6 + 16. \tag{2}$$

(2) − (1) gives

$$\gamma_4 + \gamma_5 = \alpha_{12} + \alpha_{23} + 2.$$

The constraints imply

$$\gamma_4 = \gamma_5 = 1; \ \alpha_{12} = \alpha_{23} = 0; \ \gamma_2 = \gamma_6. \tag{3}$$

Using (3), we have

$$d^{(6)}(2,6) = (1 - \alpha_{15}) + (1 - \alpha_{35}) + (3 - \alpha_{14}) + (3 - \alpha_{34}) + \gamma_2 + \gamma_6, \tag{4}$$
$$d^{(6)}(5,6) = 1 + \alpha_{15} + 1 + \alpha_{35} + (3 - \alpha_{14}) + (3 - \alpha_{34}) + (1 - \gamma_2) + 1 + \gamma_6. \tag{5}$$

$(5) - (4)$ gives
$$\alpha_{15} + \alpha_{35} + 1 = \gamma_2.$$

With the constraints and (3)
$$\gamma_2 = \gamma_6 = 1; \quad \alpha_{15} = \alpha_{35} = 0. \tag{6}$$

Then, using (3) and (6)
$$d^{(6)}(1,6) = 2 + \alpha_{14} + (3 - \alpha_{34}) + 1. \tag{7}$$

$(4) - (7)$ gives
$$\alpha_{14} = \alpha_{34} = 2. \tag{8}$$

∎

Lemma 8.6.9. *The distance $d^{(7)}$ is of L_1-type and admits the following single dichotomy decomposition*

$$\begin{aligned} d^{(7)} &= (d_{124} + d_{145} + d_{146} + d_{147} + d_{234} + d_{345} + d_{346} + d_{347}) \\ &\quad + (d_{14} + d_{34}) + (d_2 + d_4 + d_5 + d_6 + d_7). \end{aligned}$$

Proof. First we observe that the given dichotomy decomposition takes the values of $d^{(7)}$ on the pairs $(i,7)$, $i = 1, \ldots, 6$. Furthermore, its restriction to the set $\{1, \ldots, 6\}$ is the dichotomy decomposition established in the previous lemma. Thus, the given dichotomy decomposition is right. In order to prove the uniqueness, we use the previous lemma. Necessarily $d^{(7)}$ must be written as

$$\begin{aligned} d^{(7)} &= [\alpha_{124}d_{124} + (1 - \alpha_{124})\,d_{356}] + [\alpha_{125}d_{125} + (1 - \alpha_{125})\,d_{346}] \\ &\quad + [\alpha_{126}d_{126} + (1 - \alpha_{126})\,d_{345}] + [\alpha_{145}d_{145} + (1 - \alpha_{145})\,d_{236}] \\ &\quad + [\alpha_{146}d_{146} + (1 - \alpha_{146})\,d_{235}] + [\alpha_{156}d_{156} + (1 - \alpha_{156})\,d_{234}] \\ &\quad + [\beta_{14}d_{14} + (2 - \beta_{14})\,d_{147}] + [\beta_{34}d_{34} + (2 - \beta_{34})\,d_{347}] \\ &\quad + [\gamma_2 d_2 + (1 - \gamma_2)\,d_{27}] + [\gamma_4 d_4 + (1 - \gamma_4)\,d_{47}] \\ &\quad + [\gamma_5 d_5 + (1 - \gamma_5)\,d_{57}] + [\gamma_6 d_6 + (1 - \gamma_6)\,d_{67}] \\ &\quad + \gamma_7 d_7, \end{aligned}$$

with the constraints

$$\begin{aligned} \alpha_{124}, \alpha_{125}, \alpha_{126}, \alpha_{145}, \alpha_{146}, \alpha_{156} &\in [0,1], \\ \beta_{14}, \beta_{34} &\in [0,2], \\ \gamma_2, \gamma_4, \gamma_5, \gamma_6 &\in [0,1], \\ \gamma_7 &\geq 0. \end{aligned}$$

We have

$$d^{(7)}(1,7) + d^{(7)}(3,7) = 10 + 2(1 - \gamma_2) + 2(1 - \gamma_4) + 2(1 - \gamma_5) + 2(1 - \gamma_6) + 2\gamma_7, \tag{1}$$
$$d^{(7)}(4,7) + d^{(7)}(5,7) = 2(1 - \alpha_{126}) + 2\alpha_{145} + 8 + 2(1 - \gamma_2) + 2(1 - \gamma_6) + 2 + 2\gamma_7, \tag{2}$$
$$d^{(7)}(2,7) + d^{(7)}(4,7) = 2\alpha_{124} + 2(1 - \alpha_{156}) + 10 + 2(1 - \gamma_5) + 2(1 - \gamma_6) + 2\gamma_7. \tag{3}$$

(1) – (2) gives
$$(1-\gamma_4)+(1-\gamma_5)+2=(1-\alpha_{126})+\alpha_{145}.$$
With the constraints, necessarily
$$\alpha_{126}=0;\ \alpha_{145}=1;\ \gamma_4=\gamma_5=1. \qquad (4)$$
Using (4), (1) – (3) gives
$$(1-\gamma_2)+2=(1-\alpha_{156})+\alpha_{124}.$$
With the constraints, necessarily
$$\alpha_{156}=0;\ \alpha_{124}=1;\ \gamma_2=1. \qquad (5)$$
Then, using (4) and (5) we have
$$d^{(7)}(6,7)=(1-\alpha_{125})+\alpha_{146}+(2-\beta_{14})+(2-\beta_{34})+\gamma_6+\gamma_7, \qquad (6)$$
$$d^{(7)}(2,7)=1+\alpha_{125}+(1-\alpha_{146})+1+(2-\beta_{14})+(2-\beta_{34})+1+(1-\gamma_6)+\gamma_7. \qquad (7)$$
(6) – (7) gives
$$\alpha_{146}+\gamma_6=\alpha_{125}+2.$$
With the constraints, necessarily
$$\gamma_6=1;\ \alpha_{125}=0;\ \alpha_{146}=1. \qquad (8)$$
Using (4), (5) and (8), the equality (1) gives
$$\gamma_7=1. \qquad (9)$$
Then, using (4), (5), (8) and (9),
$$d^{(7)}(1,7)=3+\beta_{14}+(2-\beta_{34})+1,$$
i.e. $\beta_{14}=\beta_{34}$, so that, by (6), we obtain $\beta_{14}=\beta_{34}=1$. ∎

Remark 8.6.1. In this remark, d_1 stands for the distance of the unit regular simplex and we refer to the distance d of Remark 8.3.1. Clearly, $d^{(5)}=\frac{1}{4}(d+c_0 d_1)$ with $c_0=\frac{1}{2}$, and similar equalities hold with $d^{(6)}$ and $d^{(7)}$ for an obvious extension of d for $n>5$. Since $d^{(5)}$, $d^{(6)}$ and $d^{(7)}$ have a unique dichotomy decomposition, they lie in the boundary of $\mathcal{D}_1$ so that c_0 is the additive constant associated with them. Actually, for the first time, we solved the additive constant problem via the simplex method with a pencil for $n=5$ and a computer for $n>5$. The knowledge of the results allowed us to establish the direct and analytic proofs given in the previous lemmas. Let us note that, for $n=8$, the additive constant is still $\frac{1}{2}$. The uniqueness is not immediate but it may be proved. Moreover, the unique dichotomy decomposition contains ten 3-dichotomies which obey the conditions of Lemma 8.4.2. Thus $d^{(8)}$ is a 8-point L_1-distance which necessitates 10 axes. For $n=9$, the additive constant is 0.55.

We may compare c_0 with the Euclidean additive constant of order 2 (say c_1) associated with $\sqrt{d}$. The constant c_1 minimises

$$\min\{c \geq 0 \ : \ (d + cd_1)^{1/2} \in \mathcal{D}_e\}.$$

Since $\mathcal{D}_1 \subseteq \mathcal{D}_e^2$, necessarily we have $0 < c_1 \leq c_0 = \frac{1}{2}$. The constant c_1 is given by the smallest eigenvalue of the Torgerson's matrix associated with d. It may also be obtained by simple geometrical arguments, using the fact that $(d + c_1 d_1)^{1/2}$ belongs to the boundary of $\mathcal{D}_e$, so that there is an embedding in $\mathbb{R}^3$. Then, we obtain $0 < c_1 < \frac{1}{2}$. Consequently, for $c_1 \leq c < c_0$, the distance $(d + cd_1)^{1/2}$ is not of L_1-type and admits a Euclidean square root. Recall that such an example cannot be exhibited for a normed space. See Bretagnolle, Dacunha-Castelle and Krivine (1966). □

References

Assouad, P. (1977), Un espace hypermétrique non plongeable dans un espace L^1, C.R. Acad. Sci. Paris, t. 285, Série A, pp. 361–363.

Assouad, P., Deza, M. (1982), Metric subspaces of L^1, Publications Mathématiques d'Orsay, Dept. Mathématiques, Université de Paris-Sud, France.

Avis, D. (1977), On the Hamming Cone, Technical Report 77-5. Dept. of Operations Research, Stanford University, USA.

Ball, K. (1990), Isometric embedding in ℓ_p-spaces, Europ. J. Combinatorics, 11, pp. 305–311.

Benayade, M. , Fichet, B. (1993), Algorithms for a geometrical P.C.A. with the L_1-norm, 4th Conference of the International Federation of Classification Societies, Paris. To appear in Diday, E., et al., eds (1994), Proc. IFCS93 meeting, Springer-Verlag, New York.

Blackwell, D. , Girshick, M.A. (1966), Theory of Games, fifth printing, Wiley, New York.

Boscovich, R.J. (1757), De litteraria expeditione per pontificam ditionem, et synopsis amplioris operis, ac habentur plura eius ex exemplaria etiam sensorum impressa, Bononiensi Scientiarum et Artium Instituto Atque Academia Commentarii, 4, pp. 353–396.

Bretagnolle, J., Dacunha-Castelle, D., Krivine, J.L. (1966), Lois stables et espaces L^p, Ann. Inst. H. Poincaré Sect. B, II, n° 3, pp. 231–259.

Critchley, F.,Fichet, B. (1994), The partial order by inclusion of the principal classes of dissimilarity on a finite set, and some of their basic properties, In Van Cutsem, B., ed., Classification and Dissimilarity Analysis, Ch. 2, Lecture Notes in Statistics, Springer-Verlag, New York.

Deza, M. (Tylkin) (1960), On Hamming geometry of unitary cubes, *Dokl. Acad. Nauk SSSR*, 134, pp. 1037–1040.

Dor, L.E. (1976), Potentials and isometric embeddings in L_1, *Israel J. Math.*, 24, pp. 260–268.

Eisler, H. (1973), The algebraic and statistical tractability of the city-block metric, *British J. Math. Statist. Psych.*, 26, pp. 212–218.

Farebrother, R.W. (1987), The historical development of the L_1 and L_∞ estimation procedures 1793-1930, In Dodge, Y., ed., *Statistical Data Analysis based on the L_1-norm and Related Methods*, North-Holland, Amsterdam, pp. 37–63.

Fichet, B. (1986), Data analysis: geometric and algebraic structures, First World Congress of Bernoulli Society, Tashkent, URSS, T1, pp. 75–77.

Fichet, B. (1987), The role played by L_1 in data analysis, In Dodge, Y., ed., *Statistical Data Analysis based on the L_1-norm and Related Methods*, North-Holland, Amsterdam, pp. 185–193.

Fichet, B. (1988), L_p-spaces in data analysis, In Bock, H.H., ed., *Classification and Related Methods of Data Analysis*, North-Holland, Amsterdam, pp. 439–444.

Fichet, B. (1989), Sur la dimension des figures finies en norme L_1, 47th session of I.S.I., Paris, 1, pp. 325–326.

Fichet, B. (1992), The notion of sphericity for finite L_1-figures of data analysis, In Dodge, Y., ed., *L_1-Statistical Analysis and Related Methods*, North-Holland, Amsterdam, pp. 129–144.

Fréchet, M. (1910), Les dimensions d'un ensemble abstrait, *Math. Ann.*, 68, pp. 145–168.

Grünbaum, B. (1967), *Convex Polytopes*, Interscience Publishers, London.

Holman, W. (1972), The relation between hierarchical and Euclidean models for psychological distances, *Psychometrika*, 37, pp. 417–423.

Joly, S., Le Calvé, G. (1986), Etude des puissances d'une distance, *Statist. Anal. Données*, 11, pp. 30–50.

Laplace, P.S. (1793), Sur quelques points du système du monde, Mémoires de l'Académie Royale des Sciences de Paris, pp. 1–87. Reprinted in *Oeuvres Complètes de Laplace* (1895), Paris, Gauthier-Villars, 11, pp. 477–558.

Lawson, C.M. , Hanson, R.J. (1974), *Solving Least Squares Problems*, Englewood Cliffs, Prentice-Hall.

Le Calvé, G. (1987), L_1-embeddings of a data structure (I, D), In Dodge, Y., ed., *Statistical Data Analysis based on the L_1-norm and Related Methods*, North-Holland, Amsterdam, pp. 195–202.

Schoenberg, I.J. (1935), Remarks to Maurice Fréchet's article "Sur la définition axiomatique d'une classe d'espace distanciés vectoriellement applicable sur l'espace de Hilbert", *Ann. of Math.*, 36, pp. 724–732.

Tenenhaus, M. (1988), Canonical analysis of two convex polyhedral cones and applications, *Psychometrika*, 53, 4, pp. 503–524.

Witsenhausen, H.S. (1973), Metric inequalities and the zonoïd problem, *Proc. Amer. Math. Soc.*, 40, pp. 517–520.

Unified Reference List

References of each chapters are collated here in a single unified list. The number(s) following a reference indicates the chapter(s) where the reference is cited.

Achache, A. (1982), Galois connection of a fuzzy subset, *Fuzzy Sets and Systems*, 8, pp. 215–218. [6].

Achache, A. (1988), How to fuzzify a closure space, *J. Math.l Anal. Appl.*, 130, pp. 538–544. [6].

Al Ayoubi, B. (1991), Analyse des données de type M^1, Thèse, Université de Haute Bretagne, Rennes, France. [3].

Aschbacher, M., Baldi, P., Baum, E.B., Wilson, R.M. (1987), Embeddings of ultrametric spaces in finite dimensional structures, *SIAM J. Alg. Disc. Meth.*, 8, pp. 564–577. [2].

Assouad, P. (1977), Un espace hypermétrique non plongeable dans un espace L^1, *C.R. Acad. Sci. Paris*, t. 285, Série A, pp. 361–363. [2, 8].

Assouad, P. (1980a), Plongements isométriques dans L^1: aspect analytique, Séminaire d'Initiation à l'analyse, n° 14, Université de Paris-Sud, France, pp. 1–23. [2].

Assouad, P. (1980b), Caractérisations de sous-espaces normés de L^1 de dimension finie, Séminaire d'Analyse Fonctionnelle, Ecole Polytechnique, Palaiseau, France. [2].

Assouad, P. (1982), Sous-espaces de L^1 et inégalités hypermétriques, *C.R. Acad. Sci. Paris*, t. 294, Série 1, pp. 439–442. [2].

Assouad, P. (1984), Sur les inégalités valides dans L^1, *European J. Combin.*, 5, pp. 99–112. [2].

Assouad, P., Deza, M. (1982), Metric subspaces of L^1, Publications Mathématiques d'Orsay, Dept. Mathématiques, Université de Paris-Sud, France. [2, 8].

Avis, D. (1977), On the Hamming Cone, Technical Report 77-5. Dept. of Operations Research, Stanford University, USA. [2, 8].

Avis, D. (1978), Hamming metrics and facets of the Hamming cone, Technical Report SOCS-78.4, School of Computer Science, McGill University, Montreal, Canada. [2].

Avis, D. (1980), On the extreme rays of the metric cone, *Canad. J. Math*, 32, pp. 126–144. [2].

Avis, D. (1981), Hypermetric spaces and the Hamming Cone, *Canad. J. Math*, 33, pp. 795–802. [2].

Avis, D., Mutt (1989), All the facets of the six-point Hamming cone, *European. J. Combin.*, 10, pp. 309–312. [2].

Ball, K. (1990), Isometric embedding in ℓ_p-spaces, *Europ. J. Combinatorics*, 11, pp. 305–311. [8].

Banach, S. (1932), *Théorie des opérations linéaires*, Warsaw. [2].

Bandelt, H.J., Dress, A.W.M. (1990), A canonical decomposition theory for metrics on a finite set, Preprint 90-032, Diskrete Strukturen in der Mathematik, Universität Bielefeld 1, FRG. [2].

Bandelt, H.J., Dress, A.W.M. (1992), A canonical decomposition theory for metrics on a finite set, Adv. Math., 1, pp. 47–104. [2].

Barbut, M., Monjardet, B. (1971), *Ordre et Classification, Algèbre et Combinatoire*, Hachette, Paris. [6].

Barthélemy, J.P., Guénoche, A. (1988), *Les arbres et les représentations des proximités*, Masson, Paris. [2].

Barthélemy, J.P., Leclerc, B., Monjardet, B. (1984), Quelques aspects du consensus en classification, In Diday, E., et al., eds, *Data Analysis and Informatics*. North-Holland, Amsterdam, pp. 307–316. [2, 6].

Barthélemy, J.P., Leclerc, B., Monjardet, B. (1984), Ensembles ordonnés et taxonomie mathématique, *Ann. Discrete Mathematics*, 23, pp. 523–548. [4, 6].

Barthélemy, J.P., Leclerc, B., Monjardet, B. (1986), On the use of ordered sets in problems of comparison and consensus of classification, *J. Classification*, 3, pp. 187–224. [2, 6].

Batbedat, A. (1988), Les isomorphismes HTS et HTE (après la bijection de Benzécri/ Johnson), *Metron*, 46, pp. 47–59. [4].

Batbedat, A. (1989), Les dissimilarités Medas ou Arbas, *Statist. Anal. Données*, 14, pp. 1–18. [2, 6, 7].

Batbedat, A. (1990), *Les approches pyramidales dans la classification arborée*, Masson, Paris. [4, 6].

Batbedat, A. (1991), Ensembles ordonnés, dissimilarités, hypergraphes (finis on infinis), Personal communication. [4].

Batbedat, A. (1991), Phylogénie et dendrogrammes, Journées de Statistique, Strasbourg, France. [7].

Batbedat, A. (1992), Les distances quadrangulaires qui ont une orientation, *RAIRO-Rech. Opér.*, 26, pp. 15–29. [7].

Benayade, M. , Fichet, B. (1993), Algorithms for a geometrical P.C.A. with the L_1-norm, 4th Conference of the International Federation of Classification Societies, Paris. To appear in Diday, E., et al., eds (1994), *Proc. IFCS93 meeting*, Springer-Verlag, New York. [8].

Beninel, F. (1987), Problèmes de représentations sphériques des tableaux de dissimilarité, Thèse de 3ème cycle, Université de Haute Bretagne, Rennes, France. [3].

Benzécri, J.P. (1965), Problèmes et méthodes de la taxinomie, Rapport de recherche de l'Université de Rennes I, France. [4].

Benzécri, J.P., et al. (1973), *L'Analyse des Données. 1. La Taxinomie*, Dunod, Paris. [2, 4].

Bertrand, P., Diday, E. (1985), A visual representation of the compatibility between an order and a dissimilarity index: the pyramids, *Comput. Statist. Quart.*, 2, pp. 31–44. [6, 7].

Birkhoff, G. (1967), *Lattice Theory* (3rd ed.), Amer. Math. Soc., Providence. [6].

Blackwell, D. , Girshick, M.A. (1966), *Theory of Games*, fifth printing, Wiley, New York. [8].

Blake, I., Gilchrist, J. (1973), Addresses for graphs, *IEEE. Trans. Inform. Theory*, 19, pp. 683–688. [2].

Blumenthal, L.M. (1953), *Theory and Applications of Distance Geometry*, Clarendon Press, Oxford. [7].

Blyth, T.S., Janowitz, M.F. (1972), *Residuation theory*, Pergamon Press, Oxford. [4, 6].

Bock, H.H., ed. (1988), *Classification and Related Methods of Data Analysis*, North-Holland, Amsterdam. [7].

Boorman, S.A., Olivier, D.C. (1973), Metrics on spaces of finite trees, *J. Math. Psychol.*, 10, pp. 26–59. [6].

Boscovich, R.J. (1757), De litteraria expeditione per pontificam ditionem, et synopsis amplioris operis, ac habentur plura eius ex exemplaria etiam sensorum impressa, *Bononiensi Scientiarum et Artium Instituto Atque Academia Commentarii*, 4, pp. 353–396. [8].

Bove, G. (1989), New methods of representation of proximity data, Doctoral thesis, Università La Sapienza, Rome, Italy, (in Italian). [2].

Bretagnolle, J., Dacunha-Castelle, D., Krivine, J.L. (1966), Lois stables et espaces L^p, *Ann. Inst. H. Poincaré Sect. B*, II, n° 3, pp. 231–259. [2, 8].

Brito, P. (1994), Symbolic objects: order structure and pyramidal clustering, *IEEE Trans. Knowl. and Data Engin.*, to appear. [6].

Brossier, G. (1980), Représentation ordonnée des classifications hiérarchiques, *Statist. Anal. Données*, 5, pp. 31–44. [2].

Brossier, G. (1986), Problèmes de représentation des données par des arbres, Thèse d'Etat, Université de Haute Bretagne, Rennes 2, France. [2].

Brossier, G. (1987), Etude des matrices de proximité rectangulaires en vue de la classification, *Rev. Statist. Appl.*, 35, pp. 43–68. [2].

Buneman, P. (1971), The recovery of trees from measures of dissimilarity, In Hodson, F.R., Kendall, D.G., Taŭtu, P., eds., *Mathematics in the Archaeological and Historical Sciences*, The University Press, Edinburgh, pp. 387–395. [7].

Buneman, P. (1974), A note on metric properties of trees, *J. Combin. Theory Ser. B*, 17, pp. 48–50. [2].

Cailliez, F., Pagès, J.P. (1976), *Introduction à l'Analyse des Données*, S.M.A.S.H., Paris. [2].

Carroll, J.D. (1976), Spatial, non-spatial and hybrid models for scaling, *Psychometrika*, 41, pp. 439–463. [2].

Coppi, R., Bolasco, S. (1989), *Multiway Data Analysis*, North-Holland, Amsterdam. [2, 5].

Crippen, G., Havel, T. (1988), *Distance Geometry and Molecular Conformation*, Wiley, New York. [2].

Critchley, F. (1980), Optimal norm characterisations of multidimensional scaling methods and some related data analysis problems, In Diday, E., et al., eds., *Data Analysis and Informatics*, North-Holland, Amsterdam, pp. 209–229. [2].

Critchley, F. (1983), Ziggurats and dendrograms, Warwick Statistics Research Report n° 43, U.K. [5].

Critchley, F. (1986a), Dimensionality theorems in hierarchical cluster analysis and multidimensional scaling, In Diday, E., et al., eds., *Data Analysis and Informatics 4*, North-Holland, Amsterdam, pp. 45–70. [2].

Critchley, F. (1986b), Some observations on distance matrices, In de Leeuw, J., et al., eds., *Multidimensional Data Analysis*, DSWO Press, Leiden, pp. 53–60. [2].

Critchley, F. (1988), On certain linear mappings between inner-product and squared-distance matrices, *Linear Algebra Appl.*, 105, pp. 91–107. [2].

Critchley, F. (1988a), On certain linear mappings between inner-product and squared-distance matrices, *Linear Algebra Appl.*, 150, pp. 91–107. [7].

Critchley, F. (1988b), On exchangeability-based equivalence relations induced by strongly Robinson and, in particular, by quadripolar Robinson dissimilarity matrices, Warwick Statistics Research Report n° 152, U.K. [7].

Critchley, F. (1988c), On quadripolar Robinson dissimilarity matrices, Warwick Statistics Research Report n° 153, U.K., To appear in Diday, E., et al., eds (1994), *Proc. IFCS93 meeting*, Springer-Verlag, New York. [7].

Critchley, F. (1994), On exchangeability-based equivalence relations induced by strongly Robinson and, in particular, by quadripolar Robinson dissimilarity matrices, In Van Cutsem, B., ed., *Classification and Dissimilarity Analysis*, Ch. 7, Lecture Notes in Statistics, Springer-Verlag, New York. [2].

Critchley, F., Fichet, B. (1993), The partial order by inclusion of the principal classes of dissimilarity on a finite set, and some of their basic properties, Joint Research Report, University of Warwick, U.K. and Université d'Aix-Marseille II, France. [2].

Critchley, F., Fichet, B. (1994), The partial order by inclusion of the principal classes of dissimilarity on a finite set, and some of their basic properties, In Van Cutsem, B., ed., *Classification and Dissimilarity Analysis*, Lecture Notes in Statistics, Ch. 2, Springer-Verlag, New York. [7, 8].

Critchley, F., Marriott, P.K., Salmon, M.H. (1992), Distances in statistics, In *Proceedings of the 36th meeting of the Italian Statistical Society*, CISU, Rome, pp. 36–60. [2].

Critchley, F., Van Cutsem, B. (1989), Predissimilarities, prefilters and ultrametrics on an arbitrary set. Rapport de Recherche commun de l'Université de Warwick, U.K. et du Laboratoire TIM3-IMAG, Grenoble, France. [4].

Critchley, F., Van Cutsem, B. (1992), An order-theoretic unification and generalisation of certain fundamental bijections in mathematical classification – I, Joint Research Report, University of Warwick, U.K., and Université Joseph Fourier, Grenoble, France. [4].

Critchley, F., Van Cutsem, B. (1992), An order-theoretic unification and generalisation of certain fundamental bijections in mathematical classification – II, Joint Research Report, University of Warwick, U.K., and Université Joseph Fourier, Grenoble, France. [5].

Critchley, F., Van Cutsem, B. (1994), An order-theoretic unification and generalisation of certain fundamental bijections in mathematical classification - I, In Van Cutsem, B., ed., *Classification and Dissimilarity Analysis*, Ch. 4, Lecture Notes in Statistics, Springer-Verlag, New York. [2, 5, 6].

Critchley, F., Van Cutsem, B. (1994), An order-theoretic unification and generalisation of certain fundamental bijections in mathematical classification - II, In Van Cutsem, B., ed., *Classification and Dissimilarity Analysis*, Ch. 5, Lecture Notes in Statistics, Springer-Verlag, New York. [2, 6].

Croisot, R. (1956), Applications résiduées, *Ann. Sci. Ecole Norm. Sup. Paris*, 73, pp. 453–474. [6].

Daniel-Vatonne, M.C., and de La Higuera, C. (1993), Les termes : un modèle de représentation et structuration de données symboliques, Math. Inform. Sci. Humaines, 122, pp. 41–63. [6].

Defays, D. (1978), Analyse hiérarchique des préférences et généralisations de la transitivité, Math. Sci. Humaines, 61, pp. 5–27. [6].

Degenne, A., and Vergès, P. (1973), Introduction à l'analyse de similitude, Revue Française de Sociologie, XIV, pp. 471–512. [6].

Delclos, Th. (1987), Sur la représentation arborée en analyse des données, Mémoire D.E.A. Math. Appl. Université de Provence, Marseille, France. [2].

de Leeuw, J., Heiser, W. (1982), Theory of multidimensional scaling, In Krishnaiah, P.R., ed., Handbook of Statistics, Vol. 2, North-Holland, Amsterdam, Ch. 13. [7].

de Leeuw, J., Heiser, W., Meulman, J., Critchley, F., eds. (1986), Multidimensional Data Analysis, DSWO Press, Leiden. [2, 7].

De Soete, G., Desarbo, W.S., Furnas, G.W., Carroll, J.D. (1984), Tree representation of rectangular proximity matrices, In Degreef,E., Van Buggenhaut, J., eds., Trends in Mathematical Psychology, Elsevier Science Publishers, North-Holland, Amsterdam, pp. 377–392. [2].

Deza, M. (Tylkin) (1960), On Hamming geometry of unitary cubes, Dokl. Acad. Nauk SSSR, 134, pp. 1037–1040. [2, 8].

Deza, M. (Tylkin) (1962), On the realizibility of distance matrices in unit cubes, Problemy Kibernetiki, 7, pp. 31–45. [2].

Deza, M. (Tylkin) (1969), Linear metric properties of unitary cubes (in Russian), Proc. of the 4th Soviet Union Conference on Coding theory and transmission of information, Moscow-Tashkent, pp. 77–85. [2].

Deza, M. (1992), Hypermetrics, ℓ_1-metrics and Delauney polytopes, Proc. of Distancia 92. International Meeting on distance analysis, Rennes. pp. 7–10. [2].

Deza, M., Grishukhin, V.P., Laurent, M. (1990a), Extreme hypermetrics and L-polytopes, Report No. 90668-OR. Institut für Okonometrie und Operations Research, Universität Bonn, FRG. [2].

Deza, M., Grishukhin, V.P., Laurent, M. (1990b), Hypermetric cone is polyhedral, Combinatorica, to appear. [2].

Diday, E. (1982), Croisements, ordres et ultramétriques: application à la recherche de consensus en classification automatique, Rapport de recherche, n° 144, I.N.R.I.A., Rocquencourt, France. [2].

Diday, E. (1983), Croisements, ordres et ultramétriques, Math. Sci. Hum., 2, pp. 31–54. [2].

Diday, E. (1984), Une représentation visuelle des classes empiétantes : les pyramides, Rapport de recherche, n° 291, I.N.R.I.A., Rocquencourt, France. [2, 4, 7].

Diday, E. (1986), Une représentation visuelle des classes empiétantes : les pyramides, RAIRO Automat.-Prod. Inform. Ind., 20, pp. 475–526. [6].

Diday, E. (1986), Orders and overlapping clusters in pyramids, in In de Leeuw, J., Heiser, W., Meulman, J., Critchley, F., eds., Multidimensional Data Analysis, DSWO Press, Leiden, pp. 201–234. [7].

Diday, E. (1988), The symbolic approach in clustering and related methods of data analysis: the basic choices, In Bock, H.H., ed., Classification and Related Methods of Data Analysis, Amsterdam, North-Holland, pp. 673–683. [6].

Diday, E., Escoufier, Y., Lebart, L., Pagès, J.P., Schektman, Y., Tomassone, R., eds. (1986), Data Analysis and Informatics 4, North-Holland, Amsterdam. [7].

Dirichlet, G.L. (1850), Über die Reduction der positiven quadratischen formen mit drei unbestimmten ganzen zahlen, J. Reine Angew. Math., 50, pp. 209–227. [2].

Dobson, A.J. (1974), Unrooted trees for numerical taxonomy, J. Appl. Probab., 11, pp. 32–42. [2].

Doignon, J.-P., Monjardet, B., Roubens, M., Vincke, Ph. (1986), Biorder families, valued relations and preference modelling, J. Math. Psych. 30, pp. 435–480. [6].

Dor, L.E. (1976), Potentials and isometric embeddings in L_1, Israel J. Math., 24, pp. 260–268. [2, 8].

Dubois,D., Prade, H. (1980), Fuzzy Sets and Systems: Theory and Applications, Academic Press, New York. [6].

Dubreil,P., Croisot, R. (1954), Propriétés générales de la résiduation en liaison avec les correspondances de Galois, Collect. Math., 7, pp. 193–203. [6].

Duquenne, V. (1987), Contextual implications between attributes and some representation properties for finite lattices, In Ganter, B., Wille, R., Wolf, K.E., ed., Beiträge zur Begriffsanalyse, Wissenchaftverlag, Mannheim, pp. 213–240. [6].

Durand, C. (1988), Une approximation de Robinson inférieure maximale, Rapport de Recherche Laboratoire de Mathématiques Appliquées et Informatique, n° 88-02, Université de Provence, Marseille, France. [7].

Durand, C. (1989), Ordres et graphes pseudo-hiérarchiques: théorie et optimisation algorithmique, Thèse de l'Université de Provence, Marseille, France. [2].

Durand, C., Fichet, B. (1988), One-to-one correspondences in pyramidal representation: a unified approach, In Bock, H.H., ed., Classification and Related Methods of Data Analysis, North-Holland, Amsterdam, pp. 85–90. [2, 6, 7].

Eisler, H. (1973), The algebraic and statistical tractability of the city-block metric, *British J. Math. Statist. Psych.*, 26, pp. 212–218. [8].

Escoufier, Y. (1975), Le positionnement multidimensionnel, *Rev. Statist. Appl.*, 23, pp. 5–14. [2].

Farebrother, R.W. (1987), The historical development of the L_1 and L_∞ estimation procedures 1793-1930, In Dodge, Y., ed., *Statistical Data Analysis based on the L_1-norm and Related Methods*, North-Holland, Amsterdam, pp. 37–63. [8].

Fichet, B. (1983), Analyse factorielle sur tableaux de dissimilarité. Application aux données sur signes de présence-absence en médecine, Thèse de Biologie Humaine, Université d'Aix-Marseille II, France. [2].

Fichet, B. (1984), Sur une extension des hiérarchies et son équivalence avec certaines matrices de Robinson, Journées de Statistique, Montpellier, France. [2, 4, 7].

Fichet, B. (1986), Data analysis: geometric and algebraic structures, First World Congress of Bernoulli Society, Tashkent, URSS, T1, pp. 75–77. [2, 8].

Fichet, B. (1987), The role played by L_1 in data analysis, In Dodge, Y., ed., *Statistical Data Analysis based on the L_1-norm and Related Methods*, North-Holland, Amsterdam, pp. 185–193. [2, 8].

Fichet, B. (1988), L_p-spaces in data analysis, In Bock, H.H., ed., *Classification and Related Methods of Data Analysis*, North-Holland, Amsterdam, pp. 439–444. [2, 8].

Fichet, B. (1989), Sur la dimension des figures finies en norme L_1, 47th session of I.S.I., Paris, 1, pp. 325–326. [8].

Fichet, B. (1992), The notion of sphericity for finite L_1-figures of data analysis, In Dodge, Y., ed., *L_1-Statistical Analysis and Related Methods*, North-Holland, Amsterdam, pp. 129–144. [8].

Fichet, B., Gaud, E. (1987), On Euclidean images of a set endowed with a preordonnance, *J. Math. Psych.*, 31, pp. 24–43. [2].

Fichet, B., Le Calvé, G. (1984), Structure géométrique des principaux indices de dissimilarité sur signes de présence-absence, *Statist. Anal. Données*, 9, pp. 11–44. [2, 3].

Fraser, D.A.S., Massam, H. (1989), Mixed primal-dual bases algorithm for regression under inequality constraints: application to concave regression, *Scand. J. Statist.*, 16, pp. 65–74. [2].

Fréchet, M. (1910), Les dimensions d'un ensemble abstrait, *Math. Ann.*, 68, pp. 145–168. [2, 8].

Fréchet, M. (1935), Sur la définition axiomatique d'une classe d'espaces vectoriels distanciés applicables vectoriellement sur l'espace de Hilbert, *Ann. of Math.*, 36, pp. 705–718. [2, 3].

Ganter, B., Rindfrey, K., Skorsky, M. (1986), Software for concept analysis, In Gaul, W., Schader, M., eds., *Classification as a Tool of Research*, North-Holland, Amsterdam, pp. 161–168. [6].

Gaud, E. (1983), Représentation d'une préordonnance: étude de ses images euclidiennes, problèmes de graphes dans sa représentation hiérarchique, Thèse de 3ème cycle, Math. Appl., Université de Provence, Marseille, France. [2].

Gauss, K.F. (1831), Göttingsche gelehrte anzeigen, 2, 1075. [2].

Gelfand, A.E. (1971), Rapid seriation methods with archaeological applications, In Hodson, R.F., et al., eds., *Mathematics in the Archaeological and Historical Sciences*, Edinburgh University Press. [2].

Gentilhomme, Y. (1968), Les ensembles flous en linguistique, *Cahiers de Linguistique Théorique et Appliquée*, 5, pp. 47–65. [6].

Golumbic, M.C. (1980), *Algorithmic Graph Theory and Perfect Graphs*, Academic Press, New York. [6].

Gondran, M. (1976), La structure algébrique des classifications hiérarchiques, *Ann. INSEE*, 22-23, pp. 181–190. [6].

Gordon, A.D. (1987), A review of hierarchical classification, *J. Roy. Statist. Soc. A*, 150, pp. 119–137. [7].

Gower, J.C. (1985), Properties of Euclidean and non-Euclidean distance matrices, *Linear Algebra Appl.*, 67, pp. 81–97. [7].

Gower, J.C., Legendre, P. (1986), Metric and Euclidean properties of dissimilarity coefficients, *J. Classification*, 3, pp. 5–48. [2, 3].

Gower, J.C., Ross, G.J.S. (1969), Minimum spanning tree and single linkage analysis, *Applied Statistics*, 18, pp. 54–64. [6].

Grünbaum, B. (1967), *Convex Polytopes*, Interscience Publishers, London. [8].

Guttman, L. (1968), A general nonmetric technique for finding the smallest coordinate space for a configuration of points, *Psychometrika*, 33, pp. 469–506. [2].

Hartigan, J.A. (1967), Representations of similarity matrices by trees, *J. Amer. Statist. Assoc.*, 62, pp. 1140–1158. [4].

Hayden, T.L., Wells, J. (1988), Approximation by matrices positive semidefinite on a subspace, *Linear Algebra Appl.*, 109, pp. 115–130. [7].

Hermite, C. (1850), Sur différents objets de la théorie des nombres, *J. Reine Angew. Math.*, 40, pp. 261–315. [2].

Hodson, R.F., Kendall, D.G., Tautu, P. (1971), *Mathematics in the Archaeological and Historical Sciences*, Edinburgh University Press. [2].

Holman, W. (1972), The relation between hierarchical and Euclidean models for psychological distances, *Psychometrika*, 37, pp. 417–423. [2, 8].

Horn, R.A., Johnson, C.A. (1985), *Matrix Analysis*, Cambridge University Press. [2].

Hubert, L. (1974), Some applications of graph theory and related nonmetric techniques to problems of approximate seriation: the case of symmetric proximity measures, *British J. Math. Statist. Psych.*, 7, pp. 133–153. [2].

Janowitz, M.F. (1978), An order theoretic model for cluster analysis, *SIAM J. Appl. Math.*, 34, pp. 55–72. [2, 4, 6].

Jardine, C.J., Jardine, N., Sibson, R. (1967), The structure and construction of taxonomic hierarchies, *Math. Biosci.*, 1, 171–179. [4].

Jardine, N., Sibson, R. (1971), *Mathematical Taxonomy*, Wiley, London. [2, 4, 5, 6].

Johnson, S.C. (1967), Hierarchical clustering schemes, *Psychometrika*, 32, pp. 241–254. [2, 4, 6, 7].

Joly, S. (1989), On ternary distances $D(i,j,k)$, Second Conference of the International Federation of Classification Societies, Charlotteville, Virginia, USA. [2].

Joly, S., Le Calvé, G. (1986), Etude des puissances d'une distance, *Statist. Anal. Données*, 11, pp. 30–50. [2, 3, 8].

Joly, S., Le Calvé, G. (1992), Realisable 0-1 matrices and city-block distance, Rapport de recherche 92-1, Laboratoire Analyse des Données, Université de Haute Bretagne, Rennes, France. [3].

Joly, S., Le Calvé, G. (1992), 3-way distances, *J. Classification*, (to appear). [4].

Kaufman, A. (1973), *Introduction à la Théorie des Sous-ensembles Flous*, Masson, Paris. Translation (1975): *Introduction to the Theory of Fuzzy sets*, Academic Press, New York. [6].

Kelly, J.B. (1968), Products of zero-one matrices, *Canad. J. Math.* 20. pp. 298–329. [2, 3].

Kelly, J.B. (1970a), Metric inequalities and symmetric differences, In Shisha, O., ed, *Inequalities II*, Academic Press, New York. pp. 193–212. [2].

Kelly, J.B. (1970b), Combinatorial inequalities, In Guy, R., Hanani, H., Saver, N., Schoenberg, J., ed., *Combinatorial Structures and their applications*, Gordon and Breach, New York, pp. 201–207. [2].

Kelly, J.B. (1972), Hypermetric spaces and metric transforms, In Shisha, O., ed, *Inequalities III*, Academic Press, New York, pp. 149–158. [2, 3].

Kelly, J.B. (1975), Hypermetric spaces, In Kelly, L.M., ed, *The geometry of metric and linear spaces*, Springer-Verlag, Berlin, pp. 17–31. [2].

Krasner, M. (1944), Nombres semi-réels et espaces ultramétriques, *C.R. Acad. Sci. Paris*, 219, pp. 433–435. [2].

Kruskal, J.B. (1964), Multidimensional scaling by optimizing goodness of fit to a nonmetric hypothesis, *Psychometrika*, 29, pp. 1–27. [2].

Laplace, P.S. (1793), Sur quelques points du système du monde, Mémoires de l'Académie Royale des Sciences de Paris, pp. 1–87. Reprinted in *Oeuvres Complètes de Laplace* (1895), Paris, Gauthier-Villars, 11, pp. 477–558. [8].

Lawson, C.M. , Hanson, R.J. (1974), *Solving Least Squares Problems*, Englewood Cliffs, Prentice-Hall. [8].

Le Calvé, G. (1985), Distances à centre, *Statist. Anal. Données*, 10, pp. 29–44. [2].

Le Calvé, G. (1987), L_1-embeddings of a data structure (I, D), In Dodge, Y., ed., *Statistical Data Analysis based on the L_1-norm and Related Methods*, North-Holland, Amsterdam, pp. 195–202. [2, 3, 8].

Le Calvé, G. (1988), Similarities functions, In: Edwards, D., Raum, N.E., eds., *Proceedings of COMPSTAT 88*, Physica Verlag, Heidelberg, pp. 341–347. [3].

Leclerc, B. (1979), Semi-modularité des treillis d'ultramétriques, *C.R. Acad. Sci. Paris Sér. A*, 288, pp. 575–577. [6].

Leclerc, B. (1981), Description combinatoire des ultramétriques, *Math. Sci. Humaines*, 73, pp. 5–31. [6].

Leclerc, B. (1984a), Efficient and binary consensus functions on transitively valued relations, *Math. Social Sci.*, 8, pp. 45–61. [6].

Leclerc, B. (1984b), Indices compatibles avec une structure de treillis et fermeture résiduelle, Rapport C.M.S. P.011, Centre d'Analyse et de Mathématiques Sociales, Paris. [6].

Leclerc, B. (1990), The residuation model for ordinal construction of dissimilarities and other valued objects, Rapport C.M.S. P.063, Centre d'Analyse et de Mathématiques Sociales, Paris. [6].

Leclerc, B. (1991), Aggregation of fuzzy preferences: a theoretic Arrow-like approach, *Fuzzy Sets and Systems*, 43, pp. 291–309. [6].

Leclerc, B. (1993), Minimum spanning trees for tree metrics: abridgements and adjustments, Research Report C.M.S. P.084, Centre d'Analyse et de Mathématique Sociales, Paris, France. [7].

Lerman, I.C. (1968), Analyse du problème de la recherche d'une hiérarchie de classifications, Report 22, Maison des Sciences de l'Homme, Paris, France. [4].

Lerman, I.C. (1970), *Les bases de la classification automatique*, Gauthier-Villars, Paris. [2, 4].

Lew, J.S. (1978), Some counterexamples in multidimensional scaling, *J. Math. Psych.*, 17, pp. 247–254. [2].

Marcotorchino, F. (1991), La classification mathématique aujourd'hui, *Publications Scientifiques et Techniques d'IBM France*, 2, pp. 35–93. [1]

Mathar, R. (1985), The best Euclidean fit to a given distance matrix in prescribed dimensions, *Linear Algebra Appl.*, 67, pp. 1–6. [7].

Menger, K. (1931a), Bericht über metrische Geometrie, *Jahresber der deutschen Math.-Ver.*, 40, pp. 201–219. [2].

Menger, K. (1931b), New foundations of euclidean geometry, *Amer. J. Math.*, 53, pp. 721–745. [2].

Minkowski, H. (1891), Über die positiven quadratischen Formen und über kettenbruchahnliche Algorithm, *J. Reine Angew. Math.*, 107, pp. 278–297. [2].

Monjardet, B. (1988), A generalization of probabilistic consistency: linearity conditions for valued preference relations, In Kacprzyk, J., Roubens, M., eds., *Non-Conventional Preference Relations in Decision Making*, Lecture Notes in Econom. and Math. Syst., 301, Springer-Verlag, Berlin, pp. 36–53. [6].

Monjardet, B. (1990), Arrowian characterization of latticial federation consensus functions, *Math. Soc. Sci.*, 20, pp. 51–71. [6].

Negoita, C.V. (1981), *Fuzzy systems*, Abacus Press, Tunbridge Wells. [6].

Öre, O. (1944), Galois connections, *Trans. Amer. Math. Soc.*, 55, pp. 494–513. [6].

Oxender, D., Fox, C. (1987), *Protein Engineering*, Alan Liss Inc., New York. [2].

Patrinos, A.N., Hakimi, S.L. (1972), The distance matrix of a graph and its tree realization, *Quart. Appl. Math.*, 30, pp. 255–269. [2].

Polat, N., Flament, C. (1980), Applications galoisiennes proches d'une application entre treillis, *Math. Sci. Humaines*, 70, pp. 33–49. [6].

Robinson, W.S. (1951), A method for chronological ordering of archaeological deposits, *American Antiquity*, 16, pp. 293–301. [2, 7].

Rockafellar, R.T. (1970), *Convex Analysis*, Princeton University Press. [2].

Roubens, M. (1986), Comparison of flat fuzzy numbers, In Bandler, N., Kandel, A., eds., Proceedings of NAFIPS'86, pp. 462–476. [6].

Roubens, M., Vincke, Ph. (1988), Fuzzy possibility graphs and their application to ranking fuzzy numbers, In Kacprzyk, J., Roubens, M., eds., *Non-conventional Preference Relations in Decision Making*, Springer-Verlag, Berlin, pp. 119–128. [6].

Roux, M. (1968), Un algorithme pour construire une hiérarchie particulière, Thèse de 3ème cycle, Université Paris VI. [6].

Schoenberg, I.J. (1935), Remarks to Maurice Fréchet's article "Sur la définition axiomatique d'une classe d'espace distanciés vectoriellement applicable sur l'espace de Hilbert", Ann. of Math., 36, pp. 724–732. [2, 8].

Schoenberg, I.J. (1937), On certain metric spaces arising from Euclidean spaces by a change of metric and their imbedding in Hilbert space, Ann. of Math., 38, pp. 787–793. [2, 3].

Schoenberg, I.J. (1938), Metric spaces and positive definite functions, Trans. Amer. Math. Soc., 44, pp. 522–536. [2, 3].

Schur, J. (1911), Bemerkungen zur Theorie der beschrankter Bilinearformen mit unendlich vielen Veränderlichen, J. Reine Angew. Math., 140, pp. 1–28. [3].

Shepard, R.N. (1962), The analysis of proximities: multidimensional scaling with an unknown distance function I, II, Psychometrika, 27, pp. 125–140, pp. 219–246. [2].

Shmuely, Z. (1974), The structure of Galois connections, Pacific J. of Math., 54, pp. 209–225. [6].

Simoes-Pereira, J.M.S. (1967), A note on tree realizability of a distance matrix, J. Combin. Theory Ser. B, 6, pp. 303–310. [2].

Smolenskii, Y.A. (1969), A method for linear recording of graphs, U.S.S.R. Comput. Math. and Math. Phys., 2, pp. 396–397. [2].

Tenenhaus, M. (1988), Canonical analysis of two convex polyhedral cones and applications, Psychometrika, 53, 4, pp. 503–524. [8].

Torgerson, W.S. (1958), Theory and methods of scaling, Wiley, New York. [2, 3].

Van Cutsem, B. (1983a), Décomposition d'une ultramétrique : ultramétriques simples et semi-simples, Rapport de Recherche n°. 388, Laboratoire IMAG, Grenoble, France. [5].

Van Cutsem, B. (1983b), Ultramétriques, distances, ϕ-distances maximum dominées par une dissimilarité donnée, Statist. Anal. Données, 8, pp. 42–63. [5].

Volle, M. (1985), Conjugaison par tranches, Ann. Mat. Pura Appl., 139, pp. 279–312. [6].

Wells, J.H., Williams, L.R. (1975), Embeddings and extensions in analysis, Springer-Verlag, Berlin. [2].

Wilkeit, E. (1990), Isometric embeddings in Hamming graphs, J. Combin Theory Ser. B, 50, pp. 179–197. [2].

Wille, R. (1982), Restructuring lattice theory : an approach based on hierarchies of concepts, In Rival, I., Reidel, D., eds., *Ordered Sets*, Dordrecht, pp. 445-470. [6].

Witsenhausen, H.S. (1973), Metric inequalities and the zonoïd problem, *Proc. Amer. Math. Soc.*, 40, pp. 517–520. [2, 8].

Zaks, Y.M., Muchnik, I.B. (1989), Monotone systems for incomplete classification of a finite set of objects, *Automat. Remote Control*, 4, pp. 155–164. [4].

Zaretskii, K. (1965), Constructing a tree on the basis of a set of distances between the hanging vertices, (in Russian), *Uspekhi. Mat. Nauk*, 20, pp. 90–92. [2].

Zegers, F.E., Ten Berge, J.M.F. (1985), A family of association coefficients for metric scales, *Psychometrika*, 50, pp. 17–24. [2].

Lecture Notes in Statistics

For information about Volumes 1 to 8 please contact Springer-Verlag

Vol. 9: B. Jørgensen, Statistical Properties of the Generalized Inverse Gaussian Distribution. vi, 188 pages, 1981.

Vol. 10: A.A. McIntosh, Fitting Linear Models: An Application of Conjugate Gradient Algorithms. vi, 200 pages, 1982.

Vol. 11: D.F. Nicholls and B.G. Quinn, Random Coefficient Autoregressive Models: An Introduction. v, 154 pages, 1982.

Vol. 12: M. Jacobsen, Statistical Analysis of Counting Processes. 226 pages, 1982.

Vol. 13: J. Pfanzagl (with the assistance of W. Wefelmeyer), Contributions to a General Asymptotic Statistical Theory. vii, 315 pages, 1982.

Vol. 14: GLIM 82: Proceedings of the International Conference on Generalised Linear Models. Edited by R. Gilchrist. v, 188 pages, 1982.

Vol. 15: K.R.W. Brewer and M. Hanif, Sampling with Unequal Probabilities. ix, 164 pages, 1983.

Vol. 16: Specifying Statistical Models: From Parametric to Non-parametric, Using Bayesian or Non-Bayesian Approaches. Edited by J.P. Florens, M. Mouchart, J.P. Raoult, L. Simar, and A.F.M. Smith, xi, 204 pages, 1983.

Vol. 17: I.V. Basawa and D.J. Scott, Asymptotic Optimal Inference for Non-Ergodic Models. ix, 170 pages, 1983.

Vol. 18: W. Britton, Conjugate Duality and the Exponential Fourier Spectrum. v, 226 pages, 1983.

Vol. 19: L. Fernholz, von Mises Calculus For Statistical Functionals. 124 pages, 1983.

Vol. 20: Mathematical Learning Models — Theory and Algorithms: Proceedings of a Conference. Edited by U. Herkenrath, D. Kalin, W. Vogel. xiv, 226 pages, 1983.

Vol. 21: H. Tong, Threshold Models in Non-linear Time Series Analysis. x, 323 pages, 1983.

Vol. 22: S. Johansen, Functional Relations, Random Coefficients and Nonlinear Regression with Application to Kinetic Data, viii, pages, 1984.

Vol. 23: D.G. Saphire, Estimation of Victimization Prevalence Using Data from the National Crime Survey. v, 165 pages, 1984.

Vol. 24: T.S. Rao, M.M. Gabr, An Introduction to Bispectral Analysis and Bilinear Time Series Models. viii, 280 pages, 1984.

Vol. 25: Time Series Analysis of Irregularly Observed Data. Proceedings, 1983. Edited by E. Parzen. vii, 363 pages, 1984.

Vol. 26: Robust and Nonlinear Time Series Analysis. Proceedings, 1983. Edited by J. Franke, W. Härdle and D. Martin. ix, 286 pages, 1984.

Vol. 27: A. Janssen, H. Milbrodt, H. Strasser, Infinitely Divisible Statistical Experiments. vi, 163 pages, 1985.

Vol. 28: S. Amari, Differential-Geometrical Methods in Statistics. 290 pages, 1985.

Vol. 29: Statistics in Ornithology. Edited by B.J.T. Morgan and P.M. North. xxv, 418 pages, 1985.

Vol 30: J. Grandell, Stochastic Models of Air Pollutant Concentration. v, 110 pages, 1985.

Vol. 31: J. Pfanzagl, Asymptotic Expansions for General Statistical Models. vii, 505 pages, 1985.

Vol. 32: Generalized Linear Models. Proceedings, 1985. Edited by R. Gilchrist, B. Francis and J. Whittaker. vi, 178 pages, 1985.

Vol. 33: M. Csörgo, S. Csörgo, L. Horváth, An Asymptotic Theory for Empirical Reliability and Concentration Processes. v, 171 pages, 1986.

Vol. 34: D.E. Critchlow, Metric Methods for Analyzing Partially Ranked Data. x, 216 pages, 1985.

Vol. 35: Linear Statistical Inference. Proceedings, 1984. Edited by T. Calinski and W. Klonecki. vi, 318 pages, 1985.

Vol. 36: B. Matérn, Spatial Variation. Second Edition. 151 pages, 1986.

Vol. 37: Advances in Order Restricted Statistical Inference. Proceedings, 1985. Edited by R. Dykstra, T. Robertson and F.T. Wright. viii, 295 pages, 1986.

Vol. 38: Survey Research Designs: Towards a Better Understanding of Their Costs and Benefits. Edited by R.W. Pearson and R.F. Boruch. v, 129 pages, 1986.

Vol. 39: J.D. Malley, Optimal Unbiased Estimation of Variance Components. ix, 146 pages, 1986.

Vol. 40: H.R. Lerche, Boundary Crossing of Brownian Motion. v, 142 pages, 1986.

Vol. 41: F. Baccelli, P. Brémaud, Palm Probabilities and Stationary Queues. vii, 106 pages, 1987.

Vol. 42: S. Kullback, J.C. Keegel, J.H. Kullback, Topics in Statistical Information Theory. ix, 158 pages, 1987.

Vol. 43: B.C. Arnold, Majorization and the Lorenz Order: A Brief Introduction. vi, 122 pages, 1987.

Vol. 44: D.L. McLeish, Christopher G. Small, The Theory and Applications of Statistical Inference Functions. vi, 124 pages, 1987.

Vol. 45: J.K. Ghosh (Editor), Statistical Information and Likelihood. 384 pages, 1988.

Vol. 46: H.-G. Müller, Nonparametric Regression Analysis of Longitudinal Data. vi, 199 pages, 1988.

Vol. 47: A.J. Getson, F.C. Hsuan, {2}-Inverses and Their Statistical Application. viii, 110 pages, 1988.

Vol. 48: G.L. Bretthorst, Bayesian Spectrum Analysis and Parameter Estimation. xii, 209 pages, 1988.

Vol. 49: S.L. Lauritzen, Extremal Families and Systems of Sufficient Statistics. xv, 268 pages, 1988.

Vol. 50: O.E. Barndorff-Nielsen, Parametric Statistical Models and Likelihood. vii, 276 pages, 1988.

Vol. 51: J. Hüsler, R.-D. Reiss (Editors), Extreme Value Theory. Proceedings, 1987. x, 279 pages, 1989.

Vol. 52: P.K. Goel, T. Ramalingam, The Matching Methodology: Some Statistical Properties. viii, 152 pages, 1989.

Vol. 53: B.C. Arnold, N. Balakrishnan, Relations, Bounds and Approximations for Order Statistics. ix, 173 pages, 1989.

Vol. 54: K.R. Shah, B.K. Sinha, Theory of Optimal Designs. viii, 171 pages, 1989.

Vol. 55: L. McDonald, B. Manly, J. Lockwood, J. Logan (Editors), Estimation and Analysis of Insect Populations. Proceedings, 1988. xiv, 492 pages, 1989.

Vol. 56: J.K. Lindsey, The Analysis of Categorical Data Using GLIM. v, 168 pages, 1989.

Vol. 57: A. Decarli, B.J. Francis, R. Gilchrist, G.U.H. Seeber (Editors), Statistical Modelling. Proceedings, 1989. ix, 343 pages, 1989.

Vol. 58: O.E. Barndorff-Nielsen, P. Blæsild, P.S. Eriksen, Decomposition and Invariance of Measures, and Statistical Transformation Models. v, 147 pages, 1989.

Vol. 59: S. Gupta, R. Mukerjee, A Calculus for Factorial Arrangements. vi, 126 pages, 1989.

Vol. 60: L. Györfi, W. Härdle, P. Sarda, Ph. Vieu, Nonparametric Curve Estimation from Time Series. viii, 153 pages, 1989.

Vol. 61: J. Breckling, The Analysis of Directional Time Series: Applications to Wind Speed and Direction. viii, 238 pages, 1989.

Vol. 62: J.C. Akkerboom, Testing Problems with Linear or Angular Inequality Constraints. xii, 291 pages, 1990.

Vol. 63: J. Pfanzagl, Estimation in Semiparametric Models: Some Recent Developments. iv, 112 pages, 1990.

Vol. 64: S. Gabler, Minimax Solutions in Sampling from Finite Populations. v, 132 pages, 1990.

Vol. 65: A. Janssen, D.M. Mason, Non-Standard Rank Tests. vi, 252 pages, 1990.

Vol. 66: T. Wright, Exact Confidence Bounds when Sampling from Small Finite Universes. xvi, 431 pages, 1991.

Vol. 67: M.A. Tanner, Tools for Statistical Inference: Observed Data and Data Augmentation Methods. vi, 110 pages, 1991.

Vol. 68: M. Taniguchi, Higher Order Asymptotic Theory for Time Series Analysis. viii, 160 pages, 1991.

Vol. 69: N.J.D. Nagelkerke, Maximum Likelihood Estimation of Functional Relationships. v, 110 pages, 1992.

Vol. 70: K. Iida, Studies on the Optimal Search Plan. viii, 130 pages, 1992.

Vol. 71: E.M.R.A. Engel, A Road to Randomness in Physical Systems. ix, 155 pages, 1992.

Vol. 72: J.K. Lindsey, The Analysis of Stochastic Processes using GLIM. vi, 294 pages, 1992.

Vol. 73: B.C. Arnold, E. Castillo, J.-M. Sarabia, Conditionally Specified Distributions. xiii, 151 pages, 1992.

Vol. 74: P. Barone, A. Frigessi, M. Piccioni (Editors), Stochastic Models, Statistical Methods, and Algorithms in Image Analysis. vi, 258 pages, 1992.

Vol. 75: P.K. Goel, N.S. Iyengar (Editors), Bayesian Analysis Statistics and Econometrics. xi, 410 pages, 1992.

Vol. 76: L. Bondesson, Generalized Gamma Convolutions Related Classes of Distributions and Densities. viii, 173 pa 1992.

Vol. 77: E. Mammen, When Does Bootstrap Work? Asymp Results and Simulations. vi, 196 pages, 1992.

Vol. 78: L. Fahrmeir, B. Francis, R. Gilchrist, G. Tutz (Edit Advances in GLIM and Statistical Modelling: Proceedings o GLIM92 Conference and the 7th International Workshop Statistical Modelling, Munich, 13-17 July 1992. ix, 225 pa 1992.

Vol. 79: N. Schmitz, Optimal Sequentially Planned Dec. Procedures. xii, 209 pages, 1992.

Vol. 80: M. Fligner, J. Verducci (Editors), Probability Models Statistical Analyses for Ranking Data. xxii, 306 pages, 1992

Vol. 81: P. Spirtes, C. Glymour, R. Scheines, Causation, Pre tion, and Search. xxiii, 526 pages, 1993.

Vol. 82: A. Korostelev and A. Tsybakov, Minimax Theor Image Reconstruction. xii, 268 pages, 1993.

Vol. 83: C. Gatsonis, J. Hodges, R. Kass, N. Singpurwalla (Edit Case Studies in Bayesian Statistics. xii, 437 pages, 1993.

Vol. 84: S. Yamada, Pivotal Measures in Statistical Experim and Sufficiency. vii, 129 pages, 1994.

Vol. 85: P. Doukhan, Mixing: Properties and Examples. xi, pages, 1994.

Vol. 86: W. Vach, Logistic Regression with Missing Values i Covariates. xi, 139 pages, 1994.

Vol. 87: J. Møller, Lectures on Random Voronoi Tessellations 134 pages, 1994.

Vol. 88: J.E. Kolassa, Series Approximation Method Statistics. viii, 150 pages, 1994.

Vol. 89: P. Cheeseman, R.W. Oldford (Editors), Selecting Mo From Data: Artificial Intelligence and Statistics IV. x, 487 pa 1994.

Vol. 90: A. Csenki, Dependability for Systems with a Partitic State Space: Markov and Semi-Markov Theory and Computati Implementation. x, 241 pages, 1994.

Vol. 91: J.D. Malley, Statistical Applications of Jordan Algeb viii, 101 pages, 1994.

Vol. 92: M. Eerola, Probabilistic Causality in Longituc Studies. vii, 133 pages, 1994.

Vol. 93: Bernard Van Cutsem (Editor), Classification Dissimilarity Analysis. xiv, 238 pages, 1994.